Aquatic Biofilms

Ecology, Water Quality and Wastewater Treatment

Edited by

Anna M. Romaní

Institute of Aquatic Ecology
University of Girona
Girona
Spain

Helena Guasch

Institute of Aquatic Ecology
University of Girona
Girona
Spain

and

M. Dolors Balaguer

Laboratory of Chemical and
Environmental Engineering (LEQUiA)
Institute of the Environment
University of Girona
Girona
Spain

Caister Academic Press

Caister Academic Press
Norfolk, UK

www.caister.com

British Library Cataloguing-in-Publication Data
A catalogue record for this book is available from the British Library

ISBN: 978-1-910190-17-3 (hardback)
ISBN: 978-1-910190-18-0 (ebook)

Cover design adapted from scanning electron microscope images of aquatic biofilms. Images were obtained by V. Díaz-Villanueva (front cover) and A. Freixa (back cover) with the technical assistance of C. Carulla at the Technical Services of the University of Girona.

Contents

	Contributors	v
	Preface	xi
Part I	Biofilm Mode of Life	1
1	Limits of the Biofilm Concept and Types of Aquatic Biofilms Juanita Mora-Gómez, Anna Freixa, Núria Perujo and Laura Barral-Fraga	3
2	Laser Microscopy for the Study of Biofilms: Issues and Options Thomas R. Neu and John R. Lawrence	29
3	Interactions and Communication within Marine Biofilms Priyanka Sathe and Sergey Dobretsov	47
4	Microbial Biodiversity in Natural Biofilms Katharina Besemer	63
5	Aquatic Biofilms and Biogeochemical Processes Laura Leff, Jonathon B. Van Gray, Eugènia Martí, Stephanie N. Merbt and Anna M. Romaní	89
Part II	Biofilms and Pollution	109
6	Benthic Diatom Monitoring and Assessment of Freshwater Environments: Standard Methods and Future Challenges Soizic Morin, Nora Gómez, Elisabet Tornés, Magdalena Licursi and Juliette Rosebery	111
7	The Use of Biofilms to Assess the Effects of Chemicals on Freshwater Ecosystems Helena Guasch, Joan Artigas, Berta Bonet, Chloe Bonnineau, Oriol Canals, Natàlia Corcoll, Arnaud Foulquier, Julio López-Doval, Sandra Kim Tiam, Soizic Morin, Enrique Navarro, Stephane Pesce, Lorenzo Proia, Humbert Salvadó and Alexandra Serra	125
8	Biofilm Development in Sewer Networks Oriol Gutierrez, Guangming Jiang, Keshab Sharma and Zhiguo Yuan	145

Part III New Technologies Using Biofilms 165

9 Biofilm Biodegradation Potential 167
Freshta Akbari, Natasha Andrade, Merily Horwat and Birthe V. Kjellerup

10 Electroactive Biofilms in Water and Air Pollution Treatment 183
Anna Vilajeliu-Pons, Sebastià Puig, Alessandro Carmona-Martínez, Nicolas Bernet, Marta Coma, Federico Aulenta, Jesús Colprim and Maria Dolors Balaguer

11 Biofilms for One-stage Autotrophic Nitrogen Removal 205
José M. Carvajal-Arroyo, Tiago Rogeiro Vitor Akaboci, Maël Ruscalleda, Jesús Colprim, Emilie Courtens and Siegfried E. Vlaeminck

Glossary 223

Index 227

Contributors

Freshta Akbari
Department of Biological Sciences
Goucher College
Baltimore, MD
USA

frakb001@mail.goucher.edu

Natasha Andrade
Department of Civil and Environmental Engineering
University of Maryland at College Park
College Park, MD
USA

nandrade@umd.edu

Joan Artigas
Laboratory Microorganisms: Genome and Environment
Clermont University
University Blaise Pascal
Clermont-Ferrand
France; and
CNRS, UMR 6023
Laboratory Microorganisms: Genome and Environment
Aubière
France

joan.artigas_alejo@univ-bpclermont.fr

Federico Aulenta
Water Research Institute (IRSA)
National Research Council (CNR)
Monterotondo
Italy

aulenta@irsa.cnr.it

Maria Dolors Balaguer
Laboratory of Chemical and Environmental Engineering (LEQUiA)
Institute of the Environment
University of Girona
Girona
Spain

dolors.balaguer@udg.edu

Laura Barral-Fraga
Institute of Aquatic Ecology
University of Girona
Girona
Spain

laura.barral.fraga@gmail.com

Nicolas Bernet
INRA, UR0050
Laboratoire de Biotechnologie de l'Environnement
Narbonne
France

nicolas.bernet@supagro.inra.fr

Katharina Besemer
School of Engineering
University of Glasgow
Glasgow
United Kingdom; and
Department of Limnology and Bio-Oceanography
University of Vienna
Vienna
Austria

katharina.besemer@univie.ac.at

Berta Bonet
Institute of Aquatic Ecology
University of Girona
Girona
Spain

berta.bonet@udg.edu

Chloe Bonnineau
Institute of Life Sciences
Université Catholiqu de Louvain
Louvain-la-Neuve
Belgium

chloe.bonnineau@uclouvain.be

Oriol Canals
Laboratory of Protistology
Department of Animal Biology
Faculty of Biology
University of Barcelona
Barcelona
Spain

ocanalsd@ub.edu

Alessandro Carmona-Martínez
INRA, UR0050
Laboratoire de Biotechnologie de l'Environnement
Narbonne
France

alessandro.carmona@supagro.inra.fr

Jose M. Carvajal-Arroyo
Laboratory of Microbial Ecology and Technology (LabMET)
University of Ghent
Ghent
Belgium

josemaria.carvajalarroyo@ugent.be

Jesús Colprim
Laboratory of Chemical and Environmental Engineering (LEQUiA)
Institute of the Environment
University of Girona
Girona
Spain

j.colprim@lequia.udg.cat

Marta Coma
Laboratory of Microbial Ecology and Technology (LabMET)
Faculty of Bioscience Engineering
Ghent University
Ghent
Belgium

marta.comabech@ugent.be

Natàlia Corcoll
Department of Biological and Environmental Sciences
University of Gothenburg
Gothenburg
Sweden

natalia.corcoll@gu.se

Emilie Courtens
Laboratory of Microbial Ecology and Technology (LabMET)
Faculty of Bioscience Engineering
Ghent University
Ghent
Belgium

emilie.courtens@ugent.be

Sergey Dobretsov
Department of Marine Science and Fisheries
College of Agricultural and Marine Sciences
Sultan Qaboos University
Muscat
Oman

sergey@squ.edu.om

Arnaud Foulquier
Laboratoire d'Écologie Alpine (LECA)
UMR 5553 CNRS
Université Grenoble Alpes
Grenoble
France

arnaud.foulquier@ujf-grenoble.fr

Anna Freixa
Institute of Aquatic Ecology
University of Girona
Girona
Spain

anna.freixa@udg.edu

Nora Gómez
Laboratory of Plankton and Biofilms
Institute of Limnology "Dr. R.A. Ringuelet"
National Scientific and Technical Research Council (CONICET)
National Univeristy of La Plata (UNLP)
Buenos Aires
Argentina

nora@ilpla.edu.ar

Helena Guasch
Institute of Aquatic Ecology
University of Girona
Girona
Spain

helena.guasch@udg.edu

Oriol Gutierrez
Catalan Institute for Water Research (ICRA)
Girona
Spain; and
Advanced Water Management Centre
University of Queensland
Brisbane
Australia

ogutierrez@icra.cat; oriol@awmc.uq.edu.au

Merily Horwat
Department of Civil and Environmental Engineering
University of Maryland at College Park
College Park, MD
USA

mhorwat@umd.edu

Guangming Jiang
Advanced Water Management Centre
University of Queensland
Brisbane
Australia

g.jiang@awmc.uq.edu.au

Sandra Kim Tiam
Institut National de la Recherche Scientifique
Centre Eau Terre Environment (INRS-ETE)
Québec
Canada

Sandra.Kim_Tiam@ete.inrs.ca

Birthe V. Kjellerup
Department of Civil and Environmental Engineering
University of Maryland at College Park
College Park, MD
USA

bvk@umd.edu

John R. Lawrence
Environment Canada
National Hydrology Research Centre
Saskatoon, SK
Canada

john.lawrence@ec.gc.ca

Laura Leff
Department of Biological Sciences
Kent State University
Kent, OH
USA

lleff@kent.edu

Magdalena Licursi
Laboratory of Plankton and Biofilms
Institute of Limnology "Dr. R.A. Ringuelet"
National Scientific and Technical Research Council (CONICET)
National Univeristy of La Plata (UNLP)
Buenos Aires
Argentina

malena@ilpla.edu.ar

Julio López-Doval
Department of Ecology
Laboratory of Limnology
University of São Paulo
São Paulo
Brasil

jclopezdoval@gmail.com

Eugènia Martí
Integrative Freshwater Ecology Group
Center for Advanced Studies of Blanes (CEAB-CSIC)
Girona
Spain

eugenia@ceab.csic.es

Stephanie N. Merbt
Integrative Freshwater Ecology Group
Center for Advanced Studies of Blanes (CEAB-CSIC)
Girona
Spain

smerbt@ceab.csic.es

Juanita Mora-Gómez
Institute of Aquatic Ecology
University of Girona
Girona
Spain

juanita.mora@udg.edu
juanita.mora.gomez78@gmail.com

Soizic Morin
Aquatic Ecosystems and Global Change Research Unit (EABX)
National Research Institute of Science and Technology for Environment and Agriculture (Irstea)
Cestas
France

soizic.morin@irstea.fr

Enrique Navarro
Department of Biodiversity and Restoration
Pyrenean Institute of Ecology (CSIC)
Zaragoza
Spain

enrique.navarro@ipe.csic.es

Thomas R. Neu
Department of River Ecology
Helmholtz Center for Environmental Research – UFZ
Magdeburg
Germany

thomas.neu@ufz.de

Núria Perujo
Institute of Aquatic Ecology
University of Girona
Girona
Spain

nuria.perujo@udg.edu
nuria.perujo.buxeda@gmail.com

Stephane Pesce
Freshwater Systems, Ecology and Pollution Research Unit (MALY)
National Research Institute of Science and Technology for Environment and Agriculture (Irstea)
Villeurbanne
France

stephane.pesce@irstea.fr

Lorenzo Proia
Resources and Ecosystems Area
Catalan Institute for Water Research
Girona
Spain

proialorenzo@hotmail.it

Sebastià Puig
Laboratory of Chemical and Environmental Engineering (LEQUiA)
Institute of the Environment
University of Girona
Girona
Spain

sebastia@lequia.udg.cat

Anna M. Romaní
Institute of Aquatic Ecology
University of Girona
Girona
Spain

anna.romani@udg.edu

Juliette Rosebery
Aquatic Ecosystems and Global Change Research Unit (EABX)
National Research Institute of Science and Technology for Environment and Agriculture (Irstea)
Cestas
France

juliette.rosebery@irstea.fr

Maël Ruscalleda
Laboratory of Chemical and Environmental Engineering (LEQUiA)
Institute of the Environment
University of Girona
Girona
Spain

mael@lequia.udg.cat

Humbert Salvadó
Laboratory of Protistology
Department of Animal Biology
Faculty of Biology
University of Barcelona
Barcelona
Spain

hsalvado@ub.edu

Priyanka Sathe
Department of Marine Science and Fisheries
College of Agricultural and Marine Sciences
Sultan Qaboos University
Muscat
Oman

priyankasat@gmail.com

Alexandra Serra
Integrative Freshwater Ecology Group
Center for Advanced Studies of Blanes (CEAB-CSIC)
Girona
Spain

aserra@ceab.csic.es

Keshab Sharma
Advanced Water Management Centre
University of Queensland
Brisbane
Australia

k.sharma@awmc.uq.edu.au

Elisabet Tornés
Institute of Aquatic Ecology; and
Catalan Institute for Water Research (ICRA)
University of Girona
Girona
Spain

etornes@icra.cat

Jonathon B. Van Gray
Department of Biological Sciences
Kent State University
Kent, OH
USA

jgray39@kent.edu

Anna Vilajeliu-Pons
Laboratory of Chemical and Environmental Engineering (LEQUiA)
Institute of the Environment
University of Girona
Girona
Spain

avilajeliu@lequia.udg.cat

Tiago Rogeiro Vitor Akaboci
Laboratory of Chemical and Environmental Engineering (LEQUiA)
Institute of the Environment
University of Girona
Girona
Spain

tvitor@lequia.udg.edu

Siegfried E. Vlaeminck
Research Group of Sustainable Energy, Air and Water Technology
Department of Bioscience Engineering
University of Antwerp
Antwerp; and
Laboratory of Microbial Ecology and Technology (LabMET)
Ghent University
Ghent
Belgium

siegfried.vlaeminck@uantwerpen.be
siegfried.vlaeminck@ugent.be

Zhiguo Yuan
Advanced Water Management Centre
University of Queensland
Brisbane
Australia

z.yuan@awmc.uq.edu.au

Current Books of Interest

Climate Change and Microbial Ecology: Current Research and Future Trends	2016
Biofilms in Bioremediation: Current Research and Emerging Technologies	2016
Microalgae: Current Research and Applications	2016
Arboviruses: Molecular Biology, Evolution and Control	2016
Shigella: Molecular and Cellular Biology	2016
Alphaviruses: Current Biology	2016
Thermophilic Microorganisms	2015
Flow Cytometry in Microbiology: Technology and Applications	2015
Probiotics and Prebiotics: Current Research and Future Trends	2015
Epigenetics: Current Research and Emerging Trends	2015
Corynebacterium glutamicum: From Systems Biology to Biotechnological Applications	2015
Advanced Vaccine Research Methods for the Decade of Vaccines	2015
Antifungals: From Genomics to Resistance and the Development of Novel Agents	2015
Bacteria-Plant Interactions: Advanced Research and Future Trends	2015
Aeromonas	2015
Antibiotics: Current Innovations and Future Trends	2015
Leishmania: Current Biology and Control	2015
Acanthamoeba: Biology and Pathogenesis (2nd edition)	2015
Microarrays: Current Technology, Innovations and Applications	2014
Metagenomics of the Microbial Nitrogen Cycle: Theory, Methods and Applications	2014
Pathogenic *Neisseria*: Genomics, Molecular Biology and Disease Intervention	2014
Proteomics: Targeted Technology, Innovations and Applications	2014
Biofuels: From Microbes to Molecules	2014
Human Pathogenic Fungi: Molecular Biology and Pathogenic Mechanisms	2014
Applied RNAi: From Fundamental Research to Therapeutic Applications	2014
Halophiles: Genetics and Genomes	2014
Molecular Diagnostics: Current Research and Applications	2014
Phage Therapy: Current Research and Applications	2014
Bioinformatics and Data Analysis in Microbiology	2014
The Cell Biology of Cyanobacteria	2014
Pathogenic *Escherichia coli*: Molecular and Cellular Microbiology	2014
Campylobacter Ecology and Evolution	2014
Burkholderia: From Genomes to Function	2014
Myxobacteria: Genomics, Cellular and Molecular Biology	2014

Full details at www.caister.com

Preface

On any wet surface a biofilm is easily formed, whether it is on a building, a rock in a river, marine sediments, a decaying leaf, a sewage pipe, among others. The extensive appearance of the biofilm mode of life may be linked to its properties such as nutrient entrapment and physical protection of cells from the surrounding environment. Also, this mode of life is old, underlining its resistance. It is thought that aggregated layer-structured biofilms similar to ancient stromatolites have been relevant for the origin of first microbial cells on Earth. The unique and complex characteristics of biofilms include mechanisms and processes occurring at different scales addressed by different scientific branches. Atomic forces and chemical bonds are keys for attachment processes, development of the matrix, and chemical gradients. At the cell to organisms scale, life science analyzes cell-to-cell communication, diversity of microbial metabolisms and food web interactions in biofilms. Aquatic biofilms are also a significant component at the Earth sciences scale as shown for instance by their relevance in biogeochemical cycles.

From the first report of surface-associated bacterial cells, aquatic biofilm research have been exponentially developed in the last decades, covering the study of biofilms in marine and freshwater environments, including pristine but also those affected by pollution and anthropogenic disturbances, and of those developing in man-made systems such as water engineering processes. Although in each specific environment a distinct biofilm may develop, the drivers and gradients in biofilms show parallelisms. For instance, the oxygen gradient determining specific biogeochemical reactions is similar between naturally occurring fluvial biofilms and those developing on granules for water technology purposes. Other example is the knowledge gained from anthropogenic disturbances effects on biofilms, showing parallelisms to responses observed from biofilms growing in extreme environments and developing similar resistance strategies.

The aim of this book was to compile in a single volume the latest, up-to-date theory, methodology, and applications of aquatic biofilm's research. From the theory, a broad review of biofilm history, architecture, cell communication, biodiversity and biogeochemistry is included, updating both theory and methodology. Then, the study of biofilms developing in polluted systems as well as their use and relevance as ecotoxicological sensors is reviewed. Finally, application and profit of biofilms is shown in three examples on new technologies using biofilms. We believe the different points of view and approaches presented in the book, from theory to application, from ecology to engineering, are complementary and feed from each other contributing to our understanding of biofilm mode of life.

Anna M. Romaní, Helena Guasch, Marilós Balaguer

Part I

Biofilm Mode of Life

Limits of the Biofilm Concept and Types of Aquatic Biofilms

1

Juanita Mora-Gómez, Anna Freixa, Núria Perujo and Laura Barral-Fraga

Abstract

Nowadays, it is widely recognized that in natural aquatic settings bacterial cells are most often found in close association with wet surfaces and interfaces in the form of multicellular aggregates commonly referred to as biofilms, which also involve algae, fungi and protozoa. However, since surface-associated bacteria were first reported, the biofilm concept has been developed to include the great complexity of this microbial way of life. Biofilms in natural and anthropogenic environments are modulated by the nature of the surface in which they grow and environmental factors, which determine the biofilm composition and structure and, in consequence, its metabolism and function. As a consequence there are many types of biofilm and many terms have been used through years to try to describe particular biofilms in relation to a type of surface or environment. In the present review, we summarize the knowledge on biofilms, starting from the origin and evolution of the concept followed by a description of biofilm types based on substratum characteristics. Finally, we explain the general effects of environmental variables, contextualizing them in the wide range of natural aquatic ecosystems (including fresh and marine water) and some man-made systems (such as those associated with water distribution systems and marine environments).

Introduction

Aquatic biofilms: relevant role in the environment

Biofilm communities play a role in the environment both in maintaining and improving the ecological health of freshwater ecosystems. Biofilms are crucial in ecosystem functioning and have an excellent ability to degrade and transform pollutants, as they are involved in primary production (photosynthetic activity) (Underwood *et al.*, 2005), carbon and nutrient cycling (Davey and O'Toole, 2000), retention of inorganic and organic nutrients (Pusch *et al.*, 1998), and support of food webs (Anderson-Glena *et al.*, 2008). There are numerous benefits that a bacterial community might obtain from the formation of biofilms. For example, biofilms protect microorganisms from protozoa grazing, and host defences as in plant-associated biofilms (López *et al.*, 2010). Moreover, biofilms are also used to solve a huge problem in the treatment of wastewater, cleaning polluted water in water treatment plants by removing organic material (Lazarova and Manem, 1995). In addition to the important practical use of biofilms in biological wastewater treatment, they are also used in biodegradation and bioremediation in bioreactors (De Beer and Stoodley, 2006).

Biofilms also have negative effects in man-made systems, including biofouling and microbial-induced corrosion, which affect industrial water systems (Videla and Herrera, 2005) and drinking water distribution systems. Many problems are due to biofilm growth (Berry *et al.*, 2006) and the biodeterioration processes from the interactions between metal surfaces and biofilm. In marine systems, man-made structures such as aquaculture nets, oil and gas installations, and ship hulls are also affected by biofilm attachment and growth, the so-called microfoulers (Salta *et al.*, 2013). Marine biofouling may cause some problems such as the cost associated with increased fuel consumption by ships linked to increased frictional drag, as reported in Schultz *et al.* (2011). The survival of microorganisms on man-made surfaces is based upon interactions of many variables, including temperature, pipe surface, nutrient levels and type and concentration of disinfectants (Berry *et al.*, 2006). It is also known that attached cells have certain ecological advantages over planktonic cells such as having increased resistance to chlorine and other biocides (Emtiazi *et al.*, 2004). The resistance of biofilm cells to disinfection consists of the formation of persistent cells and protection linked to the increased production of extracellular polymeric substances.

Biofilm history and names

Over the years, a considerable amount of research has dealt with the subject of biofilm. The first official report of surface-associated microbial cells was in 1684, when Van Leeuwenhoek informed the Royal Society of London that he had observed an accumulation of microorganisms on tooth surfaces using a simple microscope. But it was not until 1936 that Zobell and Anderson introduced the term *periphytes*, referring to bacteria associated with surfaces in bottles used to store seawater (Zobell and Anderson, 1936). The examination of biofilms was promoted just after electron microscope development, which allowed high resolution at much higher magnifications than when observed under the light microscope (Donlan, 2002). In the following decades many microbiologists investigated aquatic bacteria, especially for the study and treatment of many serious diseases. Some studies reported that bacteria attached to aquatic surfaces are often 1,000 or 10,000 times greater in number than planktonic (free-floating) bacteria (Costerton *et al.*, 1978; Donland, 2002).

The first word proposed to describe microorganisms attached to aquatic surfaces was '*aufwuchs*', a German word, meaning surface growth (Ruttner, 1953). But it was not until 1975 that the term *biofilm* made its first appearance in the scientific literature (Mack *et al.*, 1975). In 1983, Wetzel described biofilms as assemblages of bacteria, algae, fungi, and protozoa within a protective matrix of extracellular polymeric substances and detritus, which colonize submerged surfaces in lakes and rivers (Wetzel, 1983). By this time scientists and engineers used developing technology to effectively study microbial communities, and biofilm research was progressively established as a relevant scientific topic. Many definitions of aquatic biofilms appeared during the eighties referring to biofilm as an assemblage of autotrophic and heterotrophic microorganisms embedded in a polymeric matrix and developing on wetted surfaces (Lock *et al.*, 1984; Costerton *et al.*, 1987; Characklis and Marshall, 1989). Later, the concept of biofilm evolved by including new knowledge from research on biofilm formation, organization, cell-to-cell communication, interaction between microorganisms, and three-dimensional structure. The great development of the research on biofilm formation in medicine and animal health (i.e. cancer processes, transplants or prosthesis) has also helped to improve the knowledge of aquatic biofilms by using relevant techniques such as confocal microscopy (see Chapter 2) and cell-to-cell signalling (see Chapter 3).

The study of biofilm in different systems, where distinct environmental variables and types of substrata are found, led to the appearance of different terms when referring to biofilm such as *periphyton* and *microphytobentos*, or other less common ones like *benthos*, *haptobenthos* and *herpobenthos*. Thus, there can be a dilemma when defining the terminology used to describe a biofilm (Wetzel, 1983), and sometimes it becomes a real problem when looking for the appropriate bibliography. In most scientific publications *periphyton* refers to freshwater ecosystems while *microphytobenthos* refers to marine ecosystems. *Periphyton* is defined as an assemblage of freshwater organisms mainly composed of photoautotrophic algae, heterotrophic and chemoautotrophic bacteria, fungi, protozoans, metazoans and viruses which grow upon a benthic substrate (Wetzel, 1983; Larned, 2010). *Microphytobenthos* are defined as populations of photoautotrophic microorganisms (diatoms, euglenids, crysophyceans, dinoflagelates) that colonize benthic substrata in marine systems, especially in intertidal and lower supra-tidal sediments reached by light (MacIntyre *et al.*, 1996; Jesus *et al.*, 2009; Pan *et al.*, 2013). The term *biofilm* was initially used in engineering and referred to attached heterotrophic communities (Wetzel, 1983), but it has been spread to both natural and anthropogenic aquatic systems as a general term referring to attached-microbial communities. Even though *biofilm* and *periphyton* are mostly used synonymously, Saikia (2001) reported slight differences between the terms *periphyton* and *biofilm*, indicating that *periphyton* is linked to nutrient dynamics in ecosystems and especially including photosynthetic organisms. Finally, *benthos* were initially described as organisms associated to the bottom or solid–liquid interfaces in aquatic systems, *haptobenthos* as adhered organisms but not penetrating a solid surface, and *herpobenthos* as organisms penetrating bottom sediments (Wetzel, 1983; Neuswanger *et al.*, 1982).

Additionally, since biofilms are formed in sand, sediment, rocks and cobbles, wood and leaves, and the surface of submerged plants, specific names for each natural substrate have also been used by adding the corresponding adjective (such as *epilithic*, *epixylic*, or *epipsammic biofilm*; see Table 1.1 and Fig. 1.1) (Lock, 1993; Romaní, 2010; Vadeboncoeur and Steinman, 2002). The term *biofilm* is also commonly used in man-made surfaces such as industrial equipment and in water management (water treatment plants and drinking water distribution systems, MacLeod *et al.*, 1990). In these artificial substrates the concept of *biofouling* (or microbial fouling) is widely used to describe the accumulation process of microorganisms on wet surfaces, especially in shipping equipment and water distribution systems (Melo and Bott, 1997).

Similar to biofilms, a special and characteristic form of attached microbial community is a *microbial mat*. Microbial mats are organo-sedimentary structures that develop on solid surfaces, formed by the trapping and binding of sediment and/or the net carbonate-precipitating activities of microorganisms, resulting in a layered structure. They are mainly formed by cyanobacteria, colourless sulfur bacteria, purple sulfur bacteria, and sulfate-reducing bacteria, and can be found in lagoons, marine intertidal and subtidal zones, hypersaline ponds, hot springs, fresh water rivers and lakes (van Gemerden, 1993; Dupraz and Vissche, 2005). Ancient microbial mats (*stromatolites*) were abundant and diverse in shallow zones of the oceans in the Proterozoic (Bottjer *et al.*, 1996). Like the first photosynthetic communities, stromatolites consumed the greenhouse gas CO_2, and produced free O_2 and H_2, playing a crucial role for the early establishment of life (Dupraz

Table 1.1 Most common names, main factors from the substratum type (support media) affecting the biofilm (intrinsic substratum factors), and dominant groups composing the biofilm for the different substrata considered

	Rocks	Sediment	Dead plant material	Living plants	Suspended aggregates (POM)	Artificial (man-made)
Most common used names	Epilithic biofilm; epilithon; periphyton	Epipsammic biofilm; epipsammon; epipelic biofilm; epipelon; hyporheic biofilm	Epixylic biofilm	Epiphytic biofilm	Snow particles, marine snow, lake snow, river snow; aggreggates, granules	Biofilm (biofouling, bioclogging)
Intrinsic substratum factors	Surface roughness; stone size and orientation	Grain size and roughness of grain; profile zonation in depth	Organic matter characteristics (chemical composition)	Biofilm–plant interaction	Particle density, size and shape; particle settling velocity; particle surface properties	Material roughness; type of material
Community composition (dominant groups)	**Algae,** cyanobacteria, bacteria, fungi and protozoa	**Bacteria,** cyanobacteria, archaea, algae, fungi and protozoa	**Fungi,** bacteria, archaea, algae, and metazoans	**Algae,** bacteria, fungi and protozoa	**Diatoms,** filamentous **cyanobacteria, bacteria,** protozoans, zooplankton	**Bacteria,** fungi and protozoa

POM, particulate organic matter.

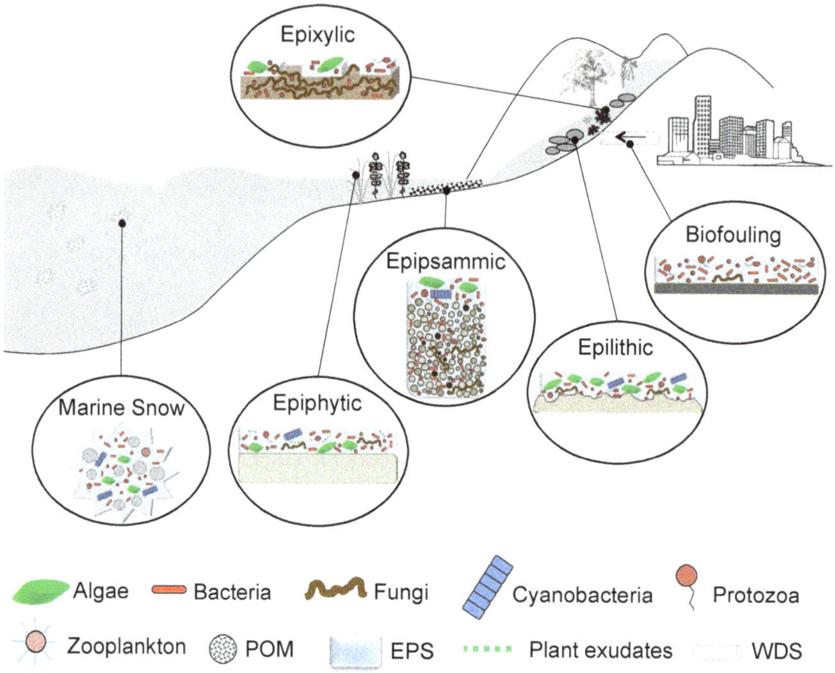

Figure 1.1 Biofilm types with regard to differential substrata find in natural and man-made environments. EPS, extracellular polymeric substances; POM, particular organic matter; WDS, water distribution system. Modified from Romaní (2010).

and Vissche, 2005). However, in the latter part of the Proterozoic they declined most likely due to metazoan diversification, increased grazing and sediment disturbance (Bottjer *et al.*, 1996).

Within the limits of the biofilm concept, *aggregates* or *granules* as suspended microbial flocs have many features in common with classical biofilms, such as the polymeric matrix and the association of diverse groups of microorganisms such as bacteria, algae and protozoans, which interact structurally and functionally (De Beer and Stoodley, 2006). Aggregates play a role in nutrient cycling and organic matter decomposition in lakes, rivers and the sea (Azam and Cho, 1987; Grossart and Simon, 1993).

What is the definition of a biofilm?

The answer to this question is not always simple due to the large variability in structure and composition of biofilm, and the different environments in which biofilm develops. The high complexity of aquatic biofilms is related to the type of substratum and the environment in which they are living (Karatan and Watnick, 2009). However, several characteristics are shared by any type of biofilm. Nowadays it is widely recognized that in natural settings bacterial cells are most often found in close association with surfaces and interfaces, in the form of multicellular aggregates commonly referred to as biofilms which also involve other microorganisms such as algae, fungi, viruses and protozoa (Branda *et al.*, 2005). A general characteristic of biofilm structure is that these microorganisms, together with extracellular enzymes and detritus, are enclosed within a polymeric matrix (Golladay and Sinsabaugh, 1991). These extracellular polymeric substances (EPS) mainly composed of polysaccharides, provide mechanical stability to biofilms, mediate their adhesion to surfaces, and form a cohesive three-dimensional polymer network that interconnects and transiently immobilizes biofilm cells (Wingender and Flemming, 2011; Decho, 2000; Gerbersdorf *et al.*, 2008). This matrix provides protection from predation, toxic substances and physical perturbations. Although the polymeric matrix is mostly constituted by water (up to 97%, Zhang and Feng, 2001) and polysaccharides, other compounds are found, such as extracellular DNA, proteins and lipids, particulate material and detritus (Costerton *et al.*, 1995; Sutherland, 2001; Lawrence and Neu, 2003).

As a consequence of this three-dimensional physical structure, tight interactions between the organisms living within the biofilm occur, building a 'micro-ecosystem' where feeding interactions, nutrient cycling, competition, synergism, and cell-to-cell signalling take place. In aquatic biofilms, diverse trophic interactions have been described, such as protozoa feeding on bacteria and algae, rotifer feeding on protozoa, bacteria and detritus, nematode feeding on algae, grazing by metazoans, etc., which might also cause structural changes to the biofilm (Augspurger *et al.*, 2008; Früh *et al.*, 2011; Majdi *et al.*, 2011; Risse-Buhl *et al.*, 2012). Thus, biofilm can be seen as the well-defined microbial loop, for example, in the pelagic ecosystem (Azam, 1998), but 'squeezed' in the physical space, enhancing the relevance of the spatial structure which facilitates cooperation (Kreft, 2004).

For the purposes of this review, we seek to give an overview of the limits of the biofilm concept by including microbial communities developing in natural aquatic environments and man-made surfaces (associated with water treatment plants and drinking water distribution systems, as well as marine surfaces). In this sense, biofilm types are described based on the substratum where they develop, and the main environmental factors influencing the different types of biofilm are also reviewed.

Types of aquatic biofilms defined by the substratum where they develop

In aquatic environments, biofilms can be highly diverse depending on the substratum where they develop. In natural environments, biofilm grows upon inert substrata such as sand, sediment, rocks and cobbles; non-living organic substrata such as wood, leaf litter or particular organic matter; and living plants such as aquatic macrophytes and macroalgae (Table 1.1). In man-made substrata, associated with drinking water distribution systems and man-made marine structures, biofilm grows upon pipes and boats. The nature of the substratum determines the composition and structure of the biofilm and, in consequence, its metabolism (Romaní, 2004a). Furthermore, some studies have demonstrated that the nature of the substratum is the most important factor for biofilm development, since substratum properties may regulate patterns of cell accumulation and distribution during the early stages of biofilm development (Terlizzi and Faimali, 2010).

Rocks

Biofilms attached to rock surfaces (also to gravel and cobbles) are referred to as *epilithic biofilms* or *epilithon* in rivers (Guasch and Sabater, 1994; Sabater *et al.*, 2006; Anderson-Glena *et al.*, 2008), marine environments (Thompson *et al.*, 2004, 2005; Firstater *et al.*, 2012; Smith *et al.*, 2010), and lakes (Lowe, 1996; Vadeboncoeur and Steinman, 2002). Compared to other biofilms, such as those that grow on sand, epilithic biofilms have a more complex structure with a higher algal biomass and they are more independent of seasonal fluctuations (Romaní and Sabater, 2001; Graba *et al.*, 2013). Data from river epilithic biofilms show a proportion of total carbon of 60–90% for algae, 10–40% for EPS, 1–5% for bacteria and less than 1% for fungi (Romaní, 2010). However, in shaded environments (i.e. forested rivers and streams), heterotrophic biomass (bacteria, fungi and protozoa) become more important (Romaní, 2010).

Stones provide a three-dimensional physical habitat for biofilm biomass (Bergey, 2008). The developing biofilms are greatly influenced by rock characteristics such as surface texture (Clifford *et al.*, 1992; Sanson *et al.*, 1995; Murdock and Dodds, 2007), stone size (Ledger and Hildrew, 1998), and stone orientation (Murdock and Dodds, 2007). On the one hand, rougher surface can cause greater biofilm accumulation on rocks, since it increases sedimentation efficiency and cell adhesion, and protects the biofilm from disturbances such as scouring and grazing (Murdock and Dodds, 2007). On the other hand, stone size and stability determine the algal resistance to scour (Peterson, 1996). Regarding stone orientation, biofilm accumulation is consistently greater on horizontal than on vertical surfaces (Baynes, 1999; Kralj *et al.*, 2006; Murdock and Dodds, 2007).

Conversely, the effect of rock chemistry on biofilms is unclear. It seems that the chemical composition of streambed stones is of little importance to algal assemblages, which may be caused by the low dissolution rate of stones (Bergey, 2008). However, some evidence points out that rock chemistry may affect early stages of biofilm colonization (Blinn *et al.*, 1980). A lack of chemical effect helps to explain the similarity of algal biomass and assemblages among such different substrates as rock and wood (e.g. Sabater *et al.*, 1998; Townsend and Gell, 2005) and the success of using artificial substrates, such as glass or clay tiles, as substitutes for natural substrates in biofilm studies (e.g. Tuchman and Stevenson, 1980; Lamberti and Resh, 1985).

Although autotrophic–heterotrophic relationships may occur in most aquatic biofilms, algal–bacterial interactions have been mainly described for epilithic biofilms. In biofilms developing on rocks, the structural stability and close spatial relationship between bacteria and algae favours the bacterial use of fresh labile organic compounds released by algae (Wetzel, 1993; Sobczak, 1996; Romaní and Sabater, 1999, 2000), affecting the whole biofilm metabolism. In addition, when a thick biofilm is developed, an anoxic layer may exist at the biofilm bottom and consequently the presence of anaerobic bacteria (Schramm *et al.*, 1999). This micro-spatial variation contributes to the intra-site differences observed in microbial communities (Anderson-Glena *et al.*, 2008).

Epilithic biofilms in aquatic environments have been defined as playing a relevant role in nutrient uptake since they can use inorganic and organic compounds from the flowing water by catalysing enzymes such as nitrogen reductase and alkaline phosphatase (Adey *et al.*, 1993). In addition, biofilm can trap particulate material from the water column and thus increase its concentration in contrast to the water column. Nutrient uptake is potentially influenced by the degree and distribution of biofilm cover, ambient macronutrient concentration, grazing, sloughing, temperature, advective transport in the river, and diffusion into the biofilm (Larned *et al.*, 2004; Saikia, 2011).

Sediment

Microorganisms attached to the particles of sandy sediments (sand and gravel) are referred to as *epipsammic*, while when developing on muddy sediments (clay or silt) they are known as *epipelic* (MacIntyre *et al.*, 1996). Sediments are a hard, inert substratum, characterized by its smaller size than other substrates. In rivers, the biofilm developing on sediment has been defined as playing a key role in organic matter decomposition, also being more heterotrophic (with higher contributions of bacteria and fungi) than the biofilm developing on rocks (Pusch *et al.*, 1998; Romaní and Sabater, 2001).

Colonization and microbial community development on sediments depends on the size, roughness and surface area of the grain. Specifically, microbial colonization on sediment grains is proportional to the grain surface area while surface area is inversely proportional to grain size. However, these assumptions could be applied only from coarse grains to fine silt (DeFlaun and Mayer, 1983) because clay particles are very rarely colonized since they are too small and smooth (Mayer and Rossi, 1982). For example, in intertidal environments, non-cohesive sediments (such as sand and gravel) have been described to exhibit greater diversity than cohesive sediments (such as clay and silt). Jesus *et al.* (2009) showed that cohesive sediments were mainly colonized by diatoms, while sandy sediments support three microphytobenthic groups (cyanobacteria, euglenids and diatoms. Besides the size of particles, the degree of roughness plays an important role, increasing microbial colonization as particle roughness increases (Meyer-Reil, 1994). In concordance, it has been reported that microbial biomass is higher on grains with more surface irregularities than smooth grains of the same size (DeFlaun and Mayer, 1983).

In aquatic environments, sediments usually show a profile zonation in depth if they are sufficiently thick, changing from an oxidized zone in the surface sediment to a reduced zone in deeper layers creating anoxic zones. In this sense, sediment thickness creates marked physical and chemical gradients that determine changes in sediment biofilm community (Froelich *et al.*, 1979). In this regard, surface sediments support heterotrophic communities that include more opportunistic species (Arnosti, 2011). Bacterial density and activity is commonly higher in surface sediments and decrease with depth (Taylor *et al.*, 2002). Hydrolysis rates of organic compounds have also been shown to decrease with depth likely due to the influx of fresh organic matter that stimulates microbial activity at the sediment surfaces (Meyer-Reil, 1986; Poremba and Hoppe, 1995; Boetius *et al.*, 2000). In rivers, the sediment hyporheic zone (interface between the river channel and groundwater) offers protection to the sediment biofilm communities against high discharge, desiccations, and extreme temperatures (Brunke and Gonser, 1997). At the hyporheic zone, grain size, sediment shape, and composition determine the porosity and hydraulic conductivity of the sediment, and influence most physical and chemical processes (Boulton *et al.*, 1998).

Nutrients are adhered to sediment particles and are available for the community living in them. In this sense, river sediment acts as a sink of nutrients in the water column (Wetzel, 1996) through nutrient uptake and deposition (Dodds, 2003). However, non-cohesive sediments (sand) have higher solute transport due to high resuspension and mixing (Ehrenhauss *et al.*, 2004), as this type of sediments have lower nutrient concentrations than cohesive sediments (Underwood, 2010).

Dead plant material

Biofilm developing on dead plant material such as wood and leaves is named *epixylic biofilm* or *epixylon* (Vadeboncoeur and Steinman, 2002). In this case, the substratum (i.e. dead organic matter surfaces) represents both a physical support medium for microbial colonization and a source of organic matter for microbes (Golladay and Sinsabaugh, 1991).

The relevance of the different organisms forming epixylic biofilms is markedly different to other biofilm types. Taxa from several fungal phyla such as Chytridiomycota, Zygomycota, Ascomycota and Basidiomycota, are the principal microbial groups growing on dead plant material, primarily due to their ability to degrade lignocellulose compounds, mainly constituent of plant tissues (Nikolcheva and Bärlocher, 2004). However, bacteria, archaea, algae and metazoans also form this kind of biofilms (Manerkar *et al.*, 2008; Gaudes *et al.*, 2009; Duarte *et al.*, 2010; Danger *et al.*, 2013).

This particular type of attached microbial community is not commonly classified as a biofilm by scientists working on decaying plant material, since colonizing microorganisms (mainly fungi) not only adhere just to the surface of dead plants but also enter and degrade the plant tissue. Fungi are able to penetrate leaves with their hyphae, whereas bacteria, archaea and algae attach to the external surfaces or colonize inside the leaves in association with hyphae growth (Baschien *et al.*, 2009). However, epixylic biofilms share two key characteristics with all aquatic biofilms: (1) microorganisms in a sessile form are aggregated and interact among them, and (2) extracellular material is produced for attachment between cells and/or to the substratum.

Microorganisms growing on dead plant material are influenced by internal and external factors (Webster and Benfield, 1986). Internal ones mainly refer to dead organic matter quality, such as cuticle toughness, nutrient content (particularly nitrogen and phosphorus), and the amount of less palatable substances such as lignin, or chemical inhibitors (Meentemeyer, 1978; Ostrofsky, 1997; Cornwell *et al.*, 2008). The quality of organic matter can vary significantly due to several reasons, such as the part of the plant being decomposed, plant taxonomic differences (Ostrofsky, 1997), intraspecies genetic variations (LeRoy *et al.*, 2007), phenological status (Kochi and Yanai, 2006) or plant condition (Lecerf and Chauvet, 2008). However, it is generally observed that slower decomposition occurs in dead plant material with high lignin content levels, low nitrogen or phosphorus concentration or greater toughness (e.g. Triska and Sedell, 1976; Gessner and Chauvet, 1994; Martínez *et al.*, 2013). External factors are environmental variables such as water temperature, nutrient concentration or flow (Webster and Benfield, 1986; Tank *et al.*, 2010). The rate of decomposition of plant organic matter is determined by the effect of interaction between the quality of plant material and the environmental variables (e.g. LeRoy and Marks, 2006). Additionally, decomposition velocity might be very important for the developing biofilm since slower decomposition enables the organic substrate to remain longer in the system so consequently greater biofilm biomass development is achieved, as occurs in wood (Golladay and Sinsabaugh, 1991).

Living plants

Biofilm growing on living plants, such as that developing on macrophytes providing a relatively steady substratum, is named *epiphytic* biofilm (Vadeboncoeur and Steinman, 2002). Submersed plants and macro algae, living in wetlands and littoral zones, are a favourable habitat for microbial growth. Aquatic plants present vast potential for microbial colonization and are frequently covered by a dense growth of algae, bacteria and other organisms favouring biofilm formation (Carpenter and Lodge, 1986). However, in deeper parts of the plant, bacteria might be favoured due to light reduction produced by the host plant, which affects algae development. Epiphytic bacteria biofilms have been found to be dominated by the bacterial group Cytophaga–Flavobacteria–Bacteroidetes and Alpha-proteobacteria in marine (Burke *et al.*, 2011) and freshwater environments (Hempel *et al.*, 2008). Moreover, high densities of nitrifying and denitrifying bacteria have been reported on submersed macrophytes (Körner, 1999).

Plant and epiphytic biofilm interact in many ways, and the interaction might be both synergic and antagonistic. Biofilm provides the macrophytes with organic compounds and carbon dioxide, and also mediates nutrient uptake and enhances nutrient recycling (Wetzel, 1993; Eriksson and Weisner, 1999). In addition, some bacterial species produce compounds against fouling organisms (Rao *et al.*, 2006), and some species enhance plant growth (Marshall *et al.*, 2006). In return, plants provide a substrate for biofilm formation and exude organic compounds and gases such as methane from the root zone, which are used by some biofilm bacteria (Heilman and Carlton, 2001). Consequently, different plant species, plant part and environmental conditions might influence epiphytic biofilms, determining changes in bacterial densities and community composition (Hempel *et al.*, 2008).

However, the interaction between the plant and the epiphytic biofilm is not always positive. Negative affectation of biofilms on submerged plants could arise from increased shading by thick biofilms and potentially also from pathogenic bacteria present in the biofilm (Underwood, 1991). Moreover, it has been shown that, in some cases, excessive biofilm growth decreases the exchange of nutrients, and reduces photosynthesis and plant growth (Sand-Jensen and Søndergaard, 1981; Asaeda *et al.*, 2004). Plants can also exert negative effects on epiphytic bacteria due to the release of secondary compounds such as polyphenols, which may inhibit bacterial growth and attachment (Van Donk and Van de Bund, 2002).

Suspended aggregates

Suspended aggregates made of microorganisms, organic and inorganic particles, are highly fragile structures suspended in fresh and sea water and are usually named *lake snow*, *river snow* or *marine snow*. Aggregates typically occur during bloom periods after an increased input of nutrients (De Beer and Stoodley, 2006)

Marine snow consists of aggregates of diatoms, filamentous cyanobacteria, bacteria, protozoans, zooplankton carcasses, abandoned larvacean houses, faecal pellets, macrophyte detritus, clay and silt minerals, calcite and other particles from the surrounding water, glued together in a polymeric matrix released from phytoplankton and bacteria. Anoxic conditions may exist within the snow particle so that diverse aerobic and anaerobic microbes colonize different niches (Alldredge, 1998). Similar aggregates are formed in some water treatment systems where they are called *granules* (see Chapter 12).

Many complex physical, chemical, biological and specific microbial processes are involved in the formation of aggregates (Simon *et al.*, 2002). Aggregation is a complex process and is controlled mainly by particle density, size and shape, settling velocity and the surface properties of the particle (Chen *et al.*, 1994; Simon *et al.*, 2002). Furthermore, porosity is an important factor controlling the sinking rate of aggregate, the flux of water through the aggregate moving relative to the surrounding water, and the flux of nutrients to and from the microorganisms colonizing the surface of the aggregate (Alldredge and Silver, 1988; Ploug, 2001).

Snow particles or suspended aggregates are ubiquitous in the water column, and sink rapidly. For this reason, suspended aggregates play a key role in transporting organic matter from surface water layers to the benthos. Due to their important role in the carbon cycle, aggregates and their associated microbial communities are a major focus of investigation (Simon *et al.*, 2002). Furthermore, aggregates are frequently identified as hotspots of carbon remineralization and microbial activities, including enzyme activities (Smith *et al.*, 1992; Grossart *et al.*, 2007). Bacterial abundance on suspended aggregates is around 10^8 bacteria/ml, which is 100-fold higher than in bulk water (Grossart *et al.*, 2007). Such aggregates are amply described in marine environments (marine snow), while few authors have studied them in river and lakes (river snow, lake snow).

Man-made surfaces

Biofilms may also form on a wide variety of artificial surfaces such as industrial or potable water system piping, where biofouling may occur by deposition and the growth of bacterial cells or flocs. In drinking water distribution systems Gram-positive bacteria and Alpha- Beta- and Gamma-proteobacteria are the dominating groups (Tokajian *et al.*, 2005). In marine environments, man-made structures such as oil and gas installations, aquaculture nets, as well as ship hulls, can also provide surfaces for attachment (Flemming, 2002). In this sense, the attachment of diatoms along with bacteria on man-made structures constitutes a major problem on artificial structures immersed in the marine environment (Salta *et al.*, 2013).

The formation of attached microbial communities to man-made surfaces follows similar sequential steps to that observed for inert natural substrates (such as rock and sediments). First, surface is conditioned with a film of polysaccharides and proteins, then pioneer planktonic cells attach to the surface and progressively other microbes arrive and biological adhesion continues to form microcolonies until the mature biofilm develops (Watnick and Kolter, 2000). Generally, biofouling can be prevented by a combination of pre-treatment (by reducing microorganism and nutrient concentrations) and preventative chemical cleaning (Guo *et al.*, 2012). In addition, there are various factors that affect biofouling formation such as solid surface material, feed water characteristics and operational conditions (mainly involving temperature, dissolved oxygen and flow velocity) (Le-Clech *et al.*, 2006).

The solid and inert man-made surface may have several characteristics that are important in the attachment process. Characklis and Marshall (1989) reported greater microbial colonization with increased surface roughness in artificial substrates. Different properties of the surface material may also exert a strong influence on the rate and extent of attachment. Most investigators have found that microorganisms attach more rapidly to hydrophobic, non-polar surfaces such as Teflon and other plastics than to hydrophilic materials such as glass or copper pipes (Lehtola *et al.*, 2005; Fletcher and Loeb, 1979; Pringle and Fletcher, 1983). In general, attachment will occur more easily on surfaces that are rougher, more hydrophobic, and coated by surface 'conditioning' films (Cooksey and Wigglesworth-Cooksey, 1995).

Biofilm responses to environmental conditions

Biofilms are present in all aquatic environments including freshwater ecosystems (streams, rivers, lakes and reservoirs), marine environments (tidal, intertidal, and deep sea), wetlands (swamps, marshes, bogs and fens), and man-made structures (such as those associated with drinking water distribution systems). However, depending on the particular characteristics of each environment, biofilms are determined by a specific set of environmental factors (Fig. 1.2). Some characteristics of the water surrounding the biofilm such as temperature, oxygen concentration, pH and nutrient levels commonly affect all aquatic environments. Other factors are only relevant in some environments and, as an example, flow is especially relevant in rivers and in drinking water pipes, while pressure and tides are especially relevant in marine environments (Fig. 1.2).

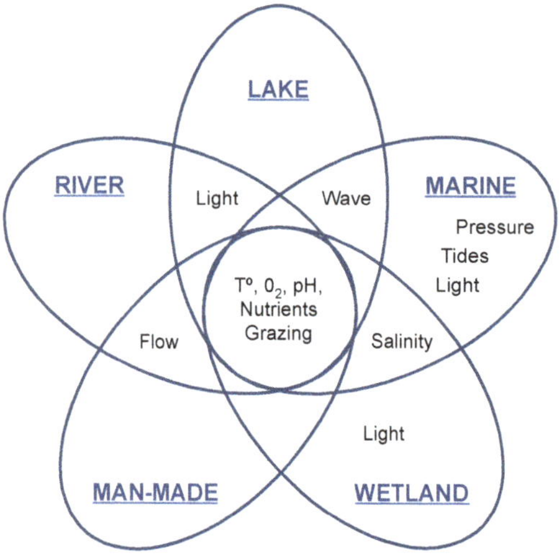

Figure 1.2 Principal factors affecting biofilm development in natural and man-made environments considered in this chapter.

Generally, rivers and streams are characterized by their unidirectional flow along a continuum, where factors such as sediment load, nutrient status, temperature, pH and water velocity, as well as the input of autochthonous and allochthonous carbon sources, vary from upstream to downstream (Koetsier *et al.*, 1997; Battin *et al.*, 2001; Winter *et al.*, 2007). Lakes and reservoirs have predictable vertical gradients of temperature, light, oxygen and pH, with cold, dark and anoxic conditions in the deepest zone (Lampert and Sommer, 2007). Wetlands are shallow environments (less than 5 m deep), which vary greatly in size and shape, and present standing water bodies with characteristic vegetation. These environments may contain water permanently, seasonally or occasionally and water might range from being fresh to salty (Hart *et al.*, 1990). Benthic marine environments can be divided in three main zones (intertidal, subtidal and deep-sea), each of them with its own habitat characteristics, which in turn influence the benthic biofilm community established (Chiu *et al.*, 2006). Specifically, the intertidal zone is highly dependent on tidal effects (Nybakken and Wallace, 1992); the subtidal zone is greatly influenced by wave action (Huzarska, 2013), and the deep-sea floor is characterized by having a disphotic zone with high pressures (Siebenaller and Somero, 1978). Finally, drinking water distribution systems and surfaces in marine environments such as boats are usually under dark and oligotrophic conditions (Kelly *et al.*, 2014).

As mentioned above, the environment plays a significant role in determining the function and structure of biofilms, all of them related with the colonizing substrata (Fig. 1.1) (Stoodley *et al.*, 2002). Some environmental factors are physical (flow, light, temperature, oxygen, salinity) and others chemical (pH, nutrient availability, organic matter composition), but biological aspects such as grazing also determine biofilm development (McIntyre *et al.*, 1996; Sabater *et al.*, 2002; Burns and Ryder, 2001). General aspects of the environmental factors affecting biofilms are reviewed below.

Light

Light is a fundamental variable affecting biofilms since light availability is a primary factor in determining the proportion of photosynthetic organisms in biofilms (Guasch and Sabater, 1995; Hill, 1996; Romaní, 2010). However, benthic light environments are rather variable, and light intensity may regulate benthic algal biomass as a function of depth, turbidity and canopy development in riparian zones (Cahoon and Nearhoof, 1999; Hill, 1996). Some species are capable of tolerating conditions of low light and resist the stresses of light deprivation (Peterson, 1996), while in some environments such as deep waters in marine environments and lakes, photosynthetic organisms are absent from biofilm communities (Lalli and Parsons, 1993).

Attenuation of light by the water column is greater in lakes, sea and deep wetlands than in running waters, but turbidity from silt and other inorganic particles substantially reduces light penetration in many rivers, reservoirs, wave-swept lakes, and estuaries. In lakes, phytoplankton abundance can control the quantity and quality of light that reaches benthic algae, resulting in an inverse relationship between phytoplankton biomass and benthic algal biomass (Hill, 1996). In small streams, light availability is mainly determined by riparian vegetation (Guasch and Sabater, 1995; Roberts *et al.*, 2004; Proia *et al.*, 2012). In a marine environment, light intensity changes in correlation to the different tidal levels.

Biofilms exposed to high light conditions usually develop a significant biomass of phototrophic organisms and thus also display photosynthetic activity. They are thicker and more structured, with a higher C/N ratio and a greater contribution of the EPS than those developing in the dark (Romaní *et al.*, 2004b). In these conditions, microbes uptake organic compounds released by algae (high-quality fresh molecules) as they are less dependent on the organic molecules available in the flowing water (Romaní *et al.*, 2004b). Photosynthetic activity promoted by light may also stimulate extracellular enzyme activity of biofilms growing on both inert and organic substrates (Francoeur *et al.*, 2006). In contrast, in low light conditions, a thinner biofilm is usually built, with lower algal biomass and, thus, microbes mainly use molecules from the dissolved organic matter (DOM) and particulate organic matter (POM) pool. These situations obviously also depend on other factors such as inorganic and organic matter available and flow velocity (Romaní, 2010).

Complementarily, some aspects related to the type of substrate also determine light intensity reaching the biofilm. In rocks, substrata texture changes light availability due to a greater area in rough surfaces, since the same amount of light energy is distributed across a larger area (Lambert's law). Moreover, direct light intensity changes with whole-substratum orientation and light direction due to microscale slope angle (Murdock and Dodds, 2007). Thus, biofilms located on the rock sides receive significantly less direct light than cells on the top when the light source is directly overhead (Hill, 1996). Likewise, light decreases rapidly in depth within sediment. In natural sediments fine particles and the detritus stored inside the cavities between larger particles reduces light penetration (Brunke and Gonser, 1997). In non-cohesive sediments (i.e. sand), light reaches a few centimetres in depth (Kühl *et al.*, 1994; Kühl and Jorgensen, 1995), while in cohesive sediments (i.e. silt), light is restricted to the top 2 mm (Kühl *et al.*, 1994).

In contrast, high light may lead to photoinhibition, defined as a reduction in the photosynthetic rate of the whole integrated biofilm (Underwood, 2002). Phototrophic organisms inhabiting high light environments (such as open sites in clear streams or tidal zones) have developed mechanisms that reduce the potentially damaging effects of high irradiances. Therefore, algae exposed to high irradiances accumulate carotenoids, which may diminish photoinhibition by absorbing photons in excess, and experience reduced quantum yields, or produce sheath pigments such as scytonemin (Hill, 1996). Similarly, light might also have negative consequences on heterotrophic organisms. It has been observed that UV radiation may limit microbial development in epixylic biofilms, as fungi are more sensitive than bacteria (Denward *et al.*, 2001). In addition, denitrification in epiphytic biofilms may be inhibited under light conditions (Eriksson, 2001). In marine ecosystems, it has been demonstrated that physiological responses to excessive insolation depress the growth of epilithic biofilms, leading to a decline in productivity during summer (Lamontagne *et al.*, 1989; Thompson *et al.*, 2004).

Temperature

Water temperature directly regulates metabolic rates, microbial growth and activity, and consequently plays an essential role in biofilm function such as organic matter processing, primary production and nutrient cycling (Brunke and Gonser, 1997; Dang *et al.*, 2009; Friberg *et al.*, 2013; Ylla *et al.*, 2014). Additionally, temperature variation may lead to alter community structure, extracellular enzyme activity, biofilm metabolism, interspecific relationships and biodiversity in all kinds of biofilms (Romaní and Sabater, 2000; Ferreira and Chauvet, 2011; Romaní *et al.*, 2014). Optimal temperatures for most biofilm range from 10 to 30°C, and higher temperatures induce heat stress and reduce growth (DeNicola and Hoagland, 1996). At the same time, species have a temperature tolerance range and below and above this range their activity is reduced or suppressed, making the response of the whole biofilm to temperature variation difficult to predict (Chauvet and Suberkropp, 1998; Rajashekhar and Kaveriappa, 2000). A rise on temperature may also increase bacterial attachment on man-made surfaces (Donlan, 2002).

Frequency and range of temperature oscillations can deeply affect the structure and function of biofilms, however, this effect varies depending on biofilm type. For example, in epixylic biofilms it was observed that diel temperature oscillations might accelerate microbial decomposer activity by up to 31%; however, under very low temperatures (~3°C) oscillations do not affect organic matter decomposition (Dang *et al.*, 2009). In contrast, in sediments, the influence of diel and seasonal temperature fluctuations are mainly observed in surface biofilms while deep sediments are more stable, with no sudden decrease/increase in temperature (Brunke and Gonser, 1997). Moreover, in intertidal areas, adaptations to temperature are really important because these areas are regularly subjected to aerial temperature variation (Nybakken and Wallace, 1992).

Studies evaluating the impact of expected temperature increase in the next decades have shown that the interaction between temperature and the quality of available organic matter may affect biofilm responses. For example, in epixylic biofilms, microorganisms living on low degradable organic matter are less affected by increasing temperature than those feeding on labile substrates (Bärlocher *et al.*, 2013; Gonçalves *et al.*, 2013). In contrast, in epilithic biofilms some evidence points out that degradation of more bioavailable organic carbon is not affected by temperature increase, while degradation of recalcitrant substances might be enhanced by warming (Ylla *et al.*, 2012).

Although less studied, a general effect of temperature on attached microbial communities is that related to biogeographic variations. In this sense, in epilithic biofilms it has been observed that at small biogeographic scales, bacterial taxon richness is greater at latitudes closer to the equator and reduced at higher altitudes, both factors (latitude and altitude) being inversely related with air temperature (Lear *et al.*, 2013).

Inorganic nutrients

Nutrients are essential resources, and their availability could determine changes in the structure and the function of microbial communities (Larned, 2010). The trophic status of aquatic ecosystems determines the microbial species dominance and microbial community composition of developing biofilms (Coleman and Burkholder, 1994; Duarte *et al.*, 2009).

In epilithic biofilms, autotrophs and heterotrophs can respond differently to nutrient supply. In biofilm heterotrophic functioning, nutrient availability can either directly (by influencing metabolic activities) or indirectly (by effect on primary producers) affect biofilm communities, and this response might change depending on light availability (e.g. Romaní and Sabater, 2000; Rier and Stevenson, 2002). High availability of nutrients can effectively enhance algal biomass and productivity when light is present (e.g. Guasch *et al.*, 1995; Ylla *et al.*, 2007).

Sediments can play an important role as sinks for inorganic nutrients by buffering their release into the water column (Wetzel, 1996; Jarvie *et al.*, 2005). As described by Dodds (2003), epipsammic biofilms may remove phosphorous from the water column, including uptake and sediment deposition. However, in eutrophic lakes, sediments can contribute to eutrophication through phosphorus release into the water column (Ribeiro *et al.*, 2008; Martins *et al.*, 2008).

In attached communities growing on plant detritus, high phosphorus and nitrogen concentrations in water promote biomass accrual, production and respiration of the microbial community, as well as its organic matter degradation activity (Gulis *et al.*, 2008; Suberkropp *et al.*, 2010). However, decomposer activity is only enhanced by nutrients up to a threshold, beyond which it decreases again (Duarte *et al.*, 2009; Woodward *et al.*, 2012). The effect of nutrient concentration on epixylic biofilm is further strongly influenced by the nutrient levels in the main organic substratum where biofilm develops and their balance with nutrients in water. For example, Royer and Minshall (2001) did not find any effect of increasing nutrient on microbial leaf decomposition, apparently due to the lack of nutrient limitation of the studied stream. Ardón and Pringle (2007) observed that biofilm growing on recalcitrant C sources did not show enhanced respiration while biofilm growing on labile C sources was stimulated by nutrients.

However, epiphytic biofilm adapted to high nutrient concentrations may increase its denitrifying capacity about a hundred times that of surface-attached microbes adapted to lower nutrient levels (Eriksson and Weisner, 1996).

Oxygen and pH

Oxygen gradients in aquatic ecosystems promote differential biochemical processes in biofilms, from aerobic respiration at the biofilm surface to methane oxidation, anaerobic respiration and fermentation occurring in zones of oxygen depletion (Brune *et al.*, 2000), commonly found in freshwater and marine sediments. Oxygen levels in water are strongly dependent on microbial activity and physical factors such as water re-aeration and depth. Additionally, human contamination providing a large amount of organic and inorganic nutrients, as observed in sewage effluent, might induce oxygen depletion and anaerobic conditions. In these cases, biofilm activity may cause diurnal fluctuation in pH, oxygen and redox. Reduced redox conditions affect the mobility of trace metals (i.e. Mn, Cu, Zn, Cd, Fe, Hg) in the interstices (Brunke and Gonser, 1997) and acidification is often accompanied by metal dissolution and deposition (Sasaki *et al.*, 2005).

Biofilm biomass production and organic matter mineralization are often considered to be highly dependent on oxygen concentration, with lower rates under anoxic conditions. In sediments, the oxygen content declines with increasing sediment depth affecting the microbial metabolism and diversity (Brunke and Gonser, 1997). In the surface, oxic sediment biofilms are composed of a high diversity of aerobic bacteria while in deeper sediment layers, with low or nil oxygen concentration, biofilms are mainly composed of nitrifiers, sulfate-reducing and methanogenic bacteria (Brune *et al.*, 2000). In contrast, the oxygen produced by photosynthetic algae also has a negative effect on chemical processes such as denitrification, nitrification, metal oxidation and sulfite oxidation in biofilms formed on rock, sediments or living plants (Eriksson, 2001; Glud, 2008). Additionally, a reduction of oxygen concentration in water can affect fungal development (i.e. slow growing and low sporulation rates), and decrease fungal diversity (Solé *et al.*, 2008; Medeiros *et al.*, 2009) reducing the microbial degradation velocity of dead plant material (Pascoal *et al.*, 2003).

Likewise, low pH may also affect biofilms in aquatic ecosystems. pH affects epilithic biofilm functioning and composition. For example, CO_2 utilization during photosynthesis in light biofilms results in an increase in the internal pH and this higher pH favours the removal of metals by precipitation and possibly adsorption (Liehr *et al.*, 1994). In heterotrophic communities, low pH may also slow down organic matter decomposition mediated by microorganisms, which effect is related to a reduced efficiency of pH-sensitive enzymes and changes in community compositions (Simon *et al.*, 2009; Clivot *et al.*, 2013a,b).

Flow

In running water, flow velocity and water level modulate biofilm growth and thickness (Augspurger and Kusel, 2010). Biofilm is disturbed by the mechanical effect of abrasion during high flow periods. For example, disturbances in river systems (such as a flood) reset biofilm succession by removing biomass and clearing substrata for colonization (Webb *et al.*, 2006). Biofilms have some mechanisms to avoid this disturbance as the thickness of extracellular polymeric substances, which contribute to biofilm stability versus flow (Costerton *et al.*, 1995; Gerbersdorf, 2008). Moreover, under flow disturbance, bacteria with higher growth rates can survive on biofilm surficial layers, while those slower growing bacteria are forced to remain at lower biofilm layers (Furumai and Rittmann, 1994).

Two types of flows are relevant to most natural and man-made environments: laminar flow and turbulent flow; both may affect biofilm microbial composition and function. Biofilms under turbulent flow commonly form filamentous 'streamers' since they can oscillate rapidly in the flow (Stoodley *et al.*, 2002). In addition, when increasing the turbulence, the boundary layer between the biofilm and the water column indirectly decreases, and consequently, increases the nutrient supply to the biofilm (De Beer *et al.*, 2006) but decreases biofilm microbial biomass and density (Biggs and Hickey, 1994). In light-grown ephilitic biofilms flow velocity has been shown to stimulate photosynthesis and increase nutrient supply (Stevenson and Glover, 1993; Augspurger and Kusel, 2010; Romaní and Marxsen, 2002). Flow velocity can also affect the biofilm structure and algal composition and its response to higher shear stresses (Graba, 2013). Flow velocity may be relevant during biofilm formation since it is reported that bacterial abundances are higher in young biofilms grown under higher flow velocities than in biofilms grown under slower flow velocities (Hunt and Parry, 1998).

In sediments, flow velocity can influence bacterial activity and solute transport (Fischer *et al.*, 2003; Battin, 2000), but the effect of water movement decreases with sediment depth. High flows affect biofilm biomass by increasing sediment mobility, which leads to the abrasion of the biofilm by suspended sediment and substrate tumbling (Biggs *et al.*, 1999).

In plant detritus, slow flowing water current might reduce microbial respiration, fungal sporulation and biomass, while high flow increases ATP, chlorophyll *a* concentration and the number of aquatic hyphomycete species (Golladay and Sinsabaugh, 1991; Schlief and Mutz, 2009). However, flow changes do not always determine differential microbial decomposer activity (Ferreira and Graça, 2006; Dewson *et al.*, 2007). Similarly, increased photosynthesis and respiration are observed for biofilms growing on living plants under high flow conditions, but denitrification capacity is favoured in stagnant waters (Eriksson, 2001).

In general, in aquatic ecosystems, flow increases biofilm colonization and metabolism up to a certain level when flow destroys the biofilm, this threshold depends on the specific environment. Similarly, in man-made systems, increased flow velocity favours biofilm attachment, if flow disturbance does not exceed critical levels (Donlan, 2002).

Waves and tides

Wave action is an important physical factor in marine shallow water habitats and in some lakes. In these habitats, turbulence produced by wave action keeps inshore water from becoming thermally stratified, and for this reason nutrients in marine shallow water habitats are rarely limited or locked up in a bottom reservoir (Nybakken and Wallace, 1992). Waves and tides determine shear stresses and influence biofilm diversity in lakes and marine systems. Specifically, biofilms that develop under high shear stresses are less diverse than those developing under lower shear stresses (Howell, 2009). This is mainly due to a reduction of biofilm maturation and maintenance in biofilms in the early stages of development occurring under high shear stress (Rochex *et al.*, 2008).

In marine systems, tides are relevant, especially in intertidal zones, because they determine the community's ability to tolerate immersion in air. Many intertidal species are well-adapted to tidal situations, so they are able to be quiescent when the tide is out and resume normal activity when the tide is in (Nybakken and Wallace, 1992). Tidal levels have been observed to cause seasonal patterns in the abundance of cyanobacteria, being maximal on the lower shore in summer and on the upper shore during winter (Thompson *et al.*, 2004). In addition, an abundance of photosynthetic microbiota can be greater on the upper shore than on the lower shore (Aleem, 1950; Castenholz, 1963; Thompson *et al.*, 2004, 2005). Additionally, microbial mats that develop in intertidal environments are exposed to water movement and wind as eroding forces. Mat formation increases the stability of sediments cementing individual sand grains to each other and increasing the critical friction velocity (Van Gemerden *et al.*, 1989).

However, waves and currents also exist in lakes and produce sediment suspension and transport from shallower to deeper areas, a process called sediment focusing. By this process, fine-grained sediments tend to concentrate in deeper parts of the lake. Wave action and currents may in part control nutrient transport and availability (Mackay *et al.*, 2011). Additionally, waves and wind as physical factors may affect the formation of lake snow aggregates which is mostly dependent on wind-induced turbulences (Grossart and Simon, 1993). In estuaries and tidal zones affected by shallow seas, macroaggregates are usually smaller and more abundant than in lakes and in open sea due to high shear rates leading to disaggregation and high resuspension rates (Chen *et al.*, 1994; Zimmermann, 1997).

Pressure

Pressure is a factor that greatly influences marine systems where it ranges from 20 to more than 1000 atm. Hydrostatic pressure is correlated to an increase in depth (Carney, 2005). Pressure increases by 10^2 kPa every 10 metres in depth. This means that pressure plays a major role in organisms' adaptation to the deep-sea environment (increased hydrostatic pressure but also low temperature) (Siebenaller and Somero, 1978). Pressure affects the morphology of bacteria (Zobell and Oppenheimer, 1950) and it may have mutagenetic effects (Zobell and Morita, 1957). Some deep-sea bacteria appear to be obligate barophiles (adapted to high pressure). However, even under optimum conditions for reproduction, barophilic bacteria in sediment found at great depths are highly specialized and are typically slower growing than shallow-water benthic species and communities (Cowie, 2010).

Salinity

Salinity variation is an important factor for biofilms in marine environments and estuaries as well as in some characteristic continental aquatic ecosystems (salty wetlands, lagoons or rivers). Marine bacteria grow optimally at a salt concentration between 3.3% and 3.5% and do not develop or develop poorly when there is no NaCl in the water. Variations in salinity influence bacterial growth, metabolism and cell attachment on artificial substrates (Chiu *et al.*, 2006). Gradients in salinity are especially important in intertidal biofilms and estuarine biofilms. Intertidal biofilms may be exposed to low/high tide alternations that influence salinity concentration due to lower/higher dilution (Nybakken and Wallace, 1992). Depending on the season, salinity has a stronger effect on marine biofilms. Chiu *et al.* (2006) observed that in summer, community composition on biofilms was affected by salinity, while in winter it was more affected by temperature than by salinity.

Estuaries are dynamic ecosystems located at the interface between marine and terrestrial environments, and are exposed to salinity gradients of between 1.0% and 3.2%, although salinity at depth might be remarkably constant (Dunn *et al.*, 2008). Salinity gradient in estuaries might regulate nitrification since the denitrifier community seems to be dominated by halotolerant bacteria (Magalhães *et al.*, 2005). Similarly, algal community composition is also influenced by the gradient of salinity. Underwood *et al.* (1998) analysed the diatom assemblages present in epipsamic biofilms in estuaries and showed *Navicula gregaria* and *N. phyllepta* to be abundant at oligo- and mesohaline sites respectively, while *Pleurosigma angulatum* and *Plagiotropis vitrea* were abundant at polyhaline sites.

Under natural conditions, some freshwater ecosystems are also influenced by increasing salinity concentrations, mainly because of evaporation combined with intrusions of groundwater and the mineral constitution of some catchments. The available data suggest that freshwater biota will be adversely affected as salinity exceeds 1 g/l, however the sensitivity of prokaryote and eukaryote biofilm components seems to be different (Nielsen *et al.*, 2003). Species replacement in bacterial communities has been observed in salinized freshwater systems, but the different communities possess similar metabolic capabilities (Hempel *et al.*, 2008; Hart *et al.*, 1990). Conversely, the majority of freshwater algal taxa do not appear to tolerate increasing salinity (Nielsen *et al.*, 2003). In addition, the breakdown of organic matter is seen to be reduced under high salinity concentration, which suggests that fungi have low tolerance of hyper saline environments (Schäfer *et al.*, 2012).

Grazing

Biofilm is strongly regulated by herbivores in aquatic ecosystems (Steinman, 1996; Thompson *et al.*, 2004). According to Lowe (1996), biotic disturbance is usually associated to the activity of grazers in the biofilm community. Grazers impact biofilm communities both by direct consumption of algal cells and by dislodgment of cells from the substratum (Lowe, 1996). Primary consumers (herbivores) in freshwater ecosystems span many taxonomic groups, but insects, molluscs (especially gastropods), crustaceans, and fish appear to be particularly important (Lamberti, 1996; Burns and Ryder, 2001). Concretely, the reduction of algal biomass due to herbivory has been demonstrated for a variety of grazer types (Steinman, 1996). Grazing maintains microbial biofilms at early successional stages and, when it is reduced, this invariably leads to an increase in microalgal and then macroalgal biomass, suggesting that top-down regulation is a key factor.

This grazing effect has been observed in marine environments (Thompson *et al.*, 2004; Firstater *et al.*, 2012), estuaries (Sullivan and Currin, 2000), rivers (Feminella and Hawkins, 1995; Lamberti, 1996; Steinman, 1996), lakes (Lowe, 1996), and in man-made systems (Esselink *et al.*, 2002).

Bacteria within biofilms may also be subjected to predation by free-living protozoa in freshwater systems (Hahn and Höfle, 2001). At the same time, because biofilms are a valuable food resource in aquatic ecosystems, their abundance may influence the physiological fitness of herbivores. Moreover, herbivore growth rates are frequently controlled by biofilm availability (Lamberti, 1996),

In plant detritus biofilms, invertebrates play a dual role, some feeding on biofilm by grazing on plant surface, but others consuming dead plant matter by shredding plant tissue. In a detrital system, synergistic and antagonistic interactions have been observed between aquatic invertebrates and microorganisms present on dead plant material (Canhoto and Graça, 2008). Microbes are responsible for increasing plant tissue nutrient content, favouring invertebrate consumption (Bärlocher and Kendrick, 1975; Foucreau *et al.*, 2013), and some insect larvae can use microbial enzymes inside their guts (Canhoto and Graça, 2008). Meanwhile, invertebrate feeding can disrupt physical barriers of dead plants, enhance their surface-to-volume ratio, and increase basal resources by egestion and excretion, favouring microbial performance (Sabetta *et al.*, 2000; Canhoto and Graça, 2008; Diaz Villanueva *et al.*, 2012). On the other hand, invertebrates and microbes can compete for plant food resources (Bärlocher, 1980); and invertebrates also predate bacterial cells, mycelia or fungal spores (Suberkropp *et al.*, 1983; Suberkropp and Wallace, 1992; Hahn and Höfle, 2001).

However, grazing pressure on biofilms is further modulated by specific environmental conditions favouring or limiting the development of grazers. Related to this, in estuarine and hypersaline environments (Wieland and Kühl, 2000), periodic drying (Lassen *et al.*, 1992) or high temperatures (Castenholz, 1984) can limit meio- and macrofauna growth, thus reducing grazing activity. Moreover, sea/air temperature regulates the foraging intensity of the principal molluscan grazers and limpets (Thompson *et al.*, 2004). A specific example of excluding grazing influence is that of microbial mats which are mainly developed under conditions that exclude or limit grazing pressure (Cohen, 1989). Since mat building depends on the suppression of invertebrate grazers, extreme salinity, temperature and drought conditions allow the development of microbial mats (Connor *et al.*, 1982; Awramik, 1984).

Conclusions

The main conclusions drawn from this review are summarized in the following points:

- Through the bibliography, biofilm is also referred to as: *periphyton, aufwuchs, microphytobentos, benthos, haptobenthos* and *herpobenthos*. Moreover, regarding to the type of substrate where the biofilm develops it is also known as *epilithic, epipsammic, epipelic, hiporheic, epixylic*, or *epiphytic* biofilm. The specific aquatic suspended aggregates behaving similarly to biofilm are called *marine snow, lake snow, river snow, aggregates* and *granules*. In man-made systems the most common term is *biofilm* and the words *biofouling* or *bioclogging* are mainly used when biofilm growth has a 'negative' effect.
- Substrata in natural and man-made environments determine biofilm structure and metabolism due to the particular characteristics of each type of surface. Therefore the roughness, size and chemical composition, as well as the possible interaction between attached community and substratum give the biofilm a particular configuration, favouring the development of either prokaryotes or eukaryotes (see Fig. 1.2 for a summary of most common biofilms).
- Biofilm characteristics are highly variable depending on the chemical, physical and biological conditions found in each particular aquatic environment. For example, photosynthetic microorganisms are promoted in biofilms growing under high light conditions, and increased temperature or nutrient conditions enhance the rate of microbial processes. The effect of most factors (light, temperature,

nutrients, oxygen, pH, flow, pressure, salinity, grazing) depends on the particular tolerance range and sensitivity of species to those factors. However, environmental variables also interact among them and it is sometimes difficult to predict accurate single responses.

- Despite the possible diverse configurations of microbial communities forming biofilms, the biofilm lifestyle is mainly characterized by: (1) multicellular association of different microbial groups (bacteria, archaea, viruses, algae, fungi, cyanobacteria, protozoa, small metazoa), (2) microorganisms are aggregated, embedded by an extracellular polymeric matrix and interact between them; and (3) they are associated to inorganic or organic substrata.

References

Adey, W., Luckett, C., and Jensen, K. (1993). Phosphorus removal from natural waters using controlled algal production. Restor. Ecol. *1*, 29–39.

Aleem, A.A. (1950). Distribution and ecology of British marine littoral diatoms. J. Ecol. *38*, 75–106.

Alldredge, A., and Silver, M.W. (1988). Characteristics, dynamics and significance of marine snow. Prog. Oceanogr. *20*, 41–82.

Alldredge, A. (1998). The carbon, nitrogen and mass content of marine snow as a function of aggregate size. Deep-sea Res. *45*, 529–541.

Anderson-Glenna, M.J., Bakkestuen, V., and Clipson, N.J.W. (2008). Spatial and temporal variability in epilithic biofilm bacterial communities along an upland river gradient. Microb. Ecol. *64*, 407–418.

Ardón, M., and Pringle, C.M. (2007). The quality of organic matter mediates the response of heterotrophic biofilms to phosphorus enrichment of the water column and substratum. Freshwater Biol. *52*, 1762–1772.

Arnosti, C. (2011). Microbial extracellular enzymes and the marine carbon cycle. Ann. Rev. Mar. Sci. 3, 401–425.

Asaeda, T., Sultana, M., Manatunge, J., and Fujino, T. (2004). The effect of epiphytic algae on the growth and production of *Potamogeton perfoliatus* L. in two light conditions. Environ. Exp. Bot. *52*, 225–238.

Augspurger, C., and Küsel, K. (2010). Flow velocity and primary production influences carbon utilization in nascent epilithic stream biofilms. Aquat. Sci. 72, 237–243.

Augspurger, C., Gleixner, G., Kramer, C., and Küsel, K. (2008). Tracking carbon flow in a 2-week-old and 6-week-old stream biofilm food web. Limnol. Oceanogr. *53*, 642–650.

Awramik, S.M. (1984). Ancient stromatolites and microbial mats. In Microbial Mats: Stromatolites, Y. Cohen, R.W. Castenholz, and H.O. Halvorson, eds. (Alan R. Liss Inc., New York, NY, USA), pp. 1–22.

Azam, F., and Cho, B.C. (1987). Bacterial utilization of organic matter in the sea. In Ecology of Microbial Communities, T. Brock, ed. (Cambridge University Press, Cambridge, UK), pp. 261–281.

Azam, F. (1998). Microbial control of oceanic carbon flux: the plot thickens. Science *280*, 696–696.

Bärlocher, F. (1980). Leaf-eating invertebrates as competitors of aquatic hyphomycetes. Oecologia *47*, 303–306.

Bärlocher, F., and Kendrick, B. (1975). Leaf-conditioning by microorganisms. Oecologia *20*, 359–362.

Bärlocher, F., Kebede, Y.K., Gonçalves, A.L., and Canhoto, C. (2013). Incubation temperature and substrate quality modulate sporulation by aquatic hyphomycetes. Microb. Ecol. *66*, 30–39.

Baschien, C., Rode, G., Böckelmann, U., Götz, P., and Szewzyk, U. (2009). Interactions between hyphosphere-associated bacteria and the fungus Cladosporium herbarum on aquatic leaf litter. Microb. Ecol. *58*, 642–650.

Battin, T.J. (2000). Hydrodynamics is a major determinant of streambed biofilm activity: from the sediment to the reach scale. Limnol. Oceanogr. *45*, 1308–1319.

Battin, T.J., Wille, A., Sattler, B., and Psenner, R. (2001). Phylogenetic and functional heterogeneity of sediment biofilms along environmental gradients in a glacial stream. Appl. Environ. Microbiol. *67*, 799–807.

Baynes, T.W. (1999). Factors structuring a subtidal encrusting community in the southern Gulf of California. Bull. Mar. Sci. *64*, 419–450.

Bergey, E.A. (2008). Does rock chemistry affect periphyton accrual in streams? Hydrobiologia *614*, 141–150.

Berry, D., Xi, C., and Raskin, L. (2006). Microbial ecology of drinking water distribution systems. Curr. Opin. Biotech. *17*, 297–302.

Biggs, B.J.F., and Hickey, C.W. (1994). Periphyton responses to a hydraulic gradient in a regulated river in New Zealand. Freshwater Biol. *32*, 49–59.

Biggs, B.J.F., Smith, R.A., and Duncan, M.J. (1999). Velocity and sediment disturbance of periphyton in headwater streams: biomass and metabolism. J. N. Am. Benthol. Soc. *18*, 222–241.

Blinn, D.W., Fredricksen, A., and Korte, V. (1980). Colonisation rates and community structure of diatoms on three dif- ferent rock substrata in a lotic system. Br. Phycol. J. *15*, 303–310.

Boetius, A., Ferdelman, T., and Lochte, K. (2000). Bacterial activity in sediments of the deep Arabian Sea in relation to vertical flux. Deep-Sea Res. PT II *47*, 2835–2875.

Bottjer, D.J., Schubert, J.K., and Droser, M.L. (1996). Comparative evolutionary palaeoecology: assessing the changing ecology of the past. Geol. Soc. London, Spec. Publ. *102*, 1–13.

Boulton, A.J., Findlay, S., Marmonier, P., Stanley, E.H., and Valett, H.M. (1998). The functional significance of the hyporheic zone in streams and rivers. Ann. Rev. Ecol. Sys. *29*, 59–81.

Branda, S.S., Vik, S., Friedman, L., and Kolter, R. (2005). Biofilms: the matrix revisited. Trends Microbiol. *13*, 20–26.

Brune, A., Frenzel, P., and Cypionka, H. (2000). Life at the oxic–anoxic interface: microbial activities and adaptations. FEMS Microbiol. Rev. *24*, 691–710.

Brunke, M., and Gonser, T. (1997). The ecological significance of exchange processes between rivers and groundwater. Freshwater Biol. *37*, 1–33.

Burke, C., Thomas, T., Lewis, M., Steinberg, P., and Kjelleberg, S. (2011). Composition, uniqueness and variability of the epiphytic bacterial community of the green alga *Ulva australis*. ISME J. *5*, 590–600.

Burns, A., and Ryder, D.S. (2001). Potential for biofilms as biological indicators in Australian riverine systems. Ecol. Man. Rest. *2*, 53–64.

Cahoon, L.B., and Nearhoof, J.E. (1999). Sediment grain size effect on benthic microalgal biomass in shallow aquatic ecosystems. Estuaries *22*, 735–741.

Canhoto, C., and Graça, M.A.S. (2008). Interactions between fungi and stream invertebrates: back to the future. In Novel Techniques and Ideas in Mycology, Sridhar, S., Barlocher, F., and Hyde, K.D., eds. (Fungal Diversity Press, Hong Kong), pp. 1–22.

Carney, R.S. (2005). Zonation of deep biota on continental margins. Oceanogr. Mar. Biol. Ann. Rev. *43*, 211–278.

Carpenter, S.R., and Lodge, D.M. (1986). Effects of submersed macrophytes on ecosystem processes. Aquat. Bot. *26*, 341–370.

Castenholz, R.W. (1963). An experimental study of the vertical distribution of littoral marine diatoms. Limnol. Oceanogr. *8*, 450–462.

Castenholz, R.W. (1984). Composition of hot spring microbial mats: a summary. In Microbial Mats: Stromatolites, Cohen, Y., Castenholz, R.W., and Halvorson, H.O., eds. (Alan R. Liss Inc., New York, NY, USA), pp. 39–58.

Characklis, W.G., and Marshall, K.C. (1989). Biofilms a basis for an interdisciplinary approach. In Biofilms, Characklis, W.G., and Marshall, K.C., eds. (John Wiley & Sons Inc., New York, NY, USA), pp. 3–15.|

Chauvet, E., and Suberkropp, K. (1998). Temperature and sporulation of aquatic hyphomycetes. Appl. Environ. Microbiol. *64*, 1522–1525.

Chen, S., Eisma, D., and Kalf, J. (1994). In situ size distribution of suspended matter during the tidal cycle in the Elbe estuary. Neth. J. Sea. Res. *32*, 37–48.

Chiu, J.M.Y., Thiyagarajan, V., Tsoy, M.M.Y., and Qian, P.Y. (2006). Qualitative and quantitative changes in marine biofilms as a function of temperature and salinity in summer and winter. Biofilms *2*, 1–13.

Clifford, H.F., Casey, R.J., and Saffran, K.A. (1992). Short-term colonization of rough and smooth tiles by benthic macroinvertebrates and algae (chlorophyll a) in two streams. J. N. Am. Benthol. Soc. *11*, 304–315.

Clivot, H., Cornut, J., Chauvet, E., Elger, A., Poupin, P., Guérold, F., and Pagnout, C. (2013a). Leaf-associated fungal diversity in acidified streams: insights from combining traditional and molecular approaches. Environ. Microbiol. *167*, 2145–2156.

Clivot, H., Danger, M., Pagnout, C., Wagner, P., Rousselle, P., Poupin, P., and Guérold, F. (2013b). Impaired leaf litter processing in acidified streams: learning from microbial enzyme activities. Microb. Ecol. *65*, 1–11.

Cohen, Y. (1989). Photosynthesis in cyanobacterial mats and its relation to the sulfur cycle: a model for microbial sulfur interactions. In Microbial Mats. Physiological Ecology of Benthic Microbial Communities, Cohen, Y., and Rosenberg, E., eds. (ASM Press, Washington, DC, USA), pp. 22–36.

Coleman, V.L., and Burkholder, J.M. (1994). Community structure and productivity of epiphytic microalgae on eelgrass (*Zostera marina* L.) under water-column nitrate enrichment. J. Exp. Mar. Bio. Ecol. *179*, 29–48.

Connor, M.S., Teal, J.M., and Valiela, I. (1982). The effect of feeding by mud snails, on the structure and metabolism of a laboratory benthic algal community. J. Exp. Mar. Biol. Ecol. *65*, 29–45.

Cooksey, K.E., and Wigglesworth-Cooksey, B. (1995). Adhesion of bacteria and diatoms to surfaces in the sea: a review. Aquat. Microb. Ecol. *9*, 87–96.

Cornwell, W.K., Cornelissen, J.H.C., Amatangelo, K., Dorrepaal, E., Eviner, V.T., Godoy, O., Hobbie, S.E., Hoorens, B., Kurokawa, H., Pérez-Harguindeguy, N., *et al.* (2008). Plant species traits are the predominant control on litter decomposition rates within biomes worldwide. Ecol. Lett. *11*, 1065–1071.

Costerton, J.W., Geesey, G.G., and Cheng, K.J. (1978). 'How bacteria stick'. Sci. Amer. *238*, 86–95.

Costerton, J.W., Cheng, K.J., Geesey, G.G., Ladd, T.I., Nickel, J.C., Dasgupta, M., and Marrie, T.J. (1987). Bacterial biofilms in nature and disease. Ann. Rev. Microbiol. *41*, 435–464.

Costerton, J.W., Lewandowski, Z., Caldwell, D.E., Korber, D.R., and Lappin-Scott, H.M. (1995). Microbial biofilms. Annu. Rev. Microbiol. *49*, 711–745.

Cowie, P.R. (2010). Biofouling Patterns with Depth. In Biofouling, S. Dürr, and J.C. Thomason, eds. (Wiley-Blackwell, New Jersey), pp. 87–122.

Dang, C.K., Schindler, M., Chauvet, E., and Gessner, M.O. (2009). Temperature oscillation coupled with fungal community shifts can modulate warming effects on litter decomposition. Ecology *90*, 122–131.

Danger, M., Cornut, J., Chauvet, E., Chavez, P., Elger, A., and Lecerf, A. (2013). Benthic algae stimulate leaf litter decomposition in detritus-based headwater streams: a case of aquatic priming effect? Ecology *94*, 1604–1613.

Davey, M.E., and O'Toole, G.A. (2000). Microbial biofilms : from ecology to molecular genetics. Microbiol. Mol. Biol. Rev. *64*, 847–867.

De Beer, D., Stoodley, P., and Lewandowski, Z. (1996). Liquid flow and mass transport in heterogeneous biofilms. Water Res. *30*, 2761–2765.

De Beer, D., and Stoodley, P. (2006). Microbial biofilms. In The Prokaryotes, Dworkin, M., Falkow, S., Rosenberg, E., Schleifer, K.H., and Stackebrandt, E., eds. (Springer, New York, NY, USA), pp. 904–937.

Decho, A.W. (2000). Microbial biofilms in intertidal systems : an overview. Cont. Shelf Res. *20*, 1257–1273.

DeFlaun, M.F., and Mayer, L.M. (1983). Relationships between bacteria and grain surfaces in intertidal sediments. Limnol. Oceanogr. *28*, 873–881.

DeNicola, D.M., and Hoagland, K.D. (1996). Effects of solar spectral irradiance (visible to UV) on a prairie stream epilithic community. J. N. Am. Benthol. Soc. *15*, 155–169.

Denward, C.M.T., Anesio, A.M., Graneli, W., and Tranvik, L.J. (2001). Solar radiation effects on decomposition of macrophyte litter in a lake littoral. Arch. Für Hydrobiol. *152*, 69–80.

Dewson, Z.S., James, A.B.W., and Death, R.G. (2007). Stream Ecosystem Functioning under reduced flow conditions. Ecol. Appl. *17*, 1797–1808.

Diaz Villanueva, V.D., Albariño, R., and Canhoto, C. (2012). Positive effect of shredders on microbial biomass and decomposition in stream microcosms. Freshwater Biol. *57*, 2504–2513.

Dodds, W.K. (2003). The role of periphyton in phosphorous retention in shallow freshwater aquatic systems. J. Phycol. *39*, 840–849.

Donlan, R.M. (2002). Biofilms: microbial life on surfaces. Emerg. Infect. Dis. *8*, 881–890.

Duarte, S., Pascoal, C., Garabétian, F., Cássio, F., and Charcosset, J.Y. (2009). Microbial decomposer communities are mainly structured by trophic status in circumneutral and alkaline streams. Appl. Environ. Microbiol. *75*, 6211–6221.

Duarte, S., Pascoal, C., Alves, A., Correia, A., and Cássio, F. (2010). Assessing the dynamic of microbial communities during leaf decomposition in a low-order stream by microscopic and molecular techniques. Microbiol. Res. *165*, 351–362.

Dunn, R.J.K., Welsh, D.T., Lee, S.Y., Lemckert, C.J., and Teasdale, P.R. (2008). Investigating the distribution and sources of organic matter in surface sediment of Coombabah Lake (Australia) using elemental, isotopic and fatty acid biomarkers. Cont. Shelf Res. *28*, 2535–2549.

Dupraz, C., and Visscher, P.T. (2005). Microbial lithification in marine stromatolites and hypersaline mats. Trends Microbiol. *13*, 429–438.

Ehrenhauss, S.,Witte, U., Buhring, S.L., and Huettel, M. (2004). Effect of advective pore water transport on distribution and degradation of diatoms in permeable North Sea sediments. Mar. Ecol. Prog. Ser. *271*, 99e111.

Emtiazi, F., Schwartz, T., Marten, S.M., Krolla-Sidenstein, P., and Obst, U. (2004). Investigation of natural biofilms formed during the production of drinking water from surface water embankment filtration. Water Res. *38*, 1197–1206.

Eriksson, P.G. (2001). Interaction effects of flow velocity and oxygen metabolism on nitrification and denitrification in biofilms on submersed macrophytes. Biogeochemistry *55*, 29–44.

Eriksson, P.G., and Weisner, S.E.B. (1996). Functional differences in epiphytic microbial communities in nutrient-rich freshwater ecosystems : an assay of denitrifying capacity. Freshwater Biol. *36*, 555–562.

Eriksson, P.G., and Weisner, S.E.B. (1999). An experimental study on effects of submersed macrophytes on nitrification and denitrification in ammonium-rich aquatic systems. Limnol. Oceanogr. *44*, 1993–1999.

Esselink, P., Fresco, L.F.M., and Dijkema, K.S. (2002). Vegetation change in a man-made salt marsh affected by a reduction in both grazing and drainage. Appl. Veg. Sci. *5*, 17–32.

Feminella, J.W., and Hawkins, C.E. (1995). Interactions between stream herbivores and periphyton: A quantitative analysis of past experiments. J. N. Am. Benthol. Soc. *14*, 465–509.

Ferreira, V., and Chauvet, E. (2011). Synergistic effects of water temperature and dissolved nutrients on litter decomposition and associated fungi. Glob. Chang. Biol. *17*, 551–564.

Ferreira, V., and Graça, M.A.S. (2006). Do Invertebrate Activity and Current Velocity Affect Fungal Assemblage Structure in Leaves? Int. Rev. Hydrobiol. *91*, 1–14.

Firstater, F.N., Hidalgo, F.J., Lomovasky, B.J., and Iribarne, O.O. (2012). Grazing and nutrient effects on a high rocky intertidal of the Peruvian central coast. J. Exp. Mar. Biol. Ecol. 422–423, 9–13.

Fischer, H., Sukhodolov, A., Wilczek, S., and Engelhardt, C. (2003). Effects of flow dynamics and sediment movement on microbial activity in a lowland river. River Res. Appl. *19*, 473–482.

Flemming, H.C. (2002). Biofouling in water systems–cases, causes and countermeasures. Appl. microbiol. Biotech. *59*, 629–640.

Fletcher, M., and Loeb, G.I. (1979). Influence of substratum characteristics on the attachment of marine pseudomonas to solid surfaces. Appl. Env. Microbiol. *37*, 67–72.

Foucreau, N., Puijalon, S., Hervant, F., and Piscart, C. (2013). Effect of leaf litter characteristics on leaf conditioning and on consumption by *Gammarus pulex*. Freshwater Biol. *58*, 1672–1681.

Furumai, H., and Rittmann, B.E. (1994). Evaluation of multiple-species biofilm and floc processes using a simplified aggregate model. Water Sci. Technol. *29*, 439–446.

Francoeur, S.N., Schaecher, M., Neely, R.K., and Kuehn, K.A. (2006). Periphytic photosynthetic stimulation of extracellular enzyme activity in aquatic microbial communities associated with decaying typha litter. Microb. Ecol. *52*, 662–669.

Friberg, N., Bergfur, J., Rasmussen, J., and Sandin, L. (2013). Changing Northern catchments: Is altered hydrology, temperature or both going to shape future stream communities and ecosystem processes? Hydrol. Process. *27*, 734–740.

Froelich, P.N., Klinkhammer, G.P., Bender, M.L., Luedtke, N.A., Heath. G.R., Cullen, D., Dauphin, P., Hammond, D., Hartmen, B., and Maynard, V. (1979). Early oxidation of organic matter in pelagic sediments of the eastern equatorial Atlantic: suboxic diagenesis. Geochim. Cosmochim. Acta *43*, 1075–1090.

Früh, D., Norf, H., and Weitere, M. (2011). Response of biofilm-dwelling ciliate communities to enrichment with algae. Aquat. Microb. Ecol. *63*, 299–309.

Gaudes, A., Artigas, J., Romaní, A.M., Sabater, S., and Muñoz, I. (2009). Contribution of microbial and invertebrate communities to leaf litter colonization in a Mediterranean stream. J. N. Am. Benthol. Soc. *28*, 34–43.

Gerbersdorf, S.U., Jancke, T., Westrich, B., and Paterson, D.M. (2008). Microbial stabilization of riverine sediments by extracellular polymeric substances. Geobiology *6*, 57–69.

Gessner, M.O., and Chauvet, E. (1994). Importance of Stream Microfungi in Controlling Breakdown Rates of Leaf Litter. Ecology *75*, 1807–1817.

Glud, R.N. (2008). Oxygen dynamics of marine sediments. Mar. Biol. Res. *4*, 243–289.

Golladay, S.W., and Sinsabaugh, R.L. (1991). Biofilm development on leaf and wood surfaces in a boreal river. Freshwater Biol. *25*, 437–450.

Gonçalves, A.L., Graça, M.A.S., and Canhoto, C. (2013). The effect of temperature on leaf decomposition and diversity of associated aquatic hyphomycetes depends on the substrate. Fungal Ecol. *6*, 546–553.

Graba, M., Sauvage, S., Moulin, F.Y., Urrea, G., Sabater, S., and Sánchez-Pérez, J.M. (2013). Interaction between local hydrodynamics and algal community in epilithic biofilm. Water Res. *47*, 2153–2163.

Grossart, H.P., and Simon, M. (1993). Limnetic macroscopic organic aggregates (lake snow): Occurrence, characteristics, and microbial dynamics in Lake Constance. Limnol. Oceanogr. *38*, 532–546.

Grossart, H.P., Tang, K.W., Kiorboe, T., and Ploug, H. (2007). Comparison of cell-specific activity between free-living and attached bacteria using isolates and natural assemblages. FEMS Microbiol. Lett. *266*, 194–200.

Guasch, H., and Sabater, S. (1994). Primary production of epilithic communities in undisturbed Mediterranean streams. Verh. Internat. Verein. Limnol. *25*, 1761–1764.

Guasch, H., Martí, E., and Sabater, S. (1995). Nutrient enrichment effects on biofilm metabolism in a Mediterranean stream. Freshwater Biol. *33*, 373–383.

Guasch, H., and Sabater, S. (1995). Seasonal variations in photosynthesis-irradiance responses by biofilms in Mediterranean streams. J. Phycol. *31*, 727–735.

Gulis, V., Suberkropp, K., and Rosemond, A.D. (2008). Comparison of fungal activities on wood and leaf litter in unaltered and nutrient-enriched headwater streams. Appl. Environ. Microbiol. *74*, 1094–1101.

Guo, W., Ngo, H.H., and Li, J. (2012). A mini-review on membrane fouling. Bioresour. Technol. *122*, 27–34.

Hahn, M.W., and Höfle, M.G. (2001). Grazing of protozoa and its effect on populations of aquatic bacteria. FEMS Microbiol. Ecol. *35*, 113–121.

Hart, B.T., Bailey, P., Edwards, R., Hortle, K., James, K., Mcmahon, A., Meredith, C., and Swadling, K. (1990). Effects of salinity on river, stream and wetland ecosystems in Victoria, Australia. Water Res. *24*, 1103–1117.

Heilman, M.A., and Carlton, R.G. (2001). Methane oxidation associated with submersed vascular macrophytes and its impact on plant diffusive methane flux. Biogeochemistry *52*, 207–224.

Hempel, M., Blume, M., Blindow, I., and Gross, E.M. (2008). Epiphytic bacterial community composition on two common submerged macrophytes in brackish water and freshwater. BMC Microbiol. *8*, 58.

Hill, W.R. (1996). Effects of light. In Algal Ecology: Freshwater Benthic Ecosystems, Stevenson, R.J., Bothwell, M.I., and Lowe, R.L., eds. (Academic Press, San Diego, CA, USA), pp. 121–148.

Howell, D. (2009). Testing the impact of biofilms on the performance of marine antifouling coatings. In Advances in Marine Antifouling Coatings and Technologies, Hellio, C., and Yebra, D., eds. (Woodhead, Cambridge, UK), pp. 422–438.

Hunt, A.P., and Parry, J.D. (1998). The effect of substratum roughness and river flow rate on the development of a freshwater biofilm community. Biofouling *12*, 287–303.

Huzarska, K. (2013). Spatial distribution of biological and physical sediment parameters in the western Gulf of Gdansk. Oceanologia *55*, 453–470.

Jarvie, H.P., Jürgens, M.D., Williams, R.J., Neal, C., Davies, J.J.L., Barrett, C., and White, J. (2005). Role of river bed sediments as sources and sinks of phosphorus across two major eutrophic UK river basins: the Hampshire Avon and Herefordshire Wye. J. Hydrol. *304*, 51–74.

Jesus, B., Brotas, V., Ribeiro, L., Mendes, C.R., Cartaxana, P., and Paterson, D.M. (2009). Adaptations of microphytobenthos assemblages to sediment type and tidal position. Cont. Shelf. Res. *29*, 1624–1634.

Karatan, E., and Watnick, P. (2009). Signals, regulatory networks, and materials that build and break bacterial biofilms. Microbiol. Mol. Biol. Rev. *73*, 310–347.

Kelly, J.J., Minalt, N., Culotti, A., Pryor, M., and Packman, A. (2014). Temporal variations in the abundance and composition of biofilm communities colonizing drinking water distribution pipes. PLoS One *9*, e98542.

Kochi, K., and Yanai, S. (2006). Shredder colonization and decomposition of green and senescent leaves during summer in a headwater stream in northern Japan. Ecol. Res. *21*, 544–550.

Koetsier, P., Vaun, MacArthur, J., and Leff, L.G. (1997). Spatial and temporal response of stream bacteria to sources of dissolved organic carbon in a blackwater stream system. Freshwater Biol. *37*, 79–89.

Körner, S. (1999). Nitrifying and denitrifying bacteria in epiphytic communities of submerged macrophytes in a treated sewage channel. Acta Hydrochim. Hydrobiol. *27*, 27–31.

Kralj, K., Plenkovic-Moraj, A., Gligora, M., Primc-Habdija, B., and Sipos, L. (2006). Structure of periphytic community on artificial substrata: influence of depth, slide orientation and colonization time in karstic Lake Visovacko, Croatia. Hydrobiologia *560*, 249–258.

Kreft, J.U. (2004). Biofilms promote altruism. Microbiology *150*, 2751–2760.

Kühl, M., Lassen, C., and Jorgensen, B.B. (1994). Optical properties of microbial mats: Light measurements with fiber-optic microprobes. NATO ASI Series *35*, 149–166.

Kühl, M., and Jorgensen, B.B. (1995). The light field of microbenthic communities: radiance distribution and microscae optics of sandy coastal sediments. Oceanogr. Literat. Rev. *42*, 560.

Lalli, C.M., and Parsons, T.R. (1993). Biological Oceanography: an Introduction (Elsevier, New York, NY, USA).

Lamberti, G.A., and Resh, V.H. (1985). Comparability of introduced tiles and natural substrates for sampling lotic bacteria, algae and macroinvertebrates. Freshwater Biol. *15*, 21–30.

Lamberti, G.A. (1996). The role of periphyton in benthic food webs. In Algal Ecology: Freshwater Benthic Ecosystems, Stevenson, R.J., Bothwell, M.I., and Lowe, R.L., eds. (Academic Press Inc., San Diego, CA, USA), pp. 533–572.

Lamontagne, I., Cardinal, A., and Fortier, L. (1989). Environ-mental forcing versus endogenous control of photosynthesis in intertidal epilithic microalgae. Mar. Ecol. Prog. Ser. *51*, 1–2.

Lampert, W., and Sommer, U. (2007). Limnoecology. The ecology of lakes and streams. (Oxford University Press, New York, NY, USA).

Larned, S.T., Nikora, V.I., and Biggs, B.J.F. (2004). Mass-transfer–limited nitrogen and phosphorus uptake by stream periphyton: A conceptual model and experimental evidence. Limnol. Oceanogr. *49*, 1992–2000.

Larned, S.T. (2010). A prospectus for periphyton: recent and future ecological research. J. N. Am. Benthol. Soc. *29*, 182–206.

Lassen, C., Ploug, H., and Jørgensen, B.B. (1992). Microalgal photosynthesis and spectral scalar irradiance in coastal marine sediments of Limfjorden, Denmark. Limnol. Oceanogr. *37*, 760–772.

Lawrence, J.R., and Neu, T.R. (2003). Microscale analyses of the formation and nature of microbial biofilm communities in river systems. Rev. Env. Sci. Biotech. *2*, 85–97.

Lazarova, V., and Manem, J. (1995). Biofilm characterization and activity analysis in water and wastewater treatment. Water Res. *29*, 2227–2245.

Le-Clech, P., Chen, V., and Fane, T.A.G. (2006). Fouling in membrane bioreactors used in wastewater treatment. J. Membrane Sci. *284*, 17–53.

Lear, G., Washington, V., Neale, M., Case, B., Buckley, H., and Lewis, G. (2013). The biogeography of stream bacteria. Global Ecol. Biogeogr. *22*, 544–554.

Lecerf, A., and Chauvet, E. (2008). Diversity and functions of leaf-decaying fungi in human-altered streams. Freshwater Biol. *53*, 1658–1672.

Ledger, M.E., and Hildrew, A.G. (1998). Temporal and spatial variation in the epilithic biofilm of an acid stream. Freshwater Biol. *40*, 655–670.

Lehtola, M.J., Miettinen, I.T., Lampola, T., Hirvonen, A., Vartiainen, T., and Martikainen, P.J. (2005). Pipeline materials modify the effectiveness of disinfectants in drinking water distribution systems. Water Res. *39*, 1962–1971.

LeRoy, C.J., and Marks, J.C. (2006). Litter quality, stream characteristics and litter diversity influence decomposition rates and macroinvertebrates. Freshwater Biol. *51*, 605–617.

LeRoy, C.J., Whitham, T.G., Wooley, S.C., and Marks, J.C. (2007). Within-species variation in foliar chemistry influences leaf-litter decomposition in a Utah river. J. N. Am. Benthol. Soc. *26*, 426–438.

Liehr, S.K., Chen, H.J., and Lin, S.H.(1994). Metals removal by algal biofilms. Water Sci. Technol. *30*, 59–68.

Lock, M.A., Wallace, R.R., Costerton, J.W., Ventullo, R.M., and Charlton, S.E.(1984). River epilithon: toward a structural-functional model. Oikos *42*, 10–22.

Lock, M.A. (1993). Attached microbial communities in rivers. In Aquatic microbiology: An Ecological Approach, Ford, T.E., ed. (Blackwell Scientific Publications, Oxford, UK), pp. 113–138.

López, D., Vlamakis, H., and Kolter, R. (2010). Biofilms. Cold Spring Harb. Perspect. Biol 2, a000398, 1–11.

Lowe, R.L. (1996). Periphyton Patterns in Lakes. In Algal Ecology: Freshwater Benthic Ecosystems, Stevenson, R.J., Bothwell, M.I., and Lowe, R.L., eds. (Academic Press Inc., San Diego, CA, USA), pp. 57–76.

Mack, W.N., Mack, J.P., and Ackerson, A.O. (1975). Microbial film development in a trickling filter. Microb. Ecol. 2, 215–226.

Mackay, E.B., Jones, I.D., Folkard, A.M., and Thackeray, S.J. (2011). Transition zones in small lakes: the importance of dilution and biological uptake on lake-wide heterogeneity. Hydrobiologia *678*, 85–97.

MacIntyre, H.L., Geider, R.J., and Miller, D.C. (1996). Microphytobenthos: the ecological role of the 'Secret Garden' of unvegetated, shallow-water marine habitats. I. Distribution, abundance and primary production. Estuaries *19*, 186–201.

MacLeod, F.A., Guiot, S.R and Costerton J.W. (1990). Layered structure of bacterial aggregates produced in an upflow anaerobic sludge bed and filter reactor. Appl. Environ. Microbiol. *56*, 1598–1607.

Magalhães, C.M., Joye, S.B., Moreira, R.M., Wiebe, W.J., and Bordalo, A.A. (2005). Effect of salinity and inorganic nitrogen concentrations on nitrification and denitrification rates in intertidal sediments and rocky biofilms of the Douro River estuary, Portugal. Water Res. *39*, 1783–1794.

Majdi, N., Traunspurger, W., Boyer, S., Mialet, B., Tackx, M., Fernandez, R., Gehner, S., Ten-Hage, L., and Buffan-Dubau, E. (2011). Response of biofilm-dwelling nematodes to habitat changes in the Garonne River, France: influence of hydrodynamics and microalgal availability. Hydrobiologia *673*, 229–244.

Manerkar, M.A., Seena, S., and Bärlocher, F. (2008). Q-RT-PCR for assessing archaea, bacteria, and fungi during leaf decomposition in a stream. Microb. Ecol. *56*, 467–473.

Marshall, K., Joint, I., Callow, M.E., and Callow, J.A. (2006). Effect of marine bacterial isolates on the growth and morphology of axenic plantlets of the green alga *Ulva linza*. Microb. Ecol. *52*, 302–310.

Martínez, A., Larrañaga, A., Pérez, J., Basaguren, A., and Pozo, J. (2013). Leaf-litter quality effects on stream ecosystem functioning: a comparison among five species. Fundam. Appl. Limnol./Arch. Für Hydrobiol. *183*, 239–248.

Martins, G., Ribeiro, D.C., Pacheco, D., Cruz, J.V., Cunha, R., Gonçalves, V., Nogueira, R., and Brito, A.G. (2008). Prospective scenarios for water quality and ecological status in Lake Sete Cidades (Portugal): the integration of mathematical modelling in decision processes. Appl. Geochem. *23*, 2171–2181.

Mayer, L., and Rossi, P. (1982). Specific surface areas in coastal sediments: Relation-ships with other textural factors. Mar. Geol. *45*, 241–252.

Medeiros, A.O., Pascoal, C., and Graça, M.A.S. (2009). Diversity and activity of aquatic fungi under low oxygen conditions. Freshwater Biol. *54*, 142–149.

Meentemeyer, V. (1978). Macroclimate and lignin control of litter decomposition rates. Ecology *59*, 465–472.

Melo, L.F., and Bott, T.R. (1997). Biofouling in water systems. Exp. Therm. Fluid. Sci. *14*, 375–381.

Meyer-Reil, L.A. (1986). Measurement of hydrolytic activity and incorporation of dissolved organic substrates by microorganisms in marine sediments. Mar. Ecol. Prog. Ser. *31*, 143–149.

Meyer-Reil, L.A. (1994). Microbial life in sedimentary biofilms – the challenge to microbial ecologists. Mar. Ecol. Prog. Ser. *112*, 303–311.

Murdock, J.N., and Dodds, W.K. (2007). Linking benthic algal biomass to stream substratum topography. J. Phycol. *43*, 449–460.

Neuswanger, D.J., Taylor, W.W., and Reynolds, J.B. (1982). Comparison of macroinvertebrate herpobenthos and haptobenthos in side channel and slough in the Upper Mississippi River. Freshwater Invertebr. Biol. *1*, 13–24.

Nielsen, D.L., Brock, M.A., Rees, G.N., and Baldwin, D.S. (2003). Effects of increasing salinity on freshwater ecosystems in Australia. Aus. J. Bot. *51*, 655–665.

Nikolcheva, L.G., and Bärlocher, F. (2004). Taxon-specific fungal primers reveal unexpectedly high diversity during leaf decomposition in a stream. Mycol. Prog. *3*, 41–50.

Nybakken, J.W., and Wallace, R.L. (1992). Marine Biology: An Ecological Approach (Addison-Wesley Longman, Chicago, IL, USA).

Ostrofsky, M.L. (1997). Relationship between chemical characteristics of autumn-shed leaves and aquatic processing rates. J. N. Am. Benthol. Soc. *16*, 750–759.

Pan, J., Bournod, C.N., Cuadrado, D.G., Vitale, A., and Piccolo, M.C. (2013). Interaction between estuarine microphytobenthos physical forcings: The role of atmospheric and sedimentary factors. Int. J. Geos. *4*, 352–361.

Pascoal, C., Pinho, M., Cássio, F., and Gomes, P. (2003). Assessing structural and functional ecosystem condition using leaf breakdown : studies on a polluted river. Freshwater Biol. *48*, 2033–2044.

Peterson, C.G. (1996). Response of Benthic Algal Communities to Natural Physical Disturbance. In Algal Ecology: Freshwater Benthic Ecosystems, Stevenson, R.J., Bothwell, M.I., and Lowe, R.L., eds. (Academic Press Inc., San Diego, CA, USA), pp. 375–402.

Ploug, H. (2001). Small-scale oxygen fluxes and remineralization in sinking aggregates. Limnol. Oceanogr. *46*, 1624–1631.

Poremba, K., and Hoppe, H.G. (1995). Spatial variation of benthic microbial production and hydrolytic enzymatic activity down the continental slope of the Celtic Sea. Mar. Ecol. Prog. Ser. *118*, 237–245.

Pringle, J.H., and Fletcher, M. (1983). Influence of substratum wettability on attachment of freshwater bacteria to solid surfaces. Appl. Env. Microbiol. *45*, 811–817.

Proia, L., Romaní, A.M., and Sabater, S. (2012). Nutrients and light effects on stream biofilms: a combined assessment with CLSM, structural and functional parameters. Hydrobiologia *695*, 281–291.

Pusch, M., Fiebig, D., Brettar, I., Eisenmann, H., Ellis, B.K., Kaplan, L.A., Lock, M.A., Naegeli, M.W., and Traunspurger, W. (1998). The role of micro-organisms in the ecological connectivity of running waters. Freshwater Biol. *40*, 453–495.

Rajashekhar, M., and Kaveriappa, K.M. (2000). Effects of temperature and light on growth and sporulation of aquatic hyphomycetes. Hydrobiologia *441*, 149–153.

Rao, D., Webb, J.S., and Kjelleberg, S. (2006). Microbial colonization and competition on the marine alga *Ulva australis*. Appl. Environ. Microbiol. *72*, 5547–5555.

Ribeiro, D.C., Martins, G., Nogueira, R., Cruz, J.V., and Brito, A.G. (2008). Phosphorus fractionation in volcanic lake sediments (Azores – Portugal). Chemosphere *70*, 1256–1263.

Rier, S.T., and Stevenson, R.J. (2002). Effects of light, dissolved organic carbon, and inorganic nutrients on the relationship between algae and heterotrophic bacteria in stream periphyton. Hydrobiologia *489*, 179–184.

Risse-Buhl, U., Karsubke, M., Schlief, J., Baschien, C., Weitere, M., and Mutz, M. (2012). Aquatic protists modulate the microbial activity associated with mineral surfaces and leaf litter. Aquat. Microb. Ecol. *66*, 133–147.

Roberts, S., Sabater, S., and Beardall, J. (2004). Benthic microalgal colonization in streams of differing riparian cover and light availability. J. Phycol. *40*, 1004–1012.

Rochex, A., Godon, J., Bernet, N., and Escudié, R. (2008). Role of shear stress on composition, diversity and dynamics of biofilm bacterial communities. Water Res. *42*, 4915– 4922.

Romaní, A.M., and Sabater, S. (1999). Effect of primary producers on the heterotrophic metabolism of a stream biofilm. Freshwater Biol. *41*, 729–736.

Romaní, A.M., and Sabater, S. (2000). Influence of algal biomass on extracellular enzyme activity in river biofilms. Microb. Ecol. *41*, 16–24.

Romaní, A.M., and Sabater, S. (2001). Structure and activity of rock and sand biofilms in a Mediterranean stream. Ecology *82*, 3232–3245.

Romaní, A.M., and Marxsen, J. (2002). Extracellular enzymatic activities in epilithic biofilms of the Breitenbach: microhabitat differences. Archiv. Für. Hydrobiologie. *155*, 541–555.

Romaní, A.M., Giorgi, A., Acuña, V., and Sabater, S. (2004a). The influence of substratum type and nutrient supply on biofilm organic matter utilization in streams. Limnol. Oceanogr. *49*, 1713–1721.

Romaní, A.M., Guasch, H., Muñoz, I., Ruana, J., and Vilalta, E. (2004b). Biofilm structure and function and possible implications for riverine DOC dynamics. Microb. Ecol. *47*, 316–328.

Romaní, A.M. (2010). Freshwater Biofilms. In Biofouling, S. Dürr and J.C. Thomason, eds. (Blackwell Publishing Ltd., Oxford, UK), pp. 137–153.

Romaní, A.M., Borrego, C.M., Díaz-Villanueva, V., Freixa, A., Gich, F., and Ylla, I. (2014). Shifts in microbial community structure and function in light- and dark-grown biofilms driven by warming. Environ. Microbiol. *16*, 2550–2567.

Royer, T.V., and Minshall, G.W. (2001). Effects of nutrient enrichment and leaf quality on the breakdown of leaves in a hardwater stream. Freshwater Biol. *46*, 603–610.

Ruttner, F. (1953). Fundamentals of Limnology (University of Toronto Press, Toronto, ON, Canada).

Sabater, S., Guasch, H., Romaní, A.M., and Muñoz, I. (2002). The effect of biological factors on the efficiency of river biofilms in improving water quality. Hydrobiologia *469*, 149–156.

Sabater, S., Gregory, S.V., and Sedell, J.R. (1998). Community dynamics and metabolism of benthic algae colonizing wood and rock substrata in a forest stream. J. Phycol. *34*, 561–567.

Sabater, S., Guasch, H., Muñoz, I., and Romaní, A.M. (2006). Hydrology, light and the use of organic and inorganic materials as structuring factors of biological communities in Mediterranean streams. Limnetica *25*(1–2), 335–348.

Sabetta, L., Costantini, M.L., Maggi, O., Persiani, A.M., and Rossi, L. (2000). Interactions between detritivores and microfungi during the leaf detritus decomposition in a volcanic lake (Lake Vico – central Italy). Hydrobiologia *439*, 49–60.

Saikia, S.K. (2011). Review on Periphyton as Mediator of nutrient transfer in aquatic ecosystems. Ecol. Balkanica *3*, 65–78.

Salta, M., Wharton, J.A., Blache, Y., Stokes, K.R., and Briand, J.F. (2013). Marine biofilms on artificial surfaces: structure and dynamics. Env. Microb. *15*, 2879–2893.

Sand-Jensen, K., and Søndergaard, M. (1981). Phytoplankton and epiphyte development and their shading effect on submerged macrophytes in lakes of different nutrient status. Int. Rev. Der Gesamten Hydrobiol. Und. Hydrogr. *66*, 529–552.

Sanson, G.D., Stolk, R., and Downes, B.J. (1995). A new method for characterizing surface-roughness and available space in biological-systems. Funct. Ecol. *9*, 127–135.

Sasaki, A., Ito, A., Aizawa, J., and Umita, T. (2005). Infuence of water and sediment quality on benthic biota in an acidified river. Water Res. *39*, 2517–2526.

Schäfer, R.B., Bundschuh, M., Rouch, D.A., Szöcs, E., von der Ohe, P.C., Pettigrove, V., Schulz, R., Nugegoda, D., and Kefford, B.J. (2012). Effects of pesticide toxicity, salinity, and other environmental variables on selected ecosystem functions in streams and the relevance for ecosystem services. Sci. Total Environ. *415*, 69–78.

Schlief, J., and Mutz, M. (2009). Effect of sudden flow reduction on the decomposition of alder leaves (*Alnus glutinosa* [L.] Gaertn.) in a temperate lowland stream: a mesocosm study. Hydrobiologia *624*, 205–217.

Schramm, A., Santegoeds, C.M., Nielsen, H.K., Ploug, H., Wagner, M., Pribyl, M., Wanner, J., Amann, R., and de Beer, D. (1999). On the occurrence of anoxic microniches, denitrification, and sulfate reduction in aerated activated sludge. Appl. Environ. Microbiol. *65*, 4189–4196.

Schultz, M.P., Bendick, J.A., Holm, E.R., and Hertel, W.M. (2011). Economic impact of biofouling on a naval surface ship. Biofouling *27*, 87–98.

Siebenaller, J., and Somero, G.N. (1978). Pressure-adaptive differences in lactate dehydrogenases of congeneric fishes living at different depths. Science *201*, 255–257.

Simon, M., Grossart, H., Schweitzer, B., and Ploug, H. (2002). Microbial ecology of organic aggregates in aquatic ecosystems. Aquat. Microb. Ecol. *28*, 175–211.

Simon, K.S., Simon, M.A., and Benfield, E.F. (2009). Variation in ecosystem function in Appalachian streams along an acidity gradient. Ecol. Appl. *19*, 1147–1160.

Smith, D.C., Simon, M., Alldredge, A.L., and Azam, F. (1992). Intense hydrolytic enzyme activity on marine aggregates and implications for rapid particle dissolution. Nature *359*, 139–142.

Smith, J.E., Hunter, C.L., and Smith, C.M. (2010). The effects of top-down versus bottom-up control on benthic coral reef community structure. Oecologia *163*, 497–507.

Sobczak, W.V. (1996). Epilithic bacterial responses to variations in algal biomass and labile dissolved organic carbon during biofilm colonization. J. N. Am. Benthol. Soc. *15*, 143–154.

Solé, M., Fetzer, I., Wennrich, R., Sridhar, K.R., Harms, H., and Krauss, G. (2008). Aquatic hyphomycete communities as potential bioindicators for assessing anthropogenic stress. Sci. Total Environ. *389*, 557–565.

Steinman, A.D. (1996). Effects of grazers on fresh water benthic algae. In Algal Ecology: Freshwater Benthic Ecosystems, Stevenson, R.J., Bothwell, M.I., and Lowe, R.L., eds. (Academic Press Inc., San Diego, CA, USA), pp. 341–373.

Stevenson, R.J., and Glover, R. (1993). Effects of algal density and current on ion transport through periphyton communities. Limnol. Oceanogra. *38*, 1276–1282.

Stoodley, P., Sauer, K., Davies, D.G., and Costerton, J.W. (2002). Biofilms as complex differentiated communities. Ann. Rev. Microbiol. *56*, 187–209.

Suberkropp, K., and Wallace, J.B. (1992). Aquatic hyphomycetes in insecticide-treated and untreated streams. J. N. Am. Benthol. Soc. *11*, 165–171.

Suberkropp, K., Arsuffi, T.L., and Anderson, J.P. (1983). Comparison of degradative ability, enzymatic activity, and palatability of aquatic hyphomycetes grown on leaf litter. Appl. Environ. Microbiol. *46*, 237–244.

Suberkropp, K., Gulis, V., Rosemond, A.D., and Benstead, J.P. (2010). Ecosystem and physiological scales of microbial responses to nutrients in a detritus-based stream: Results of a 5-year continuous enrichment. Limnol. Oceanogr. *55*, 149–160.

Sullivan, M.J., and Currin, C.A. (2000). Community structure and functional dynamics of benthic microalgae in salt marshes. In Concepts and Controversies in Tidal Marsh Ecology, Weinstein,

M.P., and Kreeger, D.A., eds. (Kluwer Academic Publishers, Dordrecht, The Netherlands), pp. 81–106.

Sutherland, I.W. (2001). The biofilm matrix – an immobilized but dynamic microbial environment. Trends Microbiol. *9*, 222–227.

Tank, J.L., Rosi-marshall, E.J., Griffiths, N.A., Entrekin, S.A., and Stephen, M.L. (2010). A review of allochthonous organic matter dynamics and metabolism in streams. J. North Am. Benthol. Soc. *29*, 118–146.

Taylor, J., Wilson, B., Mills, M., and Burns, R. (2002). Comparison of microbial numbers and enzymatic activities in surface soils and subsoils using various techniques. Soil Biol. Biochem. *34*, 387–401.

Terlizzi, A., and Faimali, M. (2010). Fouling on artificial substrata. In Biofouling, Dürr, S., and Thomason, J.C., eds. (Blackwell Publishing Ltd, Oxford, UK), pp. 170–179.

Thompson, R.C., Norton, T.A., and Hawkins, S.J. (2004). Physical stress and biological control regulate the producer–consumer balance in intertidal biofilms. Ecology *85*, 1372–1382.

Thompson, R.C., Moschella, P.S., Jenkins, S.R., Norton, T.A., and Hawkins, S.J. (2005). Differences in photosynthetic marine biofilms between sheltered and moderately exposed rocky shores. Mar. Ecol. Prog. Ser. *296*, 53–63.

Tokajian, S.T., Hashwa, F.A., Hancock, I.C., and Zalloua, P.A. (2005). Phylogenetic assessment of heterotrophic bacteria from a water distribution system using 16S rDNA sequencing. Can. J. Microbiol. *51*, 325–335.

Townsend, S.A., and Gell, P.A. (2005). The role of substrate type on benthic diatom assemblages in the Daly and Roper Rivers of the Australian wet/dry tropics. Hydrobiologia *548*, 101–115.

Triska, F., and Sedell, J.R. (1976). Decomposition of four species of leaf litter in response to nitrate manipulation. Ecology *57*, 783–792.

Tuchman, M.L., and Stevenson, R.J.(1980). Comparison of clay tile, sterilized rock, and natural substrate diatom communities in a small stream in southeastern Michigan, USA. Hydrobiologia *75*, 73–79.

Underwood, G.J.C. (1991). Note: Colonization and invasion of leaves of the aquatic macrophyte *Ceratophyllum demersum* L. by epiphytic bacteria. Microb. Ecol. *21*, 267–275.

Underwood, G.J.C., Phillips, J., and Saunders, K. (1998). Distribution of estuarine benthic diatom species along salinity and nutrient gradients. Eur. J. Phycol. *33*, 173–183.

Underwood, G.J.C. (2002). Adaptations of tropical marine microphytobenthic assemblages along a gradient of light and nutrient availability in Suva Lagoon, Fiji. Eur. J. Phycol. *37*, 449–462.

Underwood, G.J.C. (2010). Microphytobenthos and phytoplankton in the Severn estuary and possible consequences of a tidal energy barrage. Mar. Poll. Bullet. *61*, 83–91.

Underwood, G.J.C., Perkins, R.G., Consalvey, M.C., Hanlon, A.R.M., Oxborough, K., Baker, N.R., and Paterson, D.M. (2005). Patterns in microphytobenthic primary productivity: species-specific variation in migratory rhythms and photosynthetic efficiency in mixed-species biofilms. Limnol. Oceanogr. *50*, 755–767.

Vadeboncoeur, Y., and Steinman, A.D. (2002). Periphyton function in lake ecosystems. Sci. World J. *2*, 1449–1468.

Van Donk, E., and Van de Bund, W.J. (2002). Impact of submerged macrophytes including charophytes on phyto- and zooplankton communities : allelopathy versus other mechanisms. Aquat. Bot. *72*, 261–274.

Van Gemerden, H., Tughan, C.S., De Wit, R., and Herbert, R.A. (1989). Laminated microbial ecosystems on sheltered beaches in Scapa Flow, Orkney Islands. FEMS Microbiol. Ecol. *62*, 87–102.

Van Gemerden, H. (1993). Microbial mats: a joint venture. Mar. Geol. *113*, 3–25.

Videla, H.A., and Herrera, L.K. (2005). Microbiologically influenced corrosion: looking to the future. Int. Microbiol. *8*, 169–180.

Watnick, P., and Kolter, R. (2000). Biofilm, city of microbes. J. Bacteriol. *182*, 2675–2679.

Webb, J.A., Downes, B.J., Lake, P.S., and Glaister, A. (2006). Quantifying abrasion of stable substrata in streams: a new disturbance index for epilithic biota. Hydrobiologia, *559*, 443–453.

Webster, J.R., and Benfield, E.F. (1986). Vascular plant breakdown in freshwater ecosystems. Annu. Rev. Ecol. Syst. *17*, 567–594.

Wetzel, R.G. (1983). Limnology (Philadelphia, USA: Saunders College Publishing).

Wetzel, R.G. (1993). Microcommunities and microgradients: Linking nutrient regeneration, microbial mutualism, and high sustained aquatic primary production. Netherlands J. Aquat. Ecol. *27*, 3–9.

Wetzel, R.G. (1996). Benthic algae and nutrient cycling in lentic freshwater ecosystems. In Algal ecology, freshwater benthic ecosystems, Stevenson, R.J., Bothwell, M.L., and Lowe, L., eds. (Academic Press Inc., San Diego, CA, USA), pp. 641–663.

Wieland, A., and Kühl, M. (2000). Short-term temperature effects on oxygen and sulfide cycling in a hypersaline cyanobacterial mat (Solar Lake, Egypt). Mar. Ecol. Prog. Ser. *196*, 87–102.

Wingender, J., and Flemming, H.C. (2011). Biofilms in drinking water and their role as reservoir for pathogens. Int. J. Hyg. Environ. Health. *214*, 417–423.

Winter, C., Hein, T., Kavka, G., Mach, R.L., and Farnleitner, A.H. (2007). Longitudinal changes in the bacterial community composition of the Danube River: a whole-river approach. Appl. Environ. Microbiol. *73*, 421–431.

Woodward, G., Gessner, M.O., Giller, P.S., Gulis, V., Hladyz, S., Lecerf, A., Malmqvist, B., McKie, B.G., Tiegs, S.D., Cariss, H., *et al.* (2012). Continental-scale effects of nutrient pollution on stream ecosystem functioning. Science *336*, 1438–1440.

Ylla, I., Romaní, A.M., and Sabater, S. (2007). Differential effects of nutrients and light on the primary production of stream algae and mosses. Fund. Appl. Limnology *170*, 1–10.

Ylla, I., Romaní, A.M., and Sabater, S. (2012). Labile and recalcitrant organic matter utilization by river biofilm under increasing water temperature. Microb. Ecol. 64,593–604.

Ylla, I., Canhoto, C., and Romaní, A.M. 2014. Effects of warming on stream biofilm organic matter use capabilities. Microb. Ecol. *68*, 132–145.

Zhang, T., and Feng, H.H.P. (2001). Quantification of extracellular polymeric substances in biofilms by confocal laser scanning microscopy. Biotech. Letters *23*, 405–409.

Zimmermann, H. (1997). The microbial community on aggregates in the Elbe estuary. Aquat. Microb. Ecol. *13*, 37–46.

Zobell, C.E., and Anderson, D.Q. (1936). Observations on the multiplication of bacteria in different volumes of stored sea water and the influence of oxygen tension and solid surfaces. Biol. Bull. *71*, 324–342.

Zobell, C.E., and Oppenheimer, C.H. (1950). Some effects of hydrostatic pressure on the multiplication and morphology of marine bacteria. J. Bact. *60*, 771.

Zobell, C.E., and Morita, R.Y. (1957). Barophilic bacteria in some deep sea sediments. J. Bact. *73*, 563.

Laser Microscopy for the Study of Biofilms: Issues and Options

2

Thomas R. Neu and John R. Lawrence

Abstract

In this review, river biofilms and the various structural aspects of environmental biofilms are discussed. These include sample type and origin, cellular and polymeric constituents as well as examination at the micro- and meso-scale. For this purpose, selected studies are considered taking advantage of the confocal imaging approach for investigation of hydrated environmental biofilm samples. Emphasis is put on extracellular polymeric substances (EPS) as a multi-functional component of microbial biofilms. The main technical focus is on laser scanning microscopy by means of three-dimensional, multichannel imaging, specific staining techniques and digital image analysis. The advantages and limitations of the approach are critically assessed. Finally, research needs are listed in order to advance understanding of complex, environmental, real world biofilms.

Introduction

By definition, biofilm systems are a collection of microorganisms and their extracellular polymeric substances (EPS) associated with an interface. Aquatic biofilm systems comprise bio-films (stationary) and bio-aggregates (mobile), which both fulfil the definition of biofilm systems (see Chapter 1). EPS are organic polymers of microbiological origin which in biofilm systems form a matrix into which the microorganisms are embedded. At the same time EPS are responsible for the interaction with interfaces as well as with dissolved and particulate compounds. Environmental aquatic biofilm systems are heterogeneous assemblages composed of phototrophs (algae and cyanobacteria), fungi, bacteria, archaea and grazers (protists and benthic macroinvertebrates). This community is exposed to physicochemical parameters such as light, hydrodynamics, differential nutrient regimes and fluctuating conditions. The main factors shaping aquatic biofilms are twofold: (1) a complex prokaryote/eukaryote community including competitive and symbiontic interactions as well as grazing, and (2) long-term development and exposure to physicochemical environmental effects. Consequently, biofilms in natural aquatic ecosystems are rather different if compared to technical and particularly pure culture laboratory biofilms.

Aquatic biofilm systems are found in marine and freshwater habitats. The latter can be subdivided into lake and river biofilm systems. The focus in this chapter will be on river/lotic biofilm systems. In this respect the following philosophical statement might be important: 'you cannot step twice into the same river' (Heraclitus 535–475 BC). Thus heterogeneity, sampling, reproducibility and comparison represent critical issues in fluvial biofilm research. As a result, river biofilms remained for a long time a neglected field in microbiological research (Leff, 1994).

This chapter is based on two previous reviews on river biofilms discussing general aspects (Neu *et al.*, 2003) and the use of model systems (Lawrence and Neu, 2003) for the study of biofilm composition and microstructure. Furthermore, some recent reviews analyse the use of laser scanning microscopy for studying the general biofilm structure (Neu and Lawrence, 2014b), the biofilm matrix (Neu and Lawrence, 2014a) and the biofilm matrix functionality (Neu and Lawrence,

Table 2.1 Review articles on laser microscopy applications in biofilm research

Focus of review	Reference
First overview on CLSM applications	Caldwell *et al.* (1992)
Comprehensive review of CLSM	Lawrence *et al.* (1998b)
CLSM methodology	Lawrence and Neu (1999)
Short CLSM overview	Palmer, Jr. and Sternberg (1999)
Structured CLSM approach	Neu and Lawrence (2002)
One-photon LSM versus two-photon LSM	Neu and Lawrence (2005)
Spatio-temporal approaches	Palmer, Jr. *et al.* (2006)
Environmental CLSM applications	Lawrence *et al.* (2007a)
CLSM techniques and protocols	Lawrence and Neu (2007a)
CLSM of aggregates	Lawrence and Neu (2007b)
CLSM–MRI–STXM	Neu *et al.* (2010)
CLSM of hydrocarbon biofilms	Neu and Lawrence (2010)
In situ biofilm matrix analysis	Neu and Lawrence (2014a)
CLSM applications	Neu and Lawrence (2014b)

2009). Laser microscopy based methods for the study of biofilms have also been extensively reviewed (Table 2.1). The main part of this chapter has a focus on biofilm structural investigations by means of confocal laser scanning microscopy (CLSM). For this purpose the various interfaces in river habitats are differentiated (solid–liquid, liquid–liquid, liquid–gas). The challenge of analysing the biofilm matrix including its functionality will also be dealt with. Before focusing on the biofilm structure and matrix, some general issues with respect to current understanding of terms and models are discussed. Then practical considerations regarding sample type and mounting are touched upon. Finally research needs are defined for future investigations of biofilm structure and function.

Significance of sample type and origin for laser microscopy studies

With respect to sample types one has to distinguish the two major forms of river biofilm systems, bio-aggregates and bio-films. Aggregates may come directly from the water column or be developed in microcosms such as plankton columns or rolling flasks. This sample type has to be mounted in special chambers to avoid deformation and squeezing of the fragile 3-dimensional structure. In contrast, biofilms can be associated with solid environmental surfaces such as pebbles and rocks, sand and sediments or plant surfaces and wood. A common strategy in studying river biofilms is the exposure of a defined substratum which is mounted on a special device for holding it in a certain position. In many cases the material used is readily available in the lab such as glass or plastic slides. Sometimes tailor-made materials have been employed as for example clay and ceramic tiles, wood, cut rocks and metals. Often the surface roughness and texture is manipulated in order to simulate environmental surfaces. A different approach takes advantage of micro- and mesocosms which are ideally set-up next to the river in order to assure continuous once-flow-through conditions. Microcosms are usually set-up in the laboratory (Singer *et al.*, 2006; Zippel *et al.*, 2007) or in a container next to the river (http://mesocosm.eu/magdeburg). Mesocosms consisting of large outdoor flow lanes are often built parallel to the river to be as close as possible to the real environment (Battin *et al.*, 2003; Besemer *et al.*, 2009b).

With respect to specimen preparation and mounting for CLSM another aspect comes from the type of microscope available, inverted or upright. The upright version of any laser scanning microscope is often more flexible for analysing a variety of environmental samples. Aggregates can be mounted and stained in coverwell chambers with protective spacers of different heights. Defined surfaces such as glass slides can be placed in a Petri dish, stained and then examined with water immersible lenses. Similarly, pebbles, rocks, twigs or shells can be placed in a dish with higher walls for observation of larger samples. It is also possible to form a moat on larger objects (i.e. with plasticine or silicone) to create a well for staining. For further details on laser microscopy of biofilm samples the reader is referred to a number of reviews focusing on different technical and applied aspects (see Table 2.1). More recently, optical coherence tomography (OCT), a mesoscale technique, was evaluated for studying microbial biofilms. For OCT the water immersed biofilm sample can be placed directly under the imaging probe. First results of OCT imaging applied to biofilm visualization are supplied below.

The laser microscopy approach for biofilm analysis

Laser microscopy, mostly used as confocal laser scanning microscopy (CLSM), has evolved as the method of choice for examination of biofilm structure from various habitats. Most importantly, CLSM allows three-dimensional, multichannel microscopy of fully hydrated, living or fixed samples. CLSM can be used for recording the reflection (cellular or environmental) and autofluorescence signals (pigments of phototrophs) of biofilms as well as the emission signals of added fluorochromes specific for certain biochemical constituents. The fluorochromes may target nucleic acids, proteins, glycoconjugates, lipids/membranes, enzymes as well as the microhabitat (e.g. pH). The reader again is referred to a number of reviews covering staining procedures for microbial biofilms (see Table 2.1).

Despite the many advantages of CLSM, the technique has a few disadvantages to be aware of if the biofilm structure has to be examined. One of them is laser penetration into thick light scattering and absorbing samples. Thus, for thick environmental biofilms (> 150 μm) or microbial mats, the sample still has to be embedded and sectioned. Another critical point is motility for example of protists. Most CLSMs are point scanners for high resolution imaging and as a result motile or drifting objects may cause either irregular stripes or dotted tracks. However, there are instruments with different (fast) scanners available as well as spinning disc CLSMs, which can handle motility issues. Finally, large differences in signal intensity could be an issue as the signal to noise ratio can be optimized for either the bright objects or for the very faint objects but not both.

Biofilm analysis by means of CLSM usually starts with direct visual examination of the unstained sample in order to 'see' the original structure and patterns as well as possible autofluorescence using the epifluorescence mode. In a second step in the CLSM mode, the sample is optically sectioned in order to record highly resolved images. In many cases one aim is to reveal bacterial cells and their distribution mostly by using a green emitting nucleic acid-specific fluorochrome such as Syto9 or SybrGreen. In aquatic biofilms this would mean using three laser lines for excitation of the nucleic acid stain, the autofluorescence of the cyanobacteria (phycobilin and chlorophyll *a*) and the autofluorescence of algae (only chlorophyll *a*). In addition, the first line can be used for recording the reflection signal and the second line could be used for adding, for example, a lectin in order to visualize glycoconjugates. As a consequence, emission signals are recorded in four channels with the need for analysing co-localization of the phycobilin and chlorophyll *a* signal in order to isolate the cyanobacteria and lectin signal. Most laser microscopes offer the option of non-confocal transmission imaging which can be a useful addition. The variety and complexity of river biofilms examined by means of fluorescence techniques in combination with laser microscopy is presented in Figs. 2.1 and 2.2.

Laser microscopy of aquatic biofilms

CLSM has been employed for studying numerous biofilm systems from very different habitats for assessment of species specific features as well as environmental and treatment effects (Neu and Lawrence, 2014b). In the following section several selected studies with a focus on aquatic biofilms

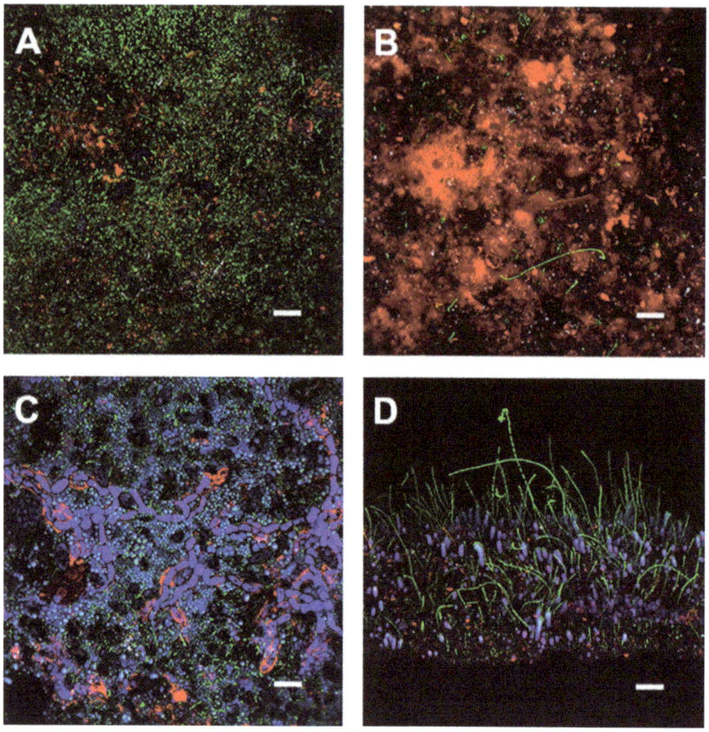

Figure 2.1 CLSM of river biofilms on pebble-sized stones illustrating heterogeneity of the biofilm landscape. Data sets are shown as maximum intensity projections. Colour allocation: white, reflection; green, bacteria (nucleic acid staining); red, glycoconjugates (lectin staining); turquoise, cyanobacteria (autofluorescence); blue, algae (autofluorescence). Scale bar = 20 μm. (A) Location dominated by bacteria of different morphology with only little glycoconjugate signal in between partly clustered. (B) Location with a bacterial base biofilm covered with large cloud-like glycoconjugate clusters forming the interface to the bulk water phase. (C) Location showing a cobblestone-like phototrophic biofilm of cyanobacteria (turquoise) and algae (blue) with bacteria in between and some glycoconjugates. (D) Location with algal and bacterial filaments oriented perpendicular to the rock surface with few glycoconjugates.

and using CLSM as a major tool will be discussed. One of the first studies demonstrating the suitability of CLSM for optical sectioning of biofilms showed the complex architecture in biofilms formed by different bacterial strains (Lawrence *et al.*, 1991). Depending on the species, cell accumulations at the bottom or top of the biofilm were found. In between the cellular biofilm constituents void spaces and channels could be identified. Similar findings were reported from experiments with contaminated groundwater in a fluidized bed reactor with biofilms developed on granular activated carbon (Massol-Deya *et al.*, 1995).

CLSM as a major tool for examining river biofilm systems was employed in a series of studies each with a different focus. In a first study the structure of river biofilms developed in rotating annular reactors was investigated (Neu and Lawrence, 1997). The turbulent flow produced a biofilm architecture with ridges oriented to flow direction. These ridges increased the surface area and created microhabitats. The bacteria were mainly colonizing the outer region of the ridges which may be composed of colloidal and particulate organic matter. Lectin staining showed large areas/volumes with lectin-specific EPS glycoconjugates. The CLSM technique employed was systematically tested and improved using the three-channel approach (Lawrence *et al.*, 1998a). In the meantime this approach was extended by adding another channel for reflection imaging as well as cyanobacteria signals via co-localization analysis of the red and far red channel as for example shown in a study on rock biofilms (Horath

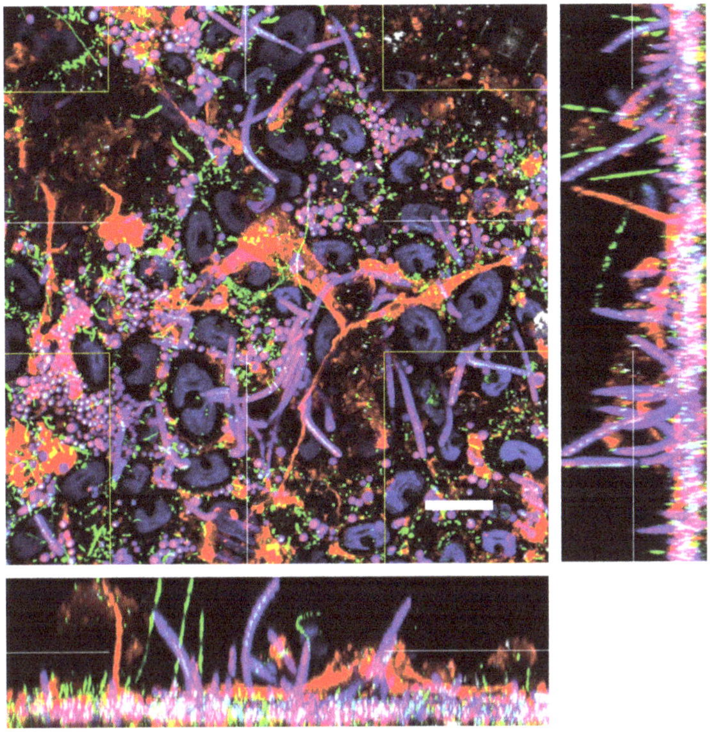

Figure 2.2 CLSM of a typical phototrophic river biofilm on a pebble stone demonstrating the 3-dimensional arrangement of biofilm constituents. The data set is shown as XY maximum intensity projection from the top (large image) as well as XZ and YZ projection from the side within the yellow cross in the XY image (rectangular images). Colour allocation: white, reflection; green, bacteria (nucleic acid staining); red, glycoconjugates (lectin staining); pink, cyanobacteria (autofluorescence); blue, algae (autofluorescence). Note that the white signal in the two side view images is a result of triple signals from the red-green-blue (RGB) overlay. The parallel lines in the side view images are caused by the user defined step size in axial direction. Scale bar = 20 µm.

et al., 2006). Thereby five different channels can be used for imaging of river biofilm constituents. In a subsequent study the river biofilms were examined using rRNA targeted probes for Alpha-, Beta- and Gamma-proteobacteria as well as the *Cytophaga–Flavobacterium* (Manz *et al.*, 1999). The river biofilms showed a distinct succession starting with Beta-proteobacteria, followed by a shift towards Alpha-proteobacteria and bacteria of the *Cytophaga–Flavobacterium* group. In terms of morphology Beta-proteobacteria showed the highest diversity. In order to run many replicate reactors, a new simple and inexpensive rotating annular reactor design was developed (Lawrence *et al.*, 2000). A comparison with the commercial system showed similar results for biofilm structure and composition. As the *in situ* analysis of the biofilm matrix remains difficult, a detailed study assessing the use of fluorescently labelled lectins was carried out (Neu *et al.*, 2001). For this purpose the lectins were applied as single, or double and triple combinations to examine glycoconjugate distribution in hydrated river biofilms.

One aspect of river biofilms is their role in sorption and degradation of contaminants which may affect biofilm structure and function. Thus, by using biofilm reactors fed with river water, the fates of atrazine and diclofop methyl were followed (Lawrence *et al.*, 2001). By using immunostaining, the sorption of atrazine to a unique microcolony was demonstrated. Both herbicides became sorbed and degraded within the biofilm which may act as a sink for both herbicides. Another study focused on the degradation of selected pharmaceuticals in river biofilms (Winkler *et al.*, 2001). It could be shown that ibuprofen was readily degraded and two degradation products could be identified. In contrast, clofibric acid was not degraded in the reactor biofilms run with river water either from the South Saskatchewan or the Elbe River. With respect to elemental contamination the effect of nickel, oxygen and nutrients on the river biofilm community was investigated (Lawrence *et al.*, 2004). Nickel concentrations at the industrial release rate had a dramatic effect on both cyanobacteria and the total phototrophic biomass. Nickel also reduced the carbohydrate utilization spectrum of the biofilm bacteria and had a negative impact on denitrification.

The response of biofilms towards nutrient additions was studied in another experimental series (Neu *et al.*, 2005). The focus was put on the presence of lectin-specific glycoconjugates produced during treatment with carbon, nitrogen, phosphorus and a combination thereof. The individual treatments showed differences in photosynthetic biomass, bacterial biomass and glycoconjugates when compared with control biofilms. It could also be shown that the reactors with single nutrient additions had a similar glycoconjugate make-up. This was also the case when the carbon, nitrogen and phosphorus treatment was compared with the control biofilms. In a detailed lectin study the microhabitat of river biofilms and microcolonies was examined (Lawrence *et al.*, 2007b). Within the microcolonies, different microdomains were identified by using a combination of lectins. It was found that some colonies had a specific glycoconjugate distribution located on the cell surface, in between the bacterial cells and around the microcolony. Thereby bacteria have the potential to design and control their immediate microenvironment. A further aspect in this study was the combination of fluorescence *in situ* hybridization (FISH) and fluorescence lectin-binding analysis (FLBA). This allowed allocation of both bacterial identity and glycoconjugate identity.

The CLSM approach was also employed by Battin and co-workers in studies on river ecosystem processes. They could show that biofilms changed the physicochemical microhabitat and thereby have a dramatic effect on transient storage of organic compounds and the retention of suspended particles (Battin *et al.*, 2003). In a later study the focus was on flow velocity in order to examine the effect of laminar, intermediate and turbulent flow on biofilm architecture and community composition (Besemer *et al.*, 2007). For this purpose, a laboratory flow lane microcosm with ceramic tiles was used. The flow velocity showed a clear effect resulting in uniform biofilms under laminar flow to ridged biofilms with streamers under turbulent flow. In addition, algae were identified as ecosystem engineers. Algae were initially part of garland-like streamers composed of bacteria and EPS decorated with diatoms. Later, filamentous algae dominated the biofilm which are, compared to bacteria, forming larger building blocks thereby changing the fluid dynamics and creating microenvironments. The effect of spatially structured landscapes on biofilm community composition was studied using large streamside flumes (Besemer *et al.*, 2009a,b). It was found that the community composition along the spatially structured landscape showed a high variation between base biofilm and streamers and suggested species sorting within different microhabitats. This heterogeneity eventually led to changes in resource use as studied by measuring carbon uptake in the form of glucose and dissolved organic carbon (Singer *et al.*, 2010).

Laser microscopy of aquatic aggregates

Similar to biofilm studies, CLSM can be employed to examine bioaggregates, aquatic flocs or reactor granules as indicated above. The first publication in this field combined *in situ* photography, laser microscopy and chemical analysis for studying fixed marine aggregates (Cowen and Holloway, 1996). Later, the group summarized the approach in a methodological manuscript in which they detailed the various fluorescence staining techniques applied to marine snow (Holloway and Cowen, 1997). CLSM has also been applied for examination of multiple parameters in non-fixed river snow (Neu, 2000). The samples were imaged in the reflection and fluorescence mode in order to visualize mineral particles, autofluorescence and specific fluorescent probes added. With a focus on glycoconjugates, the application of different fluorescently labelled lectins revealed the distribution of various glycoconjugate clusters within individual river snow aggregates. In parallel, the identity of river snow bacteria can be assessed by FISH using rRNA targeted probes (Böckelmann *et al.*, 2000). In a follow up experiment the FISH technique was combined with FLBA. By this means, the identity of bacterial species or groups can be correlated with the production of extracellular glycoconjugates (Böckelmann *et al.*, 2002). In a comparative study of aggregates from the rivers Danube and Elbe, methodological and seasonal aspects were investigated (Luef *et al.*, 2009b). Small aggregates were examined directly whereas larger ones were embedded, cryo-sectioned and post-stained. Samples were positively and negatively stained with a focus on bacteria and glycoconjugate distribution. The CLSM approach was pushed to its limit in an investigation of viruses associated with river snow (Luef *et al.*, 2009a). Although at the edge of detection, viruses could be imaged on native hydrated aggregates. The report critically discussed technical challenges which can be overcome by following certain protocols. The first combination of a full FLBA with all commercially available lectins combined with catalysed reporter deposition – fluorescence *in situ* hybridization (CARD-FISH) was applied to a >10 μm plankton fraction collected at Helgoland surface waters (Bennke *et al.*, 2013). Focus was put on the interaction of bacteroidetes with diatoms and the role of glycoconjugates. It could be demonstrated that different bacteroidetes clades interacted with diatoms using different glycoconjugates suggesting their essential role in the formation of marine aggregates.

Laser microscopy of biofilms at liquid and gaseous interfaces

Interfaces in aquatic environments are not only represented by solid surfaces or associated with mobile aggregates. Biofilms may also be found at liquid–liquid interfaces such as water–oil interfaces or at water–air interfaces. Hydrocarbon associated biofilms and their examination by CLSM has already been reviewed and discussed in methodological detail elsewhere (e.g. Neu and Lawrence, 2010). Only a few groups have actually used CLSM for looking at bacteria colonizing the oil–water interface. For example, Whyte *et al.* (1999) investigated the degradation of diesel at low temperature and showed the colonization of the droplets by *Rhodococcus*. Baldi *et al.* (2003) also studied the adhesion of bacteria to diesel droplets (Baldi *et al.*, 1999) as well as the growth of yeast on different hydrocarbons. The interaction and degradation of PCB by bacteria was shown in a study with a mixed culture isolated from contaminated soil. In a simple aquarium experiment the bacteria from the sediment colonized PCB droplets attached to floating plastic slides (Macedo *et al.*, 2005). A subsequent experiment looked at the adaptation of the microbial community to different PCB levels (Macedo *et al.*, 2007). Later, in a methodological publication CLSM was again suggested as a tool for imaging the colonization of aromatic hydrocarbons by *Mycobacteria* (Wouters et al., 2010). Nevertheless, imaging of oil–water mixtures by CLSM remains difficult with respect to mounting the samples. Challenges arise due to the specific weight (floating of hydrocarbon droplets) and hydrophobicity (sorption of hydrocarbon droplets).

An often overlooked interface in aquatic habitats is represented by the water–air interface (Norkrans, 1981). This extreme habitat, also called neuston, is characterized by an aquatic surface layer. The interface is enriched in biochemical constituents forming a kind of gelatinous layer (Wurl and Holmes, 2008; Cunliffe and Murrell, 2009). As a consequence, this nutrient-enriched layer attracts microorganisms (Franklin *et al.*, 2005). For studying sea surface microlayers, a range of different devices were employed in order to collect material for analysis (Maki, 1993). The formation and distribution of sea surface microlayers was the subject of several reviews (Wurl *et al.*, 2011; Cunliffe *et al.*, 2013). Somewhat similar are microbial surface layers forming in standing water samples (e.g. vases or puddles) or liquid microbial cultures (e.g. *Bacillus*). To the best of our knowledge these layers were examined by means of CLSM in only one survey (Krol *et al.*, 2011). The focus of this study was on plasmids transfer in *E. coli* biofilms. It turned out that, compared to biofilms at the solid–liquid interface the plasmid transfer at the water–air interface was much higher. It was suggested that apart from population density, the oxygen concentration may be the crucial factor for plasmid transfer.

Laser microscopy for studying biofilm grazing

In environmental habitats biofilms are exposed to grazing by protists and macroinvertebrates. This means there is a dynamic reciprocity between biofilm growth, biofilm detachment and sloughing as well as biofilm grazing which eventually will determine the overall biofilm architecture. There are a number of reports on biofilm grazing but only few examined biofilm structure by using CLSM. Initially, an attempt was made to visualize protists in pure culture as well as inside mature biofilms (Martin-Cereceda *et al.*, 2001; Packroff *et al.*, 2002). It turned out that it is difficult to identify a staining procedure which separates the bacterial biofilm signal from the protist signal. As protists have different feeding modes it can be expected that they affect the biofilm structure in different ways. This was followed in a flow cell study with stream biofilms (Böhme *et al.*, 2009). As a function of the protists feeding strategy, grazing led to rougher more porous biofilms and it may alter bacterial cell distribution, microcolony growth and settlement of drifting bacteria. Similarly, the effect of a free-swimming filter feeder and a surface-associated predator was investigated on pure culture biofilms (Dopheide *et al.*, 2011). It was shown that protists have a preference for certain types of biofilms based on chemical cues. Furthermore, it turned out that the filter feeder changed the biofilm structure more dramatically resulting in holes, channels and isolated grazing-resistant microcolonies. In a series of reports on grazing by the group of Kjelleberg, several different biofilms and protists were combined in order to detail their interactions. In a first study, the effect of a surface feeding flagellate was tested against biofilms of various *Pseudomonas* mutants (Matz *et al.*, 2004). It was demonstrated that *P. aeruginosa* formed microcolonies and produced toxins as a response to the grazer. A subsequent survey used protists with different feeding strategies. They could show that microcolony formation is only useful in early stages of biofilm development whereas inhibitor production was effective against a wide range of feeding types (Weitere *et al.*, 2005). In experiments with *Serratia* the role of quorum sensing in response to a suspension feeder and a surface feeder was examined (Queck *et al.*, 2006). It was suggested that differentiation into cell chains and filamentous biofilms provides protection against grazing. In a recent study with *Vibrio cholerae* WT and different mutants the focus was on bacterial polysaccharide production and quorum sensing (Sun *et al.*, 2013). It turned out that *V. cholerae* possesses various defence mechanisms against grazing pressure. In a study with macroinvertebrates, CLSM was also employed as a tool to examine biofilm structure before and after grazing. For example the effect of grazing was followed after exposing river biofilms to different benthic invertebrates (Lawrence *et al.*, 2002). It could be demonstrated that grazing by mayflies, ostracods and snails resulted in various biofilm patterns due to differential grazing strategies of the individual species.

Laser microscopy for exploring biofilm metabolism

It is also important, but difficult to determine, the metabolic status of microorganisms. CLSM however, offers a number of approaches based on fluorescent reporters. In the meantime, many combinations of fluorochromes have been suggested for measuring the viability of microorganisms. Relevant references were listed in a recent review article (see tables 7 and 8 in Neu and Lawrence, 2014b). Live–dead staining, in particular, has received a wide application and when used in a standardized protocol can provide a useful index of environmental impacts on biofilm communities. Nevertheless, a number of issues exist with *in situ* use of fluorescent reporters in 'open' and complex environmental biofilms. In agreement with Caldwell and collaborators (Caldwell *et al.*, 1997) other authors have more recently found discrepancies between fluorescence staining and culture results. These studies reinforce the dictum that all fluorescent staining should be subject to effective 'ground truth.' Several reports which critically assessed the status of viability and the application of fluorochrome combinations for measuring viability are listed for further details

Table 2.2 Critical assessments of fluorochrome combinations for bacterial viability

Focus of review	Reference
Vigour, vitality, viability	Lloyd and Hayes (1995)
Viability assessment via fluorochromes	Breeuwer and Abee (2000)
Molecular methods	Keer and Birch (2003)
Application of the BacLight kit	Stocks (2004)
BacLight kit in flow cytometry	Berney *et al.* (2007)
Mechanisms of probes for viability	Sträuber and Müller (2010)
Red but not dead? – Yeast study	Davey and Hexley (2011)
Life, death and in between: ...	Davey (2011)
Confusion over live–dead staining	Netuschil *et al.* (2014)

(Table 2.2). The timeliness of the topic 'bacterial viability' culminated in an ongoing conference series with the title 'How dead is dead?'

Large area imaging of biofilms

In order to image large biofilm areas there are two options if CLSM is employed. Firstly, a low-magnification lens with high numerical aperture (NA) (e.g. 25x, NA 0.95) can be used. Secondly, newer CLSM setups with a motorized microscope table have the option of running a so-called tile or mosaic scan. This allows recording of multiple areas at high magnification and high resolution which are finally stitched together into one large dataset. Although an attractive option, issues arise in terms of time for recording data (several hours), data size (several gigabytes) and flatness of the substratum (imaging of black volumes if substratum is tilted or structured in three dimensions). The advantage of the CLSM however lies in the multichannel mode which enables the collection of multiple biofilm parameters. Another option could be optical coherence tomography (OCT) which originated in the medical field. OCT is an interferometric technique and is based on reflection only. The advantages of OCT are its non-invasiveness, no need for adding contrasting agents and a large scan area in the millimeter range. The first application of OCT for studying biofilms was demonstrated in capillary flow cell biofilms (Xi *et al.*, 2006). Then the technique was applied to follow biofilm development in a tube reactor in order to assess the effect of different treatments (Haisch and Niessner, 2007). A direct comparison of CLSM and OCT was reported by examination of biofilms developed in a flow lane under laminar, transient and turbulent flow conditions (Wagner *et al.*, 2010). The biofilm showed different porosities depending on flow conditions. It was also demonstrated that the mesoscale structure was represented much better as compared to CLSM. Nevertheless, CLSM is needed in order to probe for different biochemical features of biofilms. The effect of biofilm grazing by eucaryotes may be also investigated by OCT (Derlon *et al.*, 2012). This was shown on ultrafiltration membranes driven by gravity and exposed to different fluxes and grazing pressure. In a follow up manuscript, a more detailed study compared OCT with CLSM, again in order to take advantage of both techniques (Derlon *et al.*, 2013). Although OCT has certain advantages in terms of depth resolution, similar to CLSM shading effects are an issue with thick samples.

Biofilm matrix

After many years of biofilm research, the biofilm matrix still represents the 'dark matter' of microbial biofilms. Although the extracellular space and its constituents, often termed EPS, is accepted as an essential part of biofilm systems, its characterization is still in its infancy. The challenge is based on matrix complexity as well as on the methods available for chemical and *in situ* analysis. Biochemically the EPS constituents comprise polysaccharides, proteins, nucleic acids, amphiphilic/lipophilic compounds and bacterial refractory compounds. In terms of constituents they may be soluble (most), insoluble (cellulose, mutan, amyloids) or structural (vesicles, nanotubes, pili). Consequently the biofilm matrix or, in other words, the extracellular space defined by the matrix constituents, initially meant to be only polysaccharides and/or proteins, becomes more and more differentiated. In this respect it is important to look at the extracellular space in general. It may comprise bacterial structures extruding from the cell surface into the extracellular space (tubes, pili), the components shed from the cell surface (membrane vesicles, amyloids), the polymers actively released (polysaccharides, proteins) but also low molecular metabolites produced and released by the bacteria. On top of that, compounds from the environment (e.g. humic compounds and colloids) will become associated and embedded into the matrix originally produced by microorganisms. Thus, the question arises: what finally makes up the biofilm matrix? Maybe the acronym EPS standing for 'extracellular polymeric substances' as a synonym for the biofilm matrix has to be questioned and discussed once again (Allison *et al.*, 2003).

A further aspect to be discussed at this point is the common model with respect to biofilm structure and matrix which has been used in many publications. The so-called 'mushroom model' seems to be quite popular but in fact represents a flow-cell artefact. The model is of course correct under laminar flow conditions in a small flow cell designed for microscopy of the developing pure or mixed culture biofilms (Crusz *et al.*, 2012). However, by looking at numerous biofilm samples from the environment as well as technical habitats it was found that the structural variation is much larger and in most cases there are no mushroom structures. Similarly, the general drawing of biofilms as surface-associated bacteria surrounded by a polymer matrix has to be questioned. Again this might be the case in many biofilms, however from environmental and biofilm reactor samples often structural features with distinct filamentous bacteria have been found. The filaments are often sticking vertically out of a basal biofilm with no obvious polymer matrix surrounding them. Another observation is the coverage of environmental biofilms by a network of very thin filamentous bacteria also showing no polymer matrix. Clearly the 'biofilm model' remains a topic for debate and discussion.

Laser microscopy of the biofilm matrix

This next section will elaborate on studies targeting the biofilm matrix. The studies will be discussed according to their focus and approach into: chemical analysis, *in situ* imaging, proteomics and the functional role of EPS.

To date the biofilm matrix has mainly been examined using two strategies, by traditional chemical techniques and by differential imaging. Chemical analysis of the matrix is usually done via extraction and subsequent biochemical analysis (Nielsen and Jahn, 1999). The critical issue is potential cell lyses and contamination of the EPS fraction by intracellular polymers. In any case, this can be done easily in pure culture studies where usually only a few polymers are produced which can be purified and characterized. Whereas from environmental samples containing numerous polymer types (Staudt *et al.*, 2003) only bulk chemical information in the form of per cent polysaccharides, proteins, etc. becomes available. References covering extraction with subsequent chemical analysis were compiled in a protocol on biofilm matrix assessment, see (Neu and Lawrence, 2014a).

The second approach takes advantage of *in situ* imaging, e.g. by CLSM employing matrix-specific probes. In comparison to electron microscopy where fixation and dehydration are necessary, CLSM does not require these artefact prone procedures. Suitable fluorochromes and probes for the biofilm matrix have been compiled in Table 2.3. Some of the fluorochromes, e.g. the cell impermeant ones may give false signals due to the staining of intracellular constituents in leaky cells. Other fluorochromes may give a double signal from the biofilm matrix as well as from biofilm cells as for example protein-specific dyes. Consequently differential imaging is required in order to separate the individual signals. The most frequently used approach is now called fluorescence lectin-binding analysis (FLBA). This comprises a full scale screening with all the commercially available lectins in order to identify a panel of useful

Table 2.3 Fluorochromes and probes suitable for biofilm matrix constituents

Fluorochrome/probe	Target	Reference
Fluorescently labelled lectins (general precedure)	Glycoconjugates	Neu and Lawrence (1999b), Neu and Lawrence (1999a), Neu *et al.* (2001)
Lectin screening (using about 70 lectins commercially available)	Selecting a panel of useful lectins for a particular sample	Staudt *et al.* (2003), Laue *et al.* (2006), Peltola *et al.* (2008), Luef *et al.* (2009b), Zippel and Neu (2011), Bennke *et al.* (2013), Weissbrodt *et al.* (2013), Zhang *et al.* (2014)
ELF 97 phosphate	Phosphatase activity	Paragas *et al.* (2002), Nedoma *et al.* (2003)
DDAO	eDNA	Allesen-Holm *et al.* (2006)
Cell impermeant dyes (TOTO-, TO-PRO-, SYTOX-series)	eDNA	Li *et al.* (1995), Roth *et al.* (1997), Lebaron *et al.* (1998)
SYPRO series (e.g. SYPRO-ORANGE)	Protein in general	Zubkov *et al.* (1999)
FM lipophilic dyes (FM 1-43, FM 4-64)	Extracellular vesicles (potentially useful)	Sharp and Pogliano (1999)

probes matching a particular biofilm system or a specific structural cue. Screening with about 70 different lectins has been applied already to a number of different biofilm systems such as reactor biofilms inoculated with river water or activated sludge (Staudt *et al.*, 2003), *Pseudomonas* biofilms (Laue *et al.*, 2006), *Deinococcus* biofilms (Peltola *et al.*, 2008), river aggregates (Luef *et al.*, 2009b), phototrophic biofilms (Zippel and Neu, 2011), marine aggregates (Bennke *et al.*, 2013), aerobic granules (Weissbrodt *et al.*, 2013) and acidophilic archaeal biofilms (Zhang *et al.*, 2014). Although potentially laborious, this approach gives rise to a targeted panel of lectins with maximal effectiveness for each system.

More recently biofilms have been subjected to the -omic approaches including proteomics of the biofilm matrix. In a very limited number of papers the extracellular proteins of pure culture biofilms were characterized with respect to their function. Examples are available for *Haemophilus* (Gallaher *et al.*, 2006), *Myxococcus* (Curtis *et al.*, 2007), *Listeria* (Dumas *et al.*, 2008) and *Shewanella* (Cao *et al.*, 2011). In most of the studies proteins with new or unknown functions were found. Even less is known for complex environmental biofilms. The few reports are from dental plaque where the effect of sucrose on matrix protein composition was investigated (Paes Leme *et al.*, 2008). Another survey focused on biofilms in acid mine drainage and compared cellular proteins with matrix proteins which were found to be different and partly unknown (Jiao *et al.*, 2011). A combination of genomics (with focus on alginate genes) and proteomics (with focus on amyloids) was employed for the analysis of activated sludge samples (Albertsen *et al.*, 2013). Interestingly most alginate related genes could be allocated to bacteroidetes and not to *Pseudomonas*.

EPS functionality represents a topic were two clusters of publications can be discussed. One of them is represented by a series of review articles compiling facts (polymer identities and properties) and references (of other reviews) finally making suggestions of what could be a function. In a first comprehensive book the many aspects of EPS identity, analysis and function were compiled (Wingender *et al.*, 1999). Several of these chapters together with other review articles were listed, assessed and critically discussed in a review on matrix functionality (Neu and Lawrence, 2009). As a result and by focusing on the nature and role of the biofilm matrix a still evolving concept of matrix functionality was suggested. The many functions maybe compiled as a list or allocated as a matrix showing multi-functionality (Table 2.4). In any case, the concept of matrix functionality may help to streamline and focus the discussion as well as future research in this direction. The other cluster is represented by pure culture studies focusing mainly on a single polymer and its function. Major studies were done with *Escherichia*,

Table 2.4 Biofilm matrix characteristics and functionality arranged to indicate multi-functionality of matrix constituents (Neu and Lawrence, 2009)

constructive polysaccharides proteins	**sorptive** polysaccharides	**adhesive** polysaccharides proteins	**anti-adhesive** amphiphiles polysaccharides
cohesive proteins (amyloids)	**active** proteins vesicles	**surface-active** amphiphiles vesicles proteins	**informative** nucleic acids vesicles proteins (lectins)
nutritive polysaccharides proteins nucleic acids	**locomotive** amphiphiles polysaccharides proteins	**conductive** proteins nucleic acids	**redox-active** bacterial refractory compounds

Pseudomonas, *Bacillus*, *Streptococcus*, among others. However, the many articles published are beyond the scope of this chapter.

Research needs

CLSM is now an established technique for *in situ* assessment of microbial biofilm systems from many different habitats. Nevertheless, in order to progress, new probes are needed especially for the complex constituents of the biofilm matrix. In terms of resolution, new techniques may help to elucidate new features of interfacial microbes. This includes, firstly, imaging at deep locations using multi-photon imaging (Neu and Lawrence, 2005; Neu *et al.*, 2004; Horton *et al.*, 2013) or by employing selective plane illumination microscopy (SPIM) (Huisken and Stainier, 2009). Secondly, the various nanoscopy techniques having a resolution across the diffraction barrier including structured illumination microscopy – SIM (Schermelleh *et al.*, 2008), stimulated emission depletion (STED) microscopy (Klar *et al.*, 2000) and blink microscopy – dSTORM/GSDIM (van de Linde *et al.*, 2008; Fölling *et al.*, 2008) will allow further progress. These demanding techniques may help to elucidate intracellular and cell surface features of hydrated microorganisms. For large area imaging optical coherence tomography (OCT) may be the method of choice, however it may need to be further improved with respect to data recording (size of area, speed) and image analysis. In addition, new tools for imaging are required at various levels. They should include fast recording of three-dimensional multichannel data sets, online deconvolution on the fly, quantification and extraction of structural information. Another area of progress will be in the field of chemical imaging. The various tools (e.g. Raman microscopy, nano-SIMS, synchrotron imaging) have already been used in a limited number of studies (Neu *et al.*, 2010). However, what is needed should comprise traditional light/laser microscopy and electron microscopy/tomography plus chemical imaging. This approach is already known and partly established as correlative microscopy. Finally, in the age of –omics, proteomics and transcriptomics are required for addressing identity and function of matrix constituents. A totally neglected -omics area is glycomics which still has to be applied in studies of microbial biofilms and especially the biofilm matrix.

Acknowledgements

The expert support by Ute Kuhlicke and George D.W. Swerhone over many years of laser microscopy on many types of biofilm systems is highly appreciated. We are very thankful to Norbert Kamjunke for collecting rock samples in the Harz mountain creeks (Figs. 2.1 and 2.2).

References

Albertsen, M., Stensballe, A., Nielsen, K.L., and Nielsen, P.-H. (2013). Digging into the extracellular matrix of a complex microbial community using a combined metagenomic and metaproteomic approach. Water Sci. Technol. 67, 1650–1656.

Allesen-Holm, M., Bundvik Barken, K., Yang, L., Klausen, M., Webb, J.S., Kjelleberg, S., Molin, S., Givskov, M., and Tolker-Nielsen, T. (2006). A characterisation of DNA release in *Pseudomonas aeruginosa* cultures and biofilms. Mol. Microbiol. 59, 1114–1128.

Allison, D.G., Sutherland, I.W., and Neu, T.R. (2003). EPS: What's in an acronym? In Biofilm Communities: Order from Chaos? McBain, A., Allison, D.G., Brading, M., Rickard, A., Verran, J., and Walker, J., eds. (BioLine, Cardiff), pp. 381–387.

Baldi, F., Ivosevic, N., Minacci, A., Pepi, M., Fani, R., Svetlicic, V., and Zutic, V. (1999). Adhesion of *Acinetobacter venetianus* to diesel fuel droplets studied with *in situ* electrochemical and molecular probes. Appl. Environ. Microbiol. 65, 2041–2048.

Baldi, F., Pepi, M., and Fava, F. (2003). Growth of Rhodosporium toruloides strain DBVPG 6662 on dibenzothiophene crystals and orimulsion. Appl. Environ. Microbiol. 69, 4689–4696.

Battin, T.J., Kaplan, L.A., Newbold, J.D., and Hansen, C.M.E. (2003). Contributions of microbial biofilms to ecosystem processes in stream mesocosms. Nature *426*, 439–442.

Bennke, C.M., Neu, T.R., Fuchs, B.M., and Amann, R. (2013). Mapping glycoconjugate-mediated interactions of marine Bacteroidetes with diatoms. Syst. Appl. Microbiol. 36, 417–425.

Berney, M., Hammes, F., Bosshard, F., Weilenmann, H.-U., and Egli, T. (2007). Assessment and interpretation of bacterial viability by using the LIVE/DEAD baclight kit in combination with flow cytometry. Appl. Environ. Microbiol. 73, 3283–3290.

Besemer, K., Hödl, I., Singer, G., and Battin, T.J. (2009a). Architectural differentiation reflects bacterial community structure in stream biofilms. The ISME Journal 3, 1318–1324.

Besemer, K., Singer, G., Hödl, I., and Battin, T.J. (2009b). Bacterial community composition of stream biofilms in spatially variable-flow environments. Appl. Environ. Microbiol. 75, 7189–7195.

Besemer, K., Singer, G., Limberger, R., Chlup, A.-K., Hochedlinger, G., Hödl, I., Baranyi, C., and Battin, T.J. (2007). Biophysical controls on community succession in stream biofilms. Appl. Environ. Microbiol. 73, 4966–4974.

Böckelmann, U., Manz, W., Neu, T.R., and Szewzyk, U. (2000). Characterization of the microbial community of lotic organic aggregates ('river snow') in the Elbe river of Germany by cultivation and molecular techniques. FEMS Microbiol. Ecol. 33, 157–170.

Böckelmann, U., Manz, W., Neu, T.R., and Szewzyk, U. (2002). A new combined technique of fluorescent *in situ* hybridization and lectin-binding-analysis (FISH-LBA) for the investigation of lotic microbial aggregates. J. Microbiol. Meth. *49*, 75–87.

Böhme, A., Risse-Buhl, U., and Küsel, K. (2009). Protists with different feeding modes change biofilm morphology. FEMS Microbiol. Ecol. *69*, 158–169.

Breeuwer, P., and Abee, T. (2000). Assessment of viability of microorganisms employing fluorescence techniques. Int. J. Food Microbiol. 55, 193–200.

Caldwell, D.E., Korber, D.R., and Lawrence, J.R. (1992). Confocal laser scanning microscopy and digital image analysis in microbial ecology. Adv. Microb. Ecol. *12*, 1–67.

Caldwell, D.E., Wolfaardt, G.M., Korber, D.R., and Lawrence, J.R. (1997). Do bacterial communities transcend Darwinism? Adv. Microb. Ecol. *15*, 105–191.

Cao, B., Shi, L., Brown, R.N., Xiong, Y., Fredrickson, J.K., Romine, M.F., Marshall, M.J., Lipton, M.S., and Beyenal, H. (2011). Extracellular polymeric substances from Shewanella sp. HRCR-1 biofilms: characterization by infrared spectroscopy and proteomics. Environ. Microbiol. *13*, 1018–1031.

Cowen, J.P., and Holloway, C.F. (1996). Structural and chemical analysis of marine aggregates: In situ macrophotography and laser confocal and electron microscopy. Mar. Biol. *126*, 163–174.

Crusz, S.A., Popat, R., Rybtke, M.T., Cámara, M., Givskov, M., Tolker-Nielsen, T., Diggle, S.P., and Williams, P. (2012). Bursting the bubble on bacterial biofilms: a flow cell methodology. Biofouling *28*, 835–842.

Cunliffe, M., and Murrell, J.C. (2009). The sea-surface microlayer is a gelatinous biofilm. The ISME Journal 3, 1001–1003.

Cunliffe, M., Engel, A., Frka, S., Gasparovic, B., Guitart, C., Murrell, J.C., Salter, M., Stolle, C., Upstill-Goddard, R., and Wurl, O. (2013). Sea surface microlayers: A unified physicochemical and biological perspective of the air–ocean interface. Prog. Oceanogr. *109*, 104–116.

Curtis, P.D., Atwood III, J., Orlando, R., and Shimkets, L.J. (2007). Proteins associated with the *Myxococcus xanthus* extracellular matrix. J. Bacteriol. *189*, 7634–7642.

Davey, H.M., and Hexley, P. (2011). Red but not dead? Membranes of stressed Saccharomyces cerevisiae are permeable to propidium iodide. Environ. Microbiol. *13*, 163–171.

Davey, H.M. (2011). Life, death, and in-between: meanings and methods in microbiology. Appl. Environ. Microbiol. *77*, 5571–5576.

Derlon, N., Koch, N., Eugster, B., Posch, T., Pernthaler, J., Pronk, W., and Morgenroth, E. (2013). Activity of metazoa governs biofilm structure formation and enhances permeate flux during gravity-driven membrane (GDM) filtration. Water Res. *47*, 2085–2095.

Derlon, N., Peter-Varbanets, M., Scheidegger, A., Pronk, W., and Morgenroth, E. (2012). Predation influences the structure of biofilm developed on ultrafiltration membranes. Water Res. *46*, 3323–3333.

Dopheide, A., Lear, G., Stott, R., and Lewis, G. (2011). Preferential feeding by the ciliates *Chilodonella* and *Tetrahymena* spp. and Effects of these protozoa on bacterial biofilm structure and composition. Appl. Environ. Microbiol. 77, 4564–4572.

Dumas, E., Meunier, B., Berdague, J.-L., Chambon, C., Desvaux, M., and Hebraud, M. (2008). Comparative analysis of extracellular and intracellular proteomes of *Listeria monocytogenes* strains reveals a correlation between protein expression and serovar. Appl. Environ. Microbiol. *74*, 7399–7409.

Fölling, J., Bossi, M., Bock, H., Medda, R., Wurm, C.A., Hein, B., Jakobs, S., Eggeling, C., and Hell, S.W. (2008). Fluorescence nanoscopy by ground-state depletion and single-molecule return. Nat. Methods 5, 943–945.

Franklin, M.P., McDonald, I.R., Bourne, D.G., Owens, N.J.P., Upstill-Goddard, R.C., and Murrell, J.C. (2005). Bacterial diversity in the bacterioneuston (sea surface microlayer): the bacterioneuston through the looking glass. Environ. Microbiol. *7*, 723–736.

Gallaher, T.K., Wu, S., Webster, P., and Aguilera, R. (2006). Identification of biofilm proteins in non-typeable *Haemophilus influenzae*. BMC Microbiol. *6*, 1–9.

Haisch, C., and Niessner, R. (2007). Visualisation of transient processes in biofilms by optical coherence tomography. Water Res. *41*, 2467–2472.

Holloway, C.F., and Cowen, J.P. (1997). Development of a scanning confocal laser microscopic technique to examine the structure and composition of marine snow. Limnol. Oceanogr. *42*, 1340–1352.

Horath, T., Neu, T.R., and Bachofen, R. (2006). Characterization of endolithic cyanobacteria in dolomite rocks in central Switzerland by reflection spectroscopy, scanning electron microscopy and laser scanning microscopy. Microb. Ecol. *51*, 353–364.

Horton, N.G., Wang, K., Kobat, D., Clark, C.G., Wise, F.W., Schaffer, C.B., and Xu, C. (2013). In vivo three-photon microscopy of subcortical structures within an intact mouse brain. Nat. Photon. *7*, 205–209.

Huisken, J., and Stainier, D.Y.R. (2009). Selective plane illumination microscopy techniques in developmental biology. Development *136*, 1963–1975.

Jiao, Y., D'haeseleer, P., Dill, B.D., Shah, M., VerBerkmoes, N.C., Hettich, R.L., Banfield, J.F., and Thelen, M.P. (2011). Identification of biofilm matrix-associated proteins from an acid mine drainage microbial community. Appl. Environ. Microbiol. *77*, 5230–5237.

Keer, J.T., and Birch, L. (2003). Molecular methods for the assessment of bacterial viability. J. Microbial. Meth. *53*, 175–183.

Klar, T.A., Jakobs, S., Dyba, M., Egner, A., and Hell, S.W. (2000). Fluorescence microscopy with diffraction resolution barrier broken by stimulated emission. Proc. Natl. Acad. Sci. U.S.A. *97*, 8206–8210.

Krol, J.E., Nguyen, H.D., Rogers, L.M., Beyenal, H., Krone, S.M., and Top, E.M. (2011). Increased transfer of a multidrug resistance plasmid in *Escherichia coli* biofilms at the air-liquid interface. Appl. Environ. Microbiol. *77*, 5079–5088.

Laue, H., Schenk, A., Hongqiao, L., Lambertsen, L., Neu, T.R., Molin, S., and Ullrich, M.S. (2006). Contribution of alginate and levan to biofilm formation by *Pseudomonas syringae*. Microbiology *152*, 2909–2918.

Lawrence, J.R., Chenier, M.R., Roy, R., Beaumier, D., Swerhone, G.D.W., Neu, T.R., and Greer, C.W. (2004). Microscale and molecular assessment of the impacts of nickel, nutrients, and oxygen level on the structure and function of river biofilm communities. Appl. Environ. Microbiol. *70*, 4326–4339.

Lawrence, J.R., Kopf, G., Headley, J.V., and Neu, T.R. (2001). Sorption and metabolism of selected herbicides in river biofilm communities. Can J. Microbiol. *47*, 634–641.

Lawrence, J.R., Korber, D., and Neu, T.R. (2007a). Analytical imaging and microscopy techniques. In Manual of Environmental Microbiology, C.J. Hurst, R.L. Crawford, J.L. Garland, D.A. Lipson, A.L. Mills, and L.D. Stetzenbach, eds. (ASM Press, Washington, DC, USA), pp. 40–68.

Lawrence, J.R., Korber, D.R., Hoyle, B.D., Costerton, J.W., and Caldwell, D.E. (1991). Optical sectioning of microbial biofilms. J. Bacteriol. *173*, 6558–6567.

Lawrence, J.R., and Neu, T.R. (1999). Confocal laser scanning microscopy for analysis of microbial biofilms. Methods Enzymol. *310*, 131–144.

Lawrence, J.R., and Neu, T.R. (2003). Microscale analyses of the formation and nature of microbial biofilm communities in river systems. Reviews in Environmental Science and Bio/Technology *2*, 85–97.

Lawrence, J.R., and Neu, T.R. (2007a). Laser scanning microscopy. In Methods for General and Molecular Microbiology, Reddy, C.A., Beveridge, T.J., Breznak, J.A., Marzluf, G.A., Schmidt, T.M., and Snyder, L.R., eds. (ASM Press, Washington, DC, USA), pp. 34–53.

Lawrence, J.R., and Neu, T.R. (2007b). Laser scanning microscopy for microbial flocs and particles. In Environmental colloids: Behavior, structure and characterisation, Wilkinson, K.J., and Lead, J.R., eds. (John Wiley & Sons Inc., Chichester, UK), pp. 469–505.

Lawrence, J.R., Neu, T.R., and Swerhone, G.D.W. (1998a). Application of multiple parameter imaging for the quantification of algal, bacterial and exopolymer components of microbial biofilms. J. Microbial. Meth. *32*, 253–261.

Lawrence, J.R., Scharf, B., Packroff, G., and Neu, T.R. (2002). Microscale evaluation of the effects of grazing by invertebrates with contrasting feeding modes on river biofilm architecture and composition. Microb. Ecol. *43*, 199–207.

Lawrence, J.R., Swerhone, G.D.W., Kuhlicke, U., and Neu, T.R. (2007b). In situ evidence for microdomains in the polymer matrix of bacterial microcolonies. Can J. Microbiol. *53*, 450–458.

Lawrence, J.R., Swerhone, G.D.W., and Neu, T.R. (2000). A simple rotating annular reactor for replicated biofilm studies. J. Microbial. Meth. *42*, 215–224.

Lawrence, J.R., Wolfaardt, G., and Neu, T.R. (1998b). The study of microbial biofilms by confocal laser scanning microscopy. In Digital image analysis of microbes, Wilkinson, M.H.F., and Shut, F., eds. (John Wiley & Sons Inc., Chichester, UK), pp. 431–465.

Lebaron, P., Catala, P., and Parthuisot, N. (1998). Effectiveness of SYTOX green stain for bacterial viability assessment. Appl. Environ. Microbiol. *64*, 2697–2700.

Leff, L.G. (1994). Stream bacterial ecology: A neglected field? ASM News *60*, 135–138.

Li, W.K.W., Jellett, J.F., and Dickie, P.M. (1995). DNA distributions in planktonic bacteria stained with TOTO or TO-PRO. Limnol. Oceanogr. *40*, 1485–1495.

Lloyd, D., and Hayes, A.J. (1995). Vigour, vitality and viability of microorganisms. FEMS Microbiol. Lett. *133*, 1–7.

Luef, B., Neu, T.R., and Peduzzi, P. (2009a). Imaging and quantifying virus fluorescence signals on aquatic aggregates: a new method and its implication for aquatic microbial ecology. FEMS Microbiol. Ecol. *68*, 372–380.

Luef, B., Neu, T.R., Zweimüller, I., and Peduzzi, P. (2009b). Structure and composition of aggregates in two large European rivers, based on confocal laser scanning microscopy and image and statistical analysis. Appl. Environ. Microbiol. *75*, 5952–5962.

Macedo, A.J., Kuhlicke, U., Neu, T.R., Timmis, K.N., and Abraham, W.-R. (2005). Three stages of a biofilm community developing at the liquid–liquid interface between polychlorinated biphenyls and water. Appl. Environ. Microbiol. *71*, 7301–7309.

Macedo, A.J., Neu, T.R., Kuhlicke, U., and Abraham, W.-R. (2007). Adaptation of microbial communities in polychlorinated biphenyls contaminated soil leading to the transformation of higher chlorinated congeners in biofilm communities. Biofilms *3*, 37–46.

Maki, J.S. (1993). The air–water interface as an extreme environment. In Aquatic Microbiology, Ford, T.E., ed. (Blackwell, Cambridge, MA, USA), pp. 409–439.

Manz, W., Wendt-Potthoff, K., Neu, T.R., Szewzyk, U., and Lawrence, J.R. (1999). Phylogenetic composition, spatial structure, and dynamics of lotic bacterial biofilms investigated by fluorescent *in situ* hybridization and confocal laser scanning microscopy. Microb. Ecol. *37*, 225–237.

Martin-Cereceda, M., Alvarez, A.M., Serrano, S., and Guinea, A. (2001). Confocal and light microscope examination of protozoa and other microorganisms in the biofilms from a rotating biological contactor wastewater treatment plant. Acta Protozool. *40*, 263–272.

Massol-Deya, A.A., Whallon, J., Hickey, R.F., and Tiedje, J.M. (1995). Channel structures in aerobic biofilms on fixed-film reactors treating contaminated groundwater. Appl. Environ. Microbiol. *61*, 769–777.

Matz, C., Bergfeld, T., Rice, S.A., and Kjelleberg, S. (2004). Microcolonies, quorum sensing and cytotoxicity determine the survival of *Pseudomonas aeruginosa* biofilms exposed to protozoan grazing. Environ. Microbiol. *6*, 218–226.

Nedoma, J., Strojsova, A., Vrba, J., Komarkova, J., and Simek, K. (2003). Extracellular phosphatase activity of natural plankton studied with ELF97 phosphate: fluorescence quantification and labelling kits. Environ. Microbiol. *5*, 462–472.

Netuschil, L., Auschill, T., Sculean, A., and Arweiler, N. (2014). Confusion over live/dead stainings for the detection of vital microorganisms in oral biofilms – which stain is suitable? BMC Oral Health *14*, 2.

Neu, T.R. (2000). In situ cell and glycoconjugate distribution in river snow studied by confocal laser scanning microscopy. Aquat. Microb. Ecol. *21*, 85–95.

Neu, T.R., Eitner, A., and Paje, M.L. (2003). Development and architecture of complex environmental biofilms – lotic biofilm systems. In Fossil and Recent Biofilms, a Natural Histrory of Life on Earth, Krumbein, W.E., Paterson, D.M., and Zavarzin, G.A., eds. (Kluwer Academic Publishers, Dordrecht), pp. 29–45.

Neu, T.R., and Lawrence, J.R. (1997). Development and structure of microbial biofilms in river water studied by confocal laser scanning microscopy. FEMS Microbiol. Ecol. *24*, 11–25.

Neu, T.R., and Lawrence, J.R. (1999a). In situ characterization of extracellular polymeric substances (EPS) in biofilm systems. In Microbial extracellular polymeric substances, Wingender, J., Neu, T.R., and Flemming, H.-C., eds. (Springer, Heidelberg), pp. 21–47.

Neu, T.R., and Lawrence, J.R. (1999b). Lectin-binding-analysis in biofilm systems. Methods Enzymol. *310*, 145–152.

Neu, T.R., and Lawrence, J.R. (2002). Laser scanning microscopy in combination with fluorescence techniques for biofilm study. In The Encyclopedia of Environmental Microbiology, Bitton, G., ed. (John Wiley and Sons Inc., New York, NY, USA), pp. 1772–1788.

Neu, T.R., and Lawrence, J.R. (2005). One-photon versus two-photon laser scanning microscopy and digital image analysis of microbial biofilms. Method. Microbiol. *34*, 87–134.

Neu, T.R., and Lawrence, J.R. (2009). Extracellular polymeric substances in microbial biofilms. In Microbial Glycobiology: Structures, Relevance and Applications, Moran, A., Brenan, P., Holst, O., and von Itzstein, M., eds. (Elsevier, San Diego, CA, USA), pp. 735–758.

Neu, T.R., and Lawrence, J.R. (2010). Examination of microbial communities on hydrocarbons by means of laser scanning microscopy. In Microbiology of Hydrocarbons, Oils, Lipids and Derived Compounds, Timmis, K.N., ed. (Springer, Heidelberg, Germany), pp. 4073–4084.

Neu, T.R., and Lawrence, J.R. (2014a). Advanced techniques for *in situ* analysis of the biofilm matrix (structure, composition, dynamics) by means of laser scanning microscopy. In Microbial Biofilms – Methods and Protocols, Donelli, G., ed. (Springer, New York, NY, USA), pp. 43–64.

Neu, T.R., Manz, B., Volke, F., Dynes, J.J., Hitchcock, A.P., and Lawrence, J.R. (2010). Advanced imaging techniques for assessment of structure, composition and function in biofilm systems. FEMS Microbiol. Ecol. 72, 1–21.

Neu, T.R., Swerhone, G.D.W., Böckelmann, U., and Lawrence, J.R. (2005). Effect of CNP on composition and structure of lotic biofilms as detected with lectin-specific glycoconjugates. Aquat. Microb. Ecol. *38*, 283–294.

Neu, T.R., Swerhone, G.D.W., and Lawrence, J.R. (2001). Assessment of lectin-binding analysis for *in situ* detection of glycoconjugates in biofilm systems. Microbiology *147*, 299–313.

Neu, T.R., Walczysko, P., and Lawrence, J.R. (2004). Two-photon imaging for studying the microbial ecology of biofilm systems. Microbes Environ. *19*, 1–6.

Neu, T., and Lawrence, J. (2014b). Investigation of Microbial Biofilm Structure by Laser Scanning

Microscopy. (Springer, Berlin/Heidelberg, Germany), pp. 1–51.

Nielsen, P.H., and Jahn, A. (1999). Extraction of EPS. In Microbial Extracellular Polymeric Substances, J. Wingender, T.R. Neu, and H.-C. Flemming, eds. (Springer, Berlin, Germany), pp. 49–72.

Norkrans, B. (1981). Surface microlayers in aquatic environments. Advances in Microb. Ecol. *4*, 51–85.

Packroff, G., Lawrence, J.R., and Neu, T.R. (2002). In situ confocal laser scanning microscopy of protozoans in culture and complex biofilm communities. Acta Protozool. *41*, 245–253.

Paes Leme, A.F., Bellato, C.M., Bedi, G., Del Bel Cury, A.A., Koo, H., and Cury, J.A. (2008). Effects of Sucrose on the extracellular matrix of plaque-like biofilm formed *in vivo*, studied by proteomic analysis. Caries Res. *42*, 435–443.

Palmer, R.J., Jr., Haagensen, J., Neu, T.R., and Sternberg, C. (2006). Confocal microscopy of biofilms – spatiotemporal approaches. In Handbook of Biological Confocal Microscopy, Pawley, J.B., ed. (Springer, New York, NY, USA), pp. 882–900.

Palmer, R.J., Jr. and Sternberg, C. (1999). Modern microcopy in biofilm research: Confocal microscopy and other approaches. Curr. Opin. Biotechnol. *10*, 263–268.

Paragas, V.B., Kramer, J.A., Fox, C., Hauglands, R.P., and Singer, V.L. (2002). The ELF-97 phosphatase substrate provides a sensitive, photostable method for labelling cytological targets. J. Microsc. *206*, 106–119.

Peltola, M., Neu, T.R., Kanto-Oqvist, L., Raulio, M., Kolari, M., and Salkinoja-Salonen, M.S. (2008). Architecture of *Deinococcus geothermalis* biofilms on glass and steel: a lectin study. Environ. Microbiol. *10*, 1752–1759.

Queck, S.Y., Weitere, M., Moreno, A.M., Rice, S.A., and Kjelleberg, S. (2006). The role of quorum sensing mediated developmental traits in the resistance of *Serratia marcescens* biofilms against protozoan grazing. Environ. Microbiol. *8*, 1017–1025.

Roth, B.L., Poot, M., Yue, S.T., and Millard, P.J. (1997). Bacterial viability and antibiotic susceptibility testing with SYTOX green nucleic acid stain. Appl. Environ. Microbiol. *63*, 2421–2431.

Schermelleh, L., Carlton, P.M., Haase, S., Shao, L., Winoto, L., Kner, P., Burke, B., Cardoso, M.C., Agard, D.A., Gustafsson, M.G.L., *et al.* (2008). Subdiffraction multicolor imaging of the nuclear periphery with 3D structured illumination microscopy. Science *320*, 1332–1336.

Sharp, M.D., and Pogliano, K. (1999). An *in vivo* membrane fusion assay implicates SpoIIIE in the final stages of engulfment during Bacillus subtilis sporulation. Proc. Natl. Acad. Sci. *96*, 14553–14558.

Singer, G., Besemer, K., Hödl, I., Chlup, A.-K., Hochedlinger, G., Stadler, P., and Battin, T. (2006). Microcosm design and evaluation to study stream microbial biofilms. Limnol. Oceanogr.: Methods *4*, 436–447.

Singer, G., Besemer, K., Schmitt-Kopplin, P., Hödl, I., and Battin, T.J. (2010). Physical heterogeneity increases biofilm resource use and its molecular diversity in stream microcosms. PLoS ONE *5*, e9988.

Staudt, C., Horn, H., Hempel, D.C., and Neu, T.R. (2003). Screening of lectins for staining lectin-specific glycoconjugates in the EPS of biofilms. In Biofilms in Medicine, Industry and Environ. Technol., Lens, P., Moran, A.P., Mahony, T., Stoodley, P., and O'Flaherty, V., eds. (IWA Publishing, UK), pp. 308–327.

Stocks, S.M. (2004). Mechanism and use of the commercially available viability stain BacLight. Cytometry A *61A*, 189–195.

Sträuber, H., and Müller, S. (2010). Viability states of bacteria – Specific mechanisms of selected probes. Cytometry A *77A*, 623–634.

Sun, S., Kjelleberg, S., and McDougald, D. (2013). Relative contributions of Vibrio polysaccharide and quorum sensing to the resistance of *Vibrio cholerae* to predation by heterotrophic protists. PLoS ONE *8*, e56338.

van de Linde, S., Kasper, R., Heilemann, M., and Sauer, M. (2008). Photoswitching microscopy with standard fluorophores. Appl. Phys. B *93*, 725–731.

Wagner, M., Taherzadeh, D., Haisch, C., and Horn, H. (2010). Investigation of the mesoscale structure and volumetric features of biofilms using optical coherence tomography. Biotechnol. Bioeng. *107*, 844–853.

Weissbrodt, D.G., Neu, T.R., Kuhlicke, U., Rappaz, Y., and Holliger, C. (2013). Assessment of bacterial and structural dynamics in aerobic granular biofilms. Front. Microbiol. *4*.

Weitere, M., Bergfeld, T., Rice, S.A., Matz, C., and Kjelleberg, S. (2005). Grazing resistance of *Pseudomonas aeruginosa* biofilms depends on type of protective mechanism, developmental stage and protozoan feeding mode. Environ. Microbiol. *7*, 1593–1601.

Whyte, L.G., Slagman, S.J., Pietrantonio, F., Bourbonniere, L., Koval, S.F., Lawrence, J.R., Inniss, W.E., and Greer, C.W. (1999). Physiological adaptations involved in alkane assimilation at a low temperature by *Rhodococcus* sp. strain Q15. Appl. Environ. Microbiol. *65*, 2961–2968.

Wingender, J., Neu, T.R., and Flemming, H.-C. (1999). Microbial extracellular polymeric substances. (Heidelberg: Springer).

Winkler, M., Lawrence, J.R., and Neu, T.R. (2001). Selective degradation of ibuprofen and clofibric acid in two model river biofilm systems. Water Res. *35*, 3197–3205.

Wouters, K., Maes, E., Spitz, J.A., Roeffaers, M.B.J., Wattiau, P., Hofkens, J., and Springael, D. (2010). A non-invasive fluorescent staining procedure allows Confocal Laser Scanning Microscopy based imaging of Mycobacterium in multispecies biofilms colonizing and degrading polycyclic aromatic hydrocarbons. J. Microbial. Meth. *83*, 317–325.

Wurl, O., Wurl, E., Miller, L., and Vagle, S. (2011). Formation and global distribution of sea-surface microlayers. Biogeoscience *8*, 121–135.

Wurl, O., and Holmes, M. (2008). The gelatinous nature of the sea-surface microlayer. Mar. Chem. *110*, 89–97.

Xi, C., Marks, D., Schlachter, S., Luo, W., and Boppart, S.A. (2006). High – resolution three – dimensional imaging of biofilm development using optical coherence tomography. J. Biomed. Opt. *11*, 1–6.

Zhang, R.Y., Neu, T.R., Bellenberg, S., Kuhlicke, U., Vera, M., and Sand, W. (2014). Use of lectin to *in situ* visualize glycoconjugates of extracellular polymeric substances in acidophilic archaeal biofilms. Microb. Biotechnol..

Zippel, B., and Neu, T.R. (2011). Characterization of glycoconjugates of extracellular polymeric substances in tufa-associated biofilms by using fluorescence lectin-binding analysis. Appl. Environ. Microbiol. 77, 506–516.

Zippel, B., Rijstenbil, J., and Neu, T.R. (2007). A flow-lane incubator for studying freshwater and marine phototrophic biofilms. J. Microbial. Meth. *70*, 336–345.

Zubkov, M.V., Fuchs, B.M., Eilers, H., Burkill, P.H., and Amann, R. (1999). Determination of total protein content of bacterial cells by SYPRO staining and flow cytometry. Appl. Environ. Microbiol. 65, 3251–3257.

Interactions and Communication within Marine Biofilms

3

Priyanka Sathe and Sergey Dobretsov

Abstract

Biofilms are the prime mode of life for many microbial species and they are omnipresent in aquatic environment. Biofilms' three-dimensional architecture provides the concomitant microbial populations with additional protection from predation and toxic substances. Marine biofilms (often termed as micro-fouling) are composed of different species of bacteria, diatoms and protozoa surrounded by a matrix of extra polymeric substances (EPS). Formation of marine biofilms depends on the species present, their microbial activity and environmental conditions. Marine biofilms attract or repel larvae of invertebrates and spores of macro-algae resulting in formation of macro-fouling communities ultimately leading to biofouling which is ever increasing threat for maritime industry. Function of any biofilm is dependent on symbiotic interactions between microbial communities. These interactions include competition, cooperation, and neutralism. The best studied chemical interaction between microbes is a cell-to-cell signalling mechanism between bacteria which is called quorum sensing (QS). This process is based on production and perception of simple chemical signals (inducers), which at concentrations higher than thresholds ones triggered differential expression of target genes and cascades of chemical reactions. In order to prevent growth of competitors, some microorganism evolved ability to interfere with QS of bacteria. This chapter reviews current understanding of the role of marine microbial communities, interactions between different microorganisms within marine biofilms and novel anti-biofilm strategies.

Introduction

Microbial communities in marine environment are abundant, diverse and complex. Metabolism of aquatic ecosystem is mediated by microbial communities forming aquatic biofilms. Aquatic microbial communities consist of two primary forms: (1) planktonic (free-floating) and (2) benthic (attached to the substratum) organisms. The equilibrium between these two states depends on environmental factors such as movement of water and substrate availability. Freshwater biofilms are composed of nutritionally diverse microorganisms, and many environmental factors, like light and temperature, can affect the biofilm structure and community composition (Flemming *et al.*, 2001). Freshwater biofilms are very well explained in literature, unlike marine biofilms (Costerton *et al.*, 1987; Characklis *et al.*, 1990; Salta *et al.*, 2013).

One litre of seawater in the surface of the ocean will typically contain 10^9 bacterial cells and the total number of bacteria species, although currently unknown, has been estimated to be as many as 1 million (Curtis *et al.*, 2002). In aqueous systems, microbial cells are found as both 'planktonic' (floating) cells and 'sessile' (attached) cells on surfaces. In marine environment the majority of microorganisms are organized in aggregates or biofilms. In biofilms microbes are enclosed in the matrix of self-produced extra polymeric substances (EPS) (Flemming *et al.*, 2001). This extracellular matrix plays a major role in holding cells together within the biofilm, while simultaneously protecting them from external insults. Biofilms developed on artificial and natural substrata in seawater contain different types of microorganisms, e.g. bacteria, archaea, protozoa, fungi and algae (Dobretsov *et al.*, 2010). In a biofilm, each group of microorganisms performs specialized metabolic functions. Biofilms have three dimensional structures (Costerton *et al.*, 1999; see also Chapter 2). Microbes form a biofilm in response to many factors, which may include cellular recognition of specific or non-specific attachment sites on the surface, nutritional cues, or in some cases, by exposure to inhibitory concentrations of antibiotics. This kind of survival strategy facilitates microorganisms to respond and adapt to physiologically limited and versatile environment and protect them from harmful environmental conditions (Jefferson *et al.*, 2004). When a cell switches to the biofilm mode of growth, it undergoes a phenotypic shift in behaviour, in which large groups of genes are differentially regulated. Biofilms are essentially multicellular organization of microbial communities, which involves physiological interaction, communication and signalling, mutual and antagonistic strategies similar to the higher forms of life (O'Toole *et al.*, 2000).

In aquatic milieu any natural and man-made substrates are quickly colonized with different species of micro- and macro-organisms (Railkin, 2004). Bacteria are considered to be the primary colonizers of substrata, forming ubiquitous biofilms or bacterial aggregates attached to abiotic and living surfaces within several hours (Zobell *et al.*, 1943). Bacterial biofilms have a complex structure and are composed of multiple bacterial species encased in EPS (Flemming *et al.*, 2001). This microbial biofilms attract or repel larvae of invertebrates and spores of macro-algae resulting in formation of macro-fouling communities (Fig. 3.1) (Yebra *et al.*, 2004; Chambers *et al.*, 2006; Magin *et al.*, 2010; Rosenhahn *et al.*, 2010).

Over the last century, marine biofilm research has been primarily focused on its adverse role in industrial biofouling. Biofouling is the term used to describe the unwanted accumulation of microscopic (micro-fouling), large multicellular (macro-fouling) organisms and deposits of their waste products (chemical fouling) on a submerged surface (Wahl *et al.*, 1989). The effects of biofouling can be felt across numerous marine industries and is a particular costly problem for desalination plants, nuclear plants, ships and other marine installations, and equipment working in sea water.

Not only macro-fouling but also biofilms can cause severe industrial problems by increasing drag force, promoting corrosion and reducing heat transfer efficiency (Yebra *et al.*, 2004, 2006; Corteson *et al.*, 1987). A biofilm 1 mm thick on a ship hull leads to the penalty of hundreds of dollars due to higher fuel consumption (Schultz *et al.*, 2011). Biofilms induce corrosion of metals by a number of mechanisms including cathodic and anodic depolarization, hydrogen production, metal reduction, and production of corrosive metabolites, such as organic acids and exopolymers (Videla and Herrera, 2004; Landousli *et al.*, 2011).

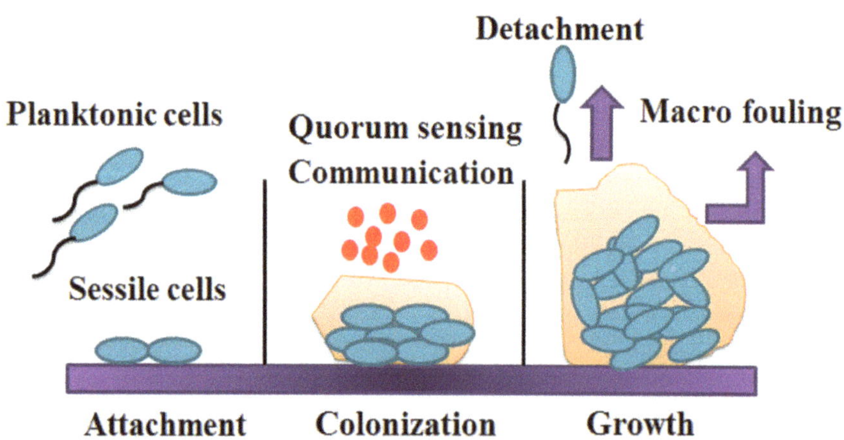

Figure 3.1 Stages in the formation of a biofilm from planktonic form of cells to community mode, these stages ultimately lead to biofouling.

Traditional way to control biofouling is based on using toxic biocides that kill fouling organisms (Yebra *et al.*, 2004). These biocides are highly toxic and kill other marine organisms as well. Antifouling biocides can accumulate in sediments, fish and marine mammals (Thomas and Brooks, 2010). Many publications have been able to demonstrate that biocidal and fouling release coatings are generally free from macrofouling but not from microbial biofilms (Cassé and Swain, 2006; Molino *et al.*, 2009; Dobretsov and Thomason, 2011; Zargiel *et al.*, 2011; Briand *et al.*, 2012). Thus, eradication of biofilms and inhibition of their growth are major industrial concerns.

Despite the recent and significant increase in the study of aquatic microbial communities, little is known about the microbial interactions within complex ecosystems such as marine water. Current knowledge of these interactions within microbial communities in marine environment is scarce whereas these communities are hotspots of exchanges between bacteria and macro-organisms.

The aims of this chapter are to discuss (1) existing knowledge of formation and composition of marine biofilms; (2) Shed light on various microbial interactions responsible for maintenance of a biofilm in marine environments; (3) Discuss novel anti-biofilm strategies that are being investigated; (4) Highlight future directions for advancing our understanding of dynamics of marine biofilms.

Adhesion and communication during biofilm formation

In order to generate a biofilm, cells switch from a planktonic to a sessile state by down-regulating the expression of flagellar genes concomitantly with an up-regulation of genes involved in production of the extracellular matrix (Fig. 3.1). During initial stages of marine biofilm production, proteins, glycoproteins, proteoglycans and polysaccharides present in the water adsorb to the substratum (Maki *et al.*, 2002). A molecular fouling can either inhibit microbial attachment or facilitate it and furnish microorganisms with a carbon source for their growth. Microbial adhesion depends on the substratum surface free energy, substratum properties and water turbulence (Zobell, 1943; Dexter *et al.*, 1975; Characklis *et al.*, 1990; Costeron *et al.*, 1995). Most microorganisms attach strongly to hydrophobic materials, such as Teflon™, than to hydrophilic materials, such as glass (Fletcher *et al.*, 1979; Railkin, 2004). Adhesion of bacteria and microalgae to surfaces of living (tissue of plants and animals) and non-living (metals, plastics, and organic particles) materials are mediated by secretion of a slimy, glue-like substance of extracellular polymers (EPS), which contain polysaccharides, lipopolysaccharides, proteins, nucleic and huminic acids (Chapter 2). Adhesion of microorganisms is not always a passive process; microbes may respond to chemical substances and swim away or towards to the substratum (Marshall *et al.*, 1971; Vandevivere *et al.*, 1993). Additionally, attached microorganisms can stimulate or suppress the co-adhesion of other planktonic microorganisms causing formation of multispecies heterogeneous biofilms (McEldowney *et al.*, 1986).

A number of investigations have shown that various physical factors (for example, flow rate, hydrodynamic forces, substrate properties, and viscosity), chemical factors (for example, nutrient availability and EPS production), and biological factors (for example, competition and predation) affect biofilm architecture both separate and in combination (Qian *et al.*, 2007). Colonization of the surface of substratum by bacteria is affected by different temperatures, motility behaviour (Kirboe and Jackson, 2001), surface chemistry and energy (Ista *et al.*, 2004). Microbial biofilms in marine environment are very heterogenic and dynamic structures and these features make them difficult to model and investigate (Dobretsov, 2010). With improvement in imaging technology and the development of electron microscopy, scanning confocal laser microscopy (SCLM), real time image capture, atomic force microscopy (AFM), nuclear magnetic resonance imaging (NMRI), various spectroscopic imaging techniques, like Raman imaging and fluorescent *in situ* hybridization (FISH) methods, our understanding of biofilm's structure and functions have been improved significantly (see Chapter 2 for more details).

Composition of marine biofilms

Competition for space and nutrition among microorganisms determines the microbial composition of biofilms. Usually, marine biofilms consist of different species of bacteria, Archaea and unicellular organisms, such as diatoms and flagellates (Dobretsov, 2010). The densities of fungi and protista are generally low in marine biofilms (Railkin, 2004; Salta *et al.*, 2013). Marine biofilms are dominated by bacteria, which are the primary colonizer of any clean substratum (Armstrong *et al.*, 2001). Viruses are possibly very abundant in marine biofilms but their presence has not been confirmed (Dobretsov, 2010). The second largest group of microorganisms in marine biofilms is diatoms or Bacillariophyta. Diatoms belonging to genes *Achnanthes*, *Amphora*, *Navicula* and *Licmophora* dominate in marine biofilms (Cookesy *et al.*, 1995; Thornton *et al.*, 2002). Protists (mainly flagellates) can be found in marine biofilms but in low densities (Underwood *et al.*, 2005; Maybruck *et al.*, 2004). Fungi, mainly *Cladosporium* species, have been isolated from marine biofilms, while their mycelia have been rarely found (Maybruck *et al.*, 2004). Marine fungi, protists and viruses in marine biofilms are rarely studied, while, probably, they play an important role as decomposers, grazers and parasites (Dobretsov, 2010).

Marine biofilms can develop even on the surface of antifouling coatings that contain toxic biocides (Callow, 1986; Dobretsov and Thomason, 2011; Chen *et al.*, 2013). However, there is only limited information about biofilms' composition and functions on antifouling coatings (Yebra *et al.*, 2004). Recent studies with culture independent molecular-based techniques suggested that the type of biocide and environmental conditions affected the composition of microbial biofilms (Briand *et al.*, 2012; Chen *et al.*, 2013). Bacteria belonging to the class Alpha-proteobacteria, Gamma-proteobacteria, firmicutes and bacteroidetes have been found on the hulls of ships coated with cuprous oxide based coatings (McNamara *et al.*, 2009). The presence of Delta-proteobacteria was reported on copper-based biocidal coatings (Chen *et al.*, 2013). The metagenomic study of marine biofilms developed on antifouling coatings demonstrated high abundance of bacteria *Acaryochloris marina* and *Maritimimonas rapanae* in 1 year old biofilms developed on copper-based antifouling coatings (Muthukrishnan *et al.*, 2014). Additionally, diatoms belonging to the genes *Nitzschia*, *Navicula*, *Amphora* and *Licmophora* were reported in biofilms on antifouling coatings (Callow, 1986; Cassé and Swain, 2006; Briand *et al.*, 2012).

In most of experiments, investigators have been studying mono-species microbial films in static conditions without the addition of nutrients (Maki, 2002; Dobretsov *et al.*, 2006). These studies give only limited information about the structure, function and the role of biofilms in the field, as natural biofilms are formed under turbulent flow conditions and composed of numerous species of microorganisms (Characklis *et al.*, 1990; Webster *et al.*, 2004). Therefore, more investigations should be done in order to investigate marine biofilms in the field conditions.

Factors affecting biofilm composition

The species composition of early marine biofilms depends mainly on two factors: (1) presence of colonizers; and (2) the physical and chemical conditions of the substratum and environment. Marine biofilms are subjected to constant dynamic changes. Any modifications in abiotic conditions (e.g. climate, depth, light regime, season, water chemistry, nutrient supply and substratum characteristics) and biotic surroundings (e.g. availability and physiology of colonizing species, competition, predation, grazing and cooperation among species) change the composition of biofilms, density, productivity, architecture, succession rate, and the production of chemical compounds by microorganisms (Salta *et al.*, 2013). For example, some microorganisms are found a whole year around, while others are seen only during particular seasons (Moss *et al.*, 2006; Forster *et al.*, 2006) and even hours (Underwood *et al.*, 2005). The diatoms *Gyrosigma balticum* and *Pleurosigma angulatum* were found at the surface of *Navicula* and *Nitzschia* dominated biofilms only during midday (Underwood *et al.*, 2005). This phenomenon can be explained by rapid (within minutes) migrations of diatoms that have tendency to adjust their position in a biofilm according to light intensity.

Additionally, biofilm sloughing and detachment can open space for re-colonization by newcomers and existing microorganisms (Underwood *et al.*, 2005).

As an example, different biofilms have been developed on chemically inert black and white substrata exposed to fouling in tropical waters (Dobretsov *et al.*, 2013). Pyrosequencing via 454 of 16S rRNA genes of bacteria from white and black substrata showed that while Alpha-proteobacteria and firmicutes dominated on both type of substrata, their quantities were different. In other words, biofilms reflected the key environmental factors of a substratum. This relationship has an important implication for the larval settlement of marine invertebrates (Hadfield *et al.*, 2001; Salta *et al.*, 2013).

Communication strategies within biofilms

Marine biofilms are usually influenced by positive, neutral or negative interactions within its counterparts. Chemical substances secreted by one species of microorganisms greatly influence the colonization of other species (Holmstro¨m *et al.*, 2002). In biofilms, microbes grow in close proximity and compete or cooperate for space and nutrients (Nadell *et al.*, 2009). These interactions between different microbial species are based on the sophisticated cell-to-cell chemical communication mechanisms, such as QS, or the secretion of secondary metabolites, like antimicrobials or siderophores (Davies *et al.*, 1998; Dong *et al.*, 2001; Dong and Zhang, 2005).

Quorum sensing

Quorum sensing (QS) is a form of bacterial population density dependent cell-to-cell communication and gene regulation that has been shown to contribute to the formation and maturation of biofilms. QS allows them to coordinate their adhesion, biofilm maturation, swarming, luminescence and chemical compound production (Waters and Bassler, 2005; Steinberg *et al.*, 2011). Overall, QS helps to control the switch from the behaviours of single cell organisms to those appropriate to cells within a biofilm (Sauer *et al.*, 2002). During this process, a class of small molecules named 'autoinducers' is produced and released into the environment (Fig. 3.2). When the concentration of these signals reaches a threshold level, these signals trigger an expression of target genes that consequently change the behaviour of bacteria (Sauer *et al.*, 2002; Steinberg *et al.*, 2011). Different species of Gram-positive and Gram-negative bacteria use a different but potentially overlapping QS signalling systems. There are three documented QS systems: the *N*-acyl homoserine lactone (AHL)-based signalling system of Gram-negative bacteria, an oligopeptide-based system in Gram-positive bacteria and a furanones-based system that is shared between both groups (Roux *et al.*, 2009; Yin *et al.*, 2012).

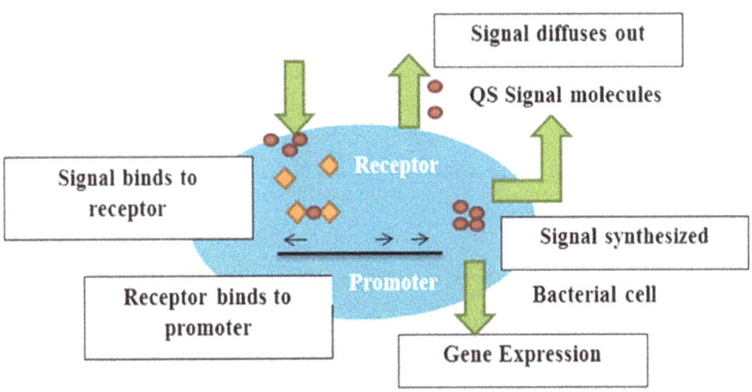

Figure 3.2 Generalized model of AHL-mediated quorum sensing in Gram-negative marine bacteria.

Mechanism of AHL-based QS signaling system

The best characterized QS signalling system is the AHL-based one (Dobretsov *et al.*, 2009). It has gained a significant interest due to its frequent role in microbial virulence mechanisms and wide distribution between marine environment (Huang *et al.*, 2009; Taylor *et al.*, 2004). This QS system is able to regulate different physiological activities of Gram-negative bacteria, which include bioluminescence, exopolysaccharides, antibiotic and enzyme production, virulence, conjugal plasmid transfer, biofilm formation, and growth (Irie and Parsek, 2008).

In AHL-based QS system, AHL signals, produced by 'LuxI' autoinducer synthase (Fig. 3.2), exit the cells after their synthesis (either by passive diffusion or active transport) and accumulate in the environment (Bassler *et al.*, 2002). Different species of bacteria can produce different AHLs, which differ from each other by the presence of functional groups, like OH, and the length of acyl-side chain (Fig. 3.3). When the extracellular concentration of AHLs reaches a certain threshold, the cognate signals bind to the cytoplasmic LuxR-type receptor, and form active dimmers (Williams *et al.*, 2007). This complex binds to specific regions (known as lux boxes) within promoter sequences and activates transcription of QS genes that change bacterial behaviour (Fig. 3.2). In the absence of AHLs, LuxR proteins are degraded quickly; conversely, binding of LuxR to the cognate AHL signal stabilizes the protein (Cho, 2009). Some bacteria and marine organisms, such as algae, are able to respond to the QS signals of other bacteria or to compromise those (Dobretsov *et al.*, 2009).

QS in marine biofilms

Information concerning the presence of AHLs and other QS molecules in the marine environment is scarce (Dobretsov *et al.*, 2007b). First evidence on the presence of QS signals in marine environment came from the study of the symbiotic bacterium *Vibrio fischeri*, which colonizes the light organ of sepiolid squid *Euprymna scolopes* (Eberhard *et al.*, 1981). Production of light by this bacterium is regulated by AHL signals (Fig. 3.3). Several studies have showed that bacteria associated with

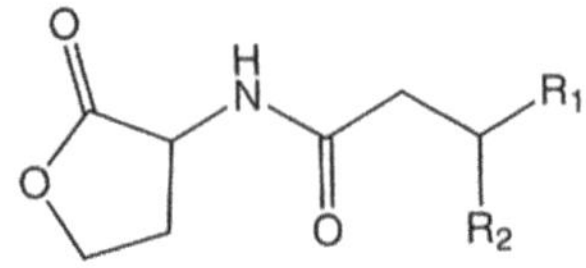

Figure 3.3 Generalized structure of acyl homoserine lactones (AHL) produced by bacteria. All of them have common lactone ring with different size (C_4–C_{18}) fatty acid acyl size chain R1. R2 = H, O or OH.

marine snow particles (Gram, 2002), and sponges (Taylor *et al.*, 2004; Mohamed *et al.*, 2008) are producing different AHL signals. AHL producing bacteria isolated from sponges belonged to *Roseobacter* sp., *Marinobacter* sp. and *Vibrio* sp. There are few publications that investigated production of QS signals in biofilms and microbial mats. Subtidal biofilms produce different size chain AHLs during formation on artificial substrata (Huang *et al.*, 2007). AHL producing culturable isolates from biofilms belonged either to Alpha-proteobacteria (all *Rhodobacteriaceae* isolates) or to Gamma-proteobacteria (genera *Pseudoalteromonas, Vibrio* and *Thalassomonas)* (Huang *et al.*, 2008). Production and degradation of AHLs by marine microbial mats (stromatolites) composed mostly of cyanobacteria and sulfate-reducing bacteria was examined (Decho, 2009). Under natural conditions microbial mats produced a wide range of AHLs (C_4 to C_{14} acyl side chain; Fig. 3.3) and most of QS signals accumulated during the night-time. *In vivo* production of QS signals and inhibitors in hot spring and desert crust microbial mats has been reported (Abed *et al.*, 2011; Dobretsov *et al.*, 2011). These studies suggest that QS signals are present in biofilms developed on artificial and natural substrata and we lack of knowledge about production and turnover of AHLs in the environment.

Interactions within biofilms and with new colonizers

In microbial biofilms different microorganisms interact with each other and new prokaryotic and eukaryotic colonizers. These intra- and inter-specific interactions between biofilm inhabiting microorganisms include helpful associations (i.e. mutualism), associations in which one microorganism benefits but other does not (commensalism), interactions beneficial for one species and harmful for another one (i.e. predation, grazing, parasitism, diseases) and various competitive interactions in which both species are adversely affected (Dobretsov, 2010). In addition, chemical compounds produced by microorganisms can inhibit or induce settlement of invertebrate larvae and spores of macroalgae (Dobretsov *et al.*, 2006; Qian *et al.*, 2007; Salta *et al.*, 2013; Marshall *et al.*, 2006). These interactions between the biofilm and higher plants and animals are of interest at several levels. Microbial biofilms provide food and vitamins for grazing macroorganisms, while both seaweeds and invertebrates supply organic nutrients to the heterotrophic bacteria in the biofilm (Thompson *et al.*, 2004). Micro-grazers, like heterotrophic flagellates, amoebas and ciliates, are among the major regulators of bacterial biomass and diversity. In marine snow biofilms the density of bacteria and their behaviour was largely controlled by the grazing pressure from flagellates (Kiørboe *et al.*, 2003). Presence of heterotrophic flagellates changed biofilm architecture and shifted the composition of bacterial communities towards less diverse with dominance of grazing resistant strains.

Some of prokaryote–eukaryote interactions have been widely investigated, others are remained undiscovered. One of interesting examples of this inter-kingdom communication is the green alga *Ulva (Enteromorpha)* sp. Zoospores of this alga use bacterial QS signals in order to detect a suitable surface for attachment (Joint *et al.*, 2007). The agglutination and attachment of diatoms *Navicula* sp. enhanced in the presence of spent culture of the bacterium *Pseudoalteromonas* sp. (Cooksey and Cooksey, 1995). Similarly, the bacterium *Pseudoalteromonas tunicata* requires the presence of diverse species of bacteria in the inoculum for its effective colonization of the surface of the alga *Ulva australis* (Rao *et al.*, 2007).

The complex architecture of a mature biofilm provides niches with distinct physico-chemical conditions, differing, e.g. in oxygen availability, in concentration of diffusible substrates and metabolic side products, in pH, and in the cell density (Costerton *et al.*, 1999). In such a mixed microbial community the strains may interact antagonistically or synergistically with each other, the latter resulting in co-colonization of distinct groups of bacteria having metabolic cooperation (O'Toole, 2000; Armstrong *et al.*, 2001; Nadell *et al.*, 2009). Microbial processes such as nitrification, anaerobic degradation of organic compounds, or bioremediation of xenobiotic compounds, have been shown to require interactions between different bacterial species within the biofilm (Paerl and Pinckney, 1996). This metabolic cooperation is advantageous to the microbial community. Nevertheless, cooperation among species is only expected under restricted conditions (Nadell *et al.*, 2009).

Under natural conditions, bacteria intra- or inter-specifically compete with their neighbours for space and resources. A surface (especially of hosts) may itself also be a trophic source where attached microorganisms catabolize organic or inorganic nutrients directly (Madigan and Martinko, 2006; Grossart, 2010). Therefore, the presence of other microorganisms on a surface reduces the availability of food and space for colonizing species (Salta *et al.*, 2013). Under such competitive selection it is not surprising that bacteria have developed special mechanisms in order to interfere with the capability of other antagonistic bacteria (Falagas *et al.*, 2008). The mechanisms of bacterial antagonism include depletion of some essential substances (e.g. substrate or a vitamin), alteration in the microenvironment (e.g. changes in the gas concentration or pH), the presentation of an obstacle or barrier, and production of antagonistic substances (Reid *et al.*, 2001).

Clearing a space to colonize by eliminating prior residents can be accomplished by production of antimicrobials, toxins, and molecules that facilitate the competitors' dispersal without actually killing those (Hibbing *et al.*, 2010). It has been shown that bacterial species that belong to the *Streptomyces, Alteromonas, Pseudoalteromonas* and *Roseobacter* genus produce a range of antibiotics (Wiese *et al.*, 2009). In the mixed species biofilms, the antibiotic produced by the bacterium *P. tunicata* remove the competing bacterial strains unless its competitor is relatively insensitive to antibacterial protein or produces strong inhibitory activity against the bacterium *P. tunicate* (James *et al.*, 1996). Antagonistic antimicrobial activity widely exists between common fouling bacteria and fungi in marine biofilms (Qian *et al.*, 2005). About 70% of tested fungal isolates (mostly *Cladosporium*) inhibited growth of marine bacterial isolates from biofilms and, in opposite, 17% of tested bacterial strains (mostly *Staphylococcus*) inhibited growth of fungi in laboratory experiments. This suggests that the competitive interactions between different components of biofilms is quite common and cannot be neglected in future investigations.

Anti-biofilm strategies

Removal of detrimental biofilms from surfaces exposed in the marine environment remains a challenge. Biofilms can be controlled by mechanical, chemical and thermal treatments (Barraud *et al.*, 2009). Conventional antifouling biocides have limited effectiveness against biofilm bacteria in industrial settings (Flemming and Ridgway, 2008). Along with the killing of bacteria, their removal from the surface is desirable (Pitts *et al.*, 2003). Controlling of bacterial biofilms is easier at the primary stages, when biofilm formation is initiated.

Antimicrobials, such as antibiotics and halogen-based treatments, have been traditionally designed to inhibit bacterial growth in water. These treatments are not effective against biofilms, as these methods require unacceptably high cost and energy inputs or lead to fatal outcomes in order to complete killing of microorganisms in the biofilms (Pitts *et al.*, 2003). To address there is a need for novel and improved anti-biofilm treatments. Several novel antimicrobial treatments against marine biofilms have been developed that are the subject of this section.

QS inhibition

Given that bacteria have evolved the ability to communicate via QS, it is reasonable that marine organisms have also evolved the ability to prevent bacterial biofilms using QS inhibitors. There are examples of interference in which microbes release enzymes that degrade autoinducers molecules (Dobretsov *et al.*, 2009). AHL-degrading microorganisms are taxonomically diverse and include true fungi (*Ascomycetes* and *Basidiomycetes*), Gram-positive and Gram-negative bacteria (Teplitski *et al.*, 2011). Two classes of bacterial AHL degrading enzymes have been identified, which are lactonases and acylases (Rajamani *et al.*, 2011). For example, *Bacillus* spp. release an AHL-lactonase that cleaves the lactone ring of AHL molecules and probably interferes with the AHL signalling of other species in the same niche (Dong *et al.*, 2002; Rajamani *et al.*, 2011). Being Gram-positive bacteria, *Bacillus* species does not produce AHLs and communicate with each other using peptide signals (Oh *et al.*, 2012). AHL-acylases cleave the acyl side chain from the lactone ring of AHL molecule (Dong *et al.*, 2002). Recently, AHL-acylase, named 'autoinducer inhibitor from cyanobacteria' (AiiC), was isolated from the marine cyanobacterium *Anabaena* sp. PCC7120 (Romero *et al.*, 2008).

Not only prokaryotes but also eukaryotes are able to control bacterial QS (Dobretsov *et al.*, 2009; Goecke *et al.*, 2010). The red alga *Delisea pulchra* produces halogenated furanones structurally similar to AHLs that interfere with bacterial AHL-mediated QS signalling (Rasmussen *et al.*, 2000). Furanones from *D. pulchra* and their synthetic analogues are potent inhibitors of QS-regulated behaviours in Gram-negative bacteria, including expression of virulence genes, production of antibiotics, bacterial motility, swarming, and biofilm formation (Steinberg and De Nys, 2002; Defoirdt *et al.*, 2004). A unicellular chlorophyte, *Chlamydomonas reinhardtii*, secretes a number of compounds that mimic the activity of AHLs and thus affect QS in bacteria (Teplitski *et al.*, 2004).

Many marine organisms are able to produce QS inhibitors but in many cases the mode of action of QS inhibitors cannot be determined (Dobretsov *et al.*, 2009; Goecke *et al.*, 2010). The sponge *Luffariella variabilis* produces manoalide, manoalide monoacetate and secomanoalide that inhibited QS of LuxR-based reporters (Skindersoe *et al.*, 2008). Demethoxy encecalin and hymenialdisin from sponges inhibited QS of the LasR-based and the LuxR-based reporters (Dobretsov *et al.*, 2011). The cyanobacterium *Lyngbya majuscula* produced malyngamide C, 8-epimalyngamide C and malyngolide that inhibited QS of LasR-based reporters (Kwan *et al.*, 2010; Dobretsov *et al.*, 2010)

Disruption of QS is considered to be an important way to prevent microbial infections (Williams *et al.*, 2007; Parsek *et al.*, 2000), as well as to prevent biofouling (Dobretsov *et al.*, 2009). QS inhibitor – kojic acid – incorporated in a non-toxic matrix was able to prevent biofouling in tropical waters (Dobretsov *et al.*, 2011). Bacteria producing AHL-degrading enzymes have been identified as promising agents in reducing biofouling (Oh *et al.*, 2012). The introduction of micro-encapsulated transgenic *E. coli* that produces AHL-degrading enzymes strongly reduced biofouling of membranes in wastewater treatment facilities over an extended period of time (Oh *et al.*, 2012).

Nitric oxide mediated biofilm dispersal

Many bacterial species, such as *Shewanella woodyi* (Liu *et al.*, 2012), *Shewanella oneidensis* (Plate *et al.*, 2012), and *Legionella pneumophila* (Carlson *et al.*, 2010) use nitric oxide (NO) in order to regulate biofilm formation through a pathway involving cyclic-di-GMP metabolism. In these bacteria, NO is sensed by H-NOX and ultimately regulated the activities of a diguanylate cyclase and/or phosphodiesterase, either directly or indirectly through a histidine kinase, to control the intracellular concentration of cyclic-di-GMP (Plate *et al.*, 2012; Carlson *et al.*, 2010). NO was identified as important factor mediating biofilm dispersal of the pathogen *P. aeruginosa* (Barraud *et al.*, 2006). Low, non-toxic concentrations of NO induced a transition from the sessile, resistant biofilm phenotypes of *P. aeruginosa* to planktonic phenotypes (Darling *et al.*, 2003; Van Alst *et al.*, 2007). Addition of antimicrobial treatment to NO treated biofilms completely removed *P. aeruginosa* biofilms (Barraud *et al.*, 2006). Cyclic-di-GMP involved in NO mediated dispersal of *P. aeruginosa* is controlled by diguanylate cyclases and phosphodiesterases (Ryan *et al.*, 2006). Genes encoding these enzymes and redox sensors are widely distributed in all bacteria (Delgado-Nixon *et al.*, 2000; Romling *et al.*, 2005). This suggests that NO dispersal of biofilms may occur in different species of bacteria. Several studies supported this hypothesis. Formation of *Staphylococcus* biofilms is inhibited by exposure to nitrite (NO_2^-), a process that involves NO (Schlag *et al.*, 2007). NO has been suggested to be an ancient and highly conserved regulator of dispersal across eukaryotic organisms (Bishop and Brandhorst, 2003). For example, dissolution and dispersal of aggregated mycelial cells of fungi, *Neurospora crassa* (Ninnemann and Maier, 1996) and *Candida tropicalis* (Wilken and Huchzermeyer, 1999), and the amoeba *Dictyostelium discoideum* (Tao *et al.*, 1997) were shown to rely on NO signalling. Taken together, these observations strongly suggest that NO can helps in biofilm dispersal and combined NO and antibiotic treatment can offer a novel strategy to control marine biofilms.

Reactive oxygen species mediated biofilm disruption

Photodynamic therapy (PDT), also known as photo radiation therapy or phototherapy, involves the use of a photoactive dye (photo sensitizer) that is activated by exposure to light of a specific wavelength in the presence of oxygen (Hashimoto *et al.*, 2005). The transfer of energy from the activated photo sensitizer to available oxygen results in the formation of toxic reactive oxygen species (ROS), such as singlet oxygen and free radicals (Hashimoto *et al.*, 2005). These very reactive ROS can damage proteins, lipids, nucleic acids, and other cellular components (Li *et al.*, 2008). It is suggested that the photocatalytic killing mechanism initially damages the surfaces weak points of the bacterial cells, before totally breakage of the cell membranes (Li *et al.*, 2008). The internal bacterial components then leak from the cells through the damaged sites. Finally, the photocatalytic reaction oxidizes the cell debris.

Photocatalysis of water create ROS using solar energy (Miller *et al.*, 2010). Titanium dioxide (TiO_2) is one of such photo-catalyst that can generate ROS when illuminated with light and thus eliminate microbes (Sapkota *et al.*, 2011). However, because of its large band *p* for excitation, only ultraviolet (UV) light irradiation can excite TiO_2, which limits its application in the living environment. However, with impurity doping, through metal coating and controlled calcinations TiO_2 has been successfully modified to expand its absorption wavelengths to the visible light region (Liu *et al.*, 2007). Previous studies have showed that the modified TiO_2 significantly reduced the numbers of surviving bacterial cells of *E. coli* in response to visible light illumination (Liu *et al.*, 2007). Several other metal oxide nanoparticles and nanocoatings are being investigated for their potent anti biofilm activity. It has been shown that ZnO nano-coatings illuminated with light inhibit growth of bacterial pathogens, such as *Escherichia coli*, *Bacillus* sp., *Salmonella* sp., *Listeria monocytogenes* and *Staphylococcus aureus* in fresh water (Jones *et al.*, 2008; Liu *et al.*, 2009; Jaisai *et al.*, 2012). Recent study demonstrated that ZnO nanorod coatings inhibited biofilm formation of the marine bacterium *Acinetobacter* sp. and settlement of the bryozoan larvae of *Bugula neritina* (Al Fori *et al.*, 2014). Vanadium pentoxide (V_2O_5 nanowires in sea water catalyse the oxidation of bromide ions and produce singlet molecular oxygen (Natalio *et al.*, 2012). *In situ* experiments showed that paints with V_2O_5 prevented biofouling when exposed to sea water for up to 60 days. These novel eco-friendly and cost-effective nano-based solutions can offer a better option for disruption of biofilms and killing microbes.

Conclusions and future directions

Recent years have witnessed a fast development in our understanding of marine biofilms, their composition, dynamics and physiology. These are derived from both laboratory and field studies and based on the development of new sensitive instruments for analysis, as well as the development of new techniques. However, the concept of interactions within marine biofilms needs further investigations for a better understanding. Unravelling the fundamentals of biofilm biology is essential for controlling marine biofilms. Metagenomics, transcriptomics and metabolomics approaches are going to be important on the prospective studies of composition and function of marine biofilms in the future.

To summarize, marine biofilms are highly complex and composed of many species of prokaryotic and eukaryotic organisms. Marine biofilms are highly dynamic and change with the changes in environmental conditions. Formation of marine biofilms depends on the species present in the environment, their microbial activity and interactions with other species present in biofilms. QS is one of the types of chemical interactions between bacterial cells, as well as between prokaryotic and eukaryotic organisms. Removal of biofilms from surfaces exposed in the marine environment remains a challenge. Various novel techniques, such as QS inhibitors, NO, and ROS can offer a better solution to get rid of these biofilms in near future. Prevention of microbial attachment and biofilm dispersal rather than killing of marine microorganisms should be future priority directions. Close collaboration between microbiologists, physicists, chemists and engineers will be necessary for the development of new non-toxic antifouling applications.

References

Adak, S., Aulak, K.S., and Stuehr, D.J. (2002). Direct evidence for nitric oxide production by a nitric-oxide synthase-like protein from *Bacillus subtilis*. J. Biol. Chem. *277*, 16167–16171.

Al-Fori, M., Dobretsov, S., Myint, M.T.Z., and Dutta, J. (2014). Antifouling properties of zinc oxide nanorod coatings. Biofouling *30*, 871–882.

Armstrong, E., Yan, L., Boyd, K.G., Wright, P.C., and Burgess, J.G. (2001). The symbiotic role of marine microbes on living surfaces. Hydrobiologia *461*, 37–40.

Barraud, N., Hassett, D.J., Hwang, S.H., Rice, S.A., Kjelleberg, S., and Webb, J.S. (2006). Involvement of nitric oxide in biofilm dispersal of *Pseudomonas aeruginosa*. J. Bacteriol. *188*, 7344–7353.

Bassler, B.L. (2002). Small talk: Cell-to-cell communication in bacteria. Cell *109*, 421–424.

Bishop, C.D., and Brandhorst, B.P. (2003). On nitric oxide signaling, metamorphosis, and the evolution of biphasic life cycles. Evol. Dev. *5*, 542–550.

Briand, J.F., Djeridi, I., Jamet, D., Coupé, S., Bressy, C., Molmeret, M., Le Berre, B., Rimet, F., Bouchez, A., and Blache, Y. (2012). Pioneer marine biofilms on artificial surfaces including antifouling coatings immersed in two contrasting French Mediterranean coast sites. Biofouling *28*, 453–463.

Callow, M.E. (1986). Fouling algae from 'in service' ships. Botanica Marina *29*, 351–358.

Cao, Y.S., and Alaerts, G.J. (1995). Influence of reactor type and shear stress on aerobic biofilm morphology, population and kinetics. Water Res. *29*, 107–118.

Carlson, H.K., Vance, R.E., and Marletta, M.A. (2010). H-NOX regulation of c-di-GMP metabolism and biofilm formation in *Legionella pneumophila*. Mol. Microbiol. *77*, 930–942.

Characklis, W., Marshall, K., and McFeters, G. (1990). The microbial cell. In Biofilm, William, G.W., and Marshall, K., eds. (John Wiley & Sons, New York, NY, USA), pp. 131–159.

Chen, C.L., Maki, J.S., Rittschof, D., and Teo, S.L.M. (2013). Early marine bacterial biofilm on a copper-based antifouling paint. Int. Biodeterior. Biodegrad. *83*, 71–76.

Cho, J.Y. (2012). Antifouling steroids isolated from red alga epiphyte filamentous bacterium *Leucothrix mucor*. Fish Sci. *78*, 683–689.

Chong, T.-M., Koh, C.-L., Sam, C.-K., Choo, Y.-M., Yin, W.-F., and Chan, K.-G. (2012). Characterization of quorum sensing and quorum quenching soil bacteria isolated from Malaysian tropical montane forest. Sensors *12*, 4846–4859.

Cooksey, K.E., and Wigglesworth-Cooksey, B. (1995). Adhesion of bacteria and diatoms to surfaces in the sea: a review. Aquat. Microb. Ecol. *9*, 87–96.

Costerton, J.W. (1995). Overview of microbial biofilms. J. Ind. Microbiol. *15*, 137–140.

Costerton, J.W., Cheng, K.J., Geesey, G.G., Ladd, T.I., Nickel, J.C., Dasgupta, M., and Marrie, T.J. (1987). Bacterial biofilms in nature and disease. Annu. Rev. Microbiol. *41*, 435–464.

Costerton, J.W., Lewandowski, Z., Caldwell, D.E., Korber, D.R., and Lappin-Scott, H.M. (1995). Microbial biofilms. Annu. Rev. Microbiol. *49*, 711–745.

Costerton, J.W., Stewart, P.S., and Greenberg, E.P. (1999). Bacterial biofilms: A common cause of persistent infections. Science *284*, 1318–1322.

Curtis, T.P., Sloan, W.T., and Scannell, J.W. (2002). Estimating prokaryotic diversity and its limits. Proc. Natl. Acad. Sci. U.S.A. *99*, 10494–10499.

Dahms, H.-U., and Qian, P.-Y. (2005). Exposure of biofilms to meiofaunal copepods affects the larval settlement of *Hydroides elegans* (Polychaeta). Mar. Ecol. Prog. Ser. *297*, 203–214.

Dang, H., and Lovell, C.R. (2000). Bacterial primary colonization and early succession on surfaces in marine waters as determined by amplified rRNA gene restriction analysis and sequence analysis of 16S rRNA genes. Appl. Environ. Microbiol. *66*, 467–475.

Darling, K.E., and Evans, T.J. (2003). Effects of nitric oxide on Pseudomonas aeruginosa infection of epithelial cells from a human respiratory cell line derived from a patient with cystic fibrosis. Infect. Immun. *71*, 2341–2349.

Davies, D.G., Parsek, M.R., Pearson, J.P., Iglewski, B.H., Costerton, J., and Greenberg, E. (1998). The involvement of cell-to-cell signals in the development of a bacterial biofilm. Science *280*, 295–298.

Decho, A.W., Visscher, P.T., Ferry, J., Kawaguchi, T., He, L., Przekop, K.M., Norman, R.S., and Reid, R.P. (2009). Autoinducers extracted from microbial mats reveal a surprising diversity of *N*-acylhomoserine lactones (AHLs) and abundance changes that may relate to diel pH. Environ. Microbiol. *11*, 409–420.

Delgado-Nixon, V.M., Gonzalez, G., and Gilles-Gonzalez, M.-A. (2000). Dos, a heme-binding PAS protein from Escherichia coli, is a direct oxygen sensor. Biochemistry *39*, 2685–2691.

Dexter, S., Sullivan, J., Williams, J., and Watson, S. (1975). Influence of substrate wettability on the attachment of marine bacteria to various surfaces. Appl. Microbiol. *30*, 298–308.

Dobretsov, S.V., and Qian, P.-Y. (2002). Effect of bacteria associated with the green alga *Ulva reticulata* on marine micro-and macrofouling. Biofouling *18*, 217–228.

Dobretsov, S., and Qian, P.-Y. (2004). The role of epibotic bacteria from the surface of the soft coral *Dendronephthya* sp. in the inhibition of larval settlement. J. Exp. Marine Biol. Ecol. *299*, 35–50.

Dobretsov, S., and Qian, P.-Y. (2006). Facilitation and inhibition of larval attachment of the bryozoan *Bugula neritina* in association with mono-species and multi-species biofilms. J. Exp. Marine Biol. Ecol. *333*, 263–274.

Dobretsov, S., and Thomason, J.C. (2011). The development of marine biofilms on two commercial non-biocidal coatings: a comparison between silicone and fluoropolymer technologies. Biofouling *27*, 869–880.

Dobretsov, S., Dahms, H.-U., and Qian, P.-Y. (2006). Inhibition of biofouling by marine microorganisms and their metabolites. Biofouling *22*, 43–54.

Dobretsov, S., Dahms, H.-U., YiLi, H., Wahl, M., and Qian, P.-Y. (2007). The effect of quorum-sensing blockers on the formation of marine microbial communities and larval attachment. FEMS Microbiol. Ecol. *60*, 177–188.

Dobretsov, S., Teplitski, M., and Paul, V. (2009). Mini-review: quorum sensing in the marine environment and its relationship to biofouling. Biofouling *25*, 413–427.

Dobretsov, S., Dürr, S., and Thomason, J. (2010a). Marine biofilms. Biofouling, 123–136.

Dobretsov, S., Teplitski, M., Alagely, A., Gunasekera, S.P., and Paul, V.J. (2010b). Malyngolide from the cyanobacterium *Lyngbya majuscula* interferes with quorum sensing circuitry. Environ. Microbiol. Rep. *2*, 739–744.

Dobretsov, S., Teplitski, M., Bayer, M., Gunasekera, S., Proksch, P., and Paul, V.J. (2011). Inhibition of marine biofouling by bacterial quorum sensing inhibitors. Biofouling *27*, 893–905.

Dobretsov, S., Abed, R.M., and Voolstra, C.R. (2013). The effect of surface colour on the formation of marine micro and macrofouling communities. Biofouling *29*, 617–627.

Dong, Y.-H., and Zhang, L.-H. (2005). Quorum sensing and quorum-quenching enzymes. J. Microbiol. *43*, 101–109.

Dong, Y.-H., Wang, L.-H., Xu, J.-L., Zhang, H.-B., Zhang, X.-F., and Zhang, L.-H. (2001). Quenching quorum-sensing-dependent bacterial infection by an N-acyl homoserine lactonase. Nature *411*, 813–817.

Eberhard, A., Burlingame, A., Eberhard, C., Kenyon, G., Nealson, K., and Oppenheimer, N. (1981). Structural identification of autoinducer of Photobacterium fischeri luciferase. Biochemistry *20*, 2444–2449.

Fenchel, T. (1986). The ecology of heterotrophic microflagellates. In Advances in Microb. Ecol., Marshall, K.C., ed. (Springer, New York, NY, USA), pp. 57–97.

Flemming, H.-C., and Ridgway, H. (2008). Biofilm control: conventional and alternative approaches. In

Springer Series on Biofilms (Springer-Verlag, Berlin/Heidelberg, Germany), pp. 103–117.

Flemming, H., and Wingender, J. (2001). Relevance of microbial extracellular polymeric substances (EPSs)—Part II: Technical aspects. Water Sci. Technol. *43*, 9–16.

Flemming, H.-C., and Wingender, J. (2010). The biofilm matrix. Nat. Rev. Microbiol. *8*, 623–633.

Fletcher, M., and Loeb, G. (1979). Influence of substratum characteristics on the attachment of a marine pseudomonad to solid surfaces. Appl. Environ. Microbiol. *37*, 67–72.

Forster, R., Créach, V., Sabbe, K., Vyverman, W., and Stal, L. (2006). Biodiversity-ecosystem function relationship in microphytobenthic diatoms of the Westerschelde estuary. Mar. Ecol. Prog. Ser. *311*, 191–201.

Franck Cassé, G.W.S. (2006). The development of microfouling on four commercial antifouling coatings under static and dynamic immersion. Int. Biodeterior. Biodegradation *57*, 179–185.

Geesey, G., Beech, I., Bremer, P., Webster, B., and Wells, D. (2000). Biocorrosion. In Biofilms II: Process Analysis and Applications, Bryers, J.D., ed. (Wiley-Liss, New York, NY, USA), pp. 281–325.

Goecke, F.R., Labes, A., Wiese, J., and Imhoff, J.F. (2010). Chemical interactions between marine macroalgae and bacteria. Mar. Ecol. Prog. Ser. *409*, 267–299.

Gram, L., Grossart, H.-P., Schlingloff, A., and Kiørboe, T. (2002). Possible quorum sensing in marine snow bacteria: production of acylated homoserine lactones by Roseobacter strains isolated from marine snow. Appl. Environ. Microbiol. *68*, 4111–4116.

Grossart, H.P. (2010). Ecological consequences of bacterioplankton lifestyles: changes in concepts are needed. Environ. Microbiol. Rep. *2*, 706–714.

Gusarov, I., Starodubtseva, M., Wang, Z.-Q., McQuade, L., Lippard, S.J., Stuehr, D.J., and Nudler, E. (2008). Bacterial nitric-oxide synthases operate without a dedicated redox partner. J. Biol. Chem. *283*, 13140–13147.

Harder, T., Lam, C., and Qian, P.-Y. (2002). Induction of larval settlement in the polychaete *Hydroides elegans* by marine biofilms: an investigation of monospecific diatom films as settlement cues. Mar. Ecol. Prog. Ser. *229*, 105–112.

Hashimoto, K., Irie, H., and Fujishima, A. (2005). TiO2 photocatalysis: a historical overview and future prospects. Jpn. J. Appl. Phys. *44*, 8269.

Hentzer, M., Riedel, K., Rasmussen, T.B., Heydorn, A., Andersen, J.B., Parsek, M.R., Rice, S.A., Eberl, L., Molin, S., and Høiby, N. (2002). Inhibition of quorum sensing in *Pseudomonas aeruginosa* biofilm bacteria by a halogenated furanone compound. Microbiology *148*, 87–102.

Holmström, C., and Kjelleberg, S. (1999). Marine Pseudoalteromonas species are associated with higher organisms and produce biologically active extracellular agents. FEMS Microbiol. Ecol. *30*, 285–293.

Holmström, C., Egan, S., Franks, A., McCloy, S., and Kjelleberg, S. (2002). Antifouling activities expressed by marine surface associated Pseudoalteromonas species. FEMS Microbiol. Ecol. *41*, 47–58.

Huang, Y.-L., Dobretsov, S., Ki, J.-S., Yang, L.-H., and Qian, P.-Y. (2007). Presence of acyl-homoserine lactone in subtidal biofilm and the implication in larval behavioral response in the polychaete *Hydroides elegans*. Microb. Ecol. *54*, 384–392.

Huang, Y.-L., Ki, J.-S., Case, R.J., and Qian, P.-Y. (2008). Diversity and acyl-homoserine lactone production among subtidal biofilm-forming bacteria. Aquat. Microb. Ecol. *52*, 185.

Irie, Y., and Parsek, M. (2008). Quorum sensing and microbial biofilms. In Bacterial Biofilms, Romeo, T., ed. (Springer-Verlag, Berlin/Heidelberg, Germany), pp. 67–84.

Ista, L.K., Fan, H., Baca, O., and López, G.P. (1996). Attachment of bacteria to model solid surfaces: oligo (ethylene glycol) surfaces inhibit bacterial attachment. FEMS Microbiol. Lett. *142*, 59–63.

Jaisai, M., Baruah, S., and Dutta, J. (2012). Paper modified with ZnO nanorods–antimicrobial studies. Beilstein J. Nanotechnol. *3*, 684–691.

James, S.G., Holmström, C., and Kjelleberg, S. (1996). Purification and characterization of a novel antibacterial protein from the marine bacterium D2. Appl. Environ. Microbiol. *62*, 2783–2788.

Jefferson, K.K. (2004). What drives bacteria to produce a biofilm? FEMS Microbiol. Lett. *236*, 163–173.

Jenal, U. (2004). Cyclic di-guanosine-monophosphate comes of age: a novel secondary messenger involved in modulating cell surface structures in bacteria? Curr. Opin. Microbiol. *7*, 185–191.

Joint, I., Tait, K., Callow, M.E., Callow, J.A., Milton, D., Williams, P., and Cámara, M. (2002). Cell-to-cell communication across the prokaryote-eukaryote boundary. Science *298*, 1207–1207.

Joint, I., Tait, K., and Wheeler, G. (2007). Cross-kingdom signalling: exploitation of bacterial quorum sensing molecules by the green seaweed Ulva. Philos. Trans. R. Soc. B Biol. Sci. *362*, 1223–1233.

Jones, N., Ray, B., Ranjit, K.T., and Manna, A.C. (2008). Antibacterial activity of ZnO nanoparticle suspensions on a broad spectrum of microorganisms. FEMS Microbiol. Lett. *279*, 71–76.

Jones, P.R., Cottrell, M.T., Kirchman, D.L., and Dexter, S.C. (2007). Bacterial community structure of biofilms on artificial surfaces in an estuary. Microb. Ecol. *53*, 153–162.

Joubert, L.-M., Wolfaardt, G., and Botha, A. (2006). Microbial exopolymers link predator and prey in a model yeast biofilm system. Microb. Ecol. *52*, 187–197.

Kiørboe, T., Tang, K., Grossart, H.-P., and Ploug, H. (2003). Dynamics of microbial communities on marine snow aggregates: colonization, growth, detachment, and grazing mortality of attached bacteria. Appl. Environ. Microbiol. *69*, 3036–3047.

Kjellerup, B.V., Thomsen, T.R., Nielsen, J.L., Olesen, B., Frølund, B., and Nielsen, P.H. (2005). Microbial diversity in biofilms from corroding heating systems. Biofouling *21*, 19–29.

Kwan, J.C., Meickle, T., Ladwa, D., Teplitski, M., Paul, V., and Luesch, H. (2011). Lyngbyoic acid, a "tagged" fatty acid from a marine cyanobacterium, disrupts quorum sensing in *Pseudomonas aeruginosa*. Mol. Biosyst. *7*, 1205–1216.

Lau, S.C., Thiyagarajan, V., Cheung, S.C., and Qian, P.-Y. (2005). Roles of bacterial community composition in biofilms as a mediator for larval settlement of three marine invertebrates. Aquat. Microb. Ecol. *38*, 41–51.

Le Roux, F., Zouine, M., Chakroun, N., Binesse, J., Saulnier, D., Bouchier, C., Zidane, N., Ma, L., Rusniok, C., and Lajus, A. (2009). Genome sequence of Vibrio splendidus: an abundant planctonic marine species with a large genotypic diversity. Environ. Microbiol. *11*, 1959–1970.

Lee, S.J., Park, S.-Y., Lee, J.-J., Yum, D.-Y., Koo, B.-T., and Lee, J.-K. (2002). Genes encoding the *N*-acyl homoserine lactone-degrading enzyme are widespread in many subspecies of *Bacillus thuringiensis*. Appl. Environ. Microbiol. *68*, 3919–3924.

Lewandowski, Z. (2000). Structure and function of biofilms. In Biofilms: Recent Advances in Their Study and Control, Evans, L.V., ed. (Harwood Academic Publishers, Newark, NJ, USA), pp. 1–17.

Li, Q., Mahendra, S., Lyon, D.Y., Brunet, L., Liga, M.V., Li, D., and Alvarez, P.J. (2008). Antimicrobial nanomaterials for water disinfection and microbial control: potential applications and implications. Water Res. *42*, 4591–4602.

Liu, N., Xu, Y., Hossain, S., Huang, N., Coursolle, D., Gralnick, J.A., and Boon, E.M. (2012). Nitric oxide regulation of cyclic di-GMP synthesis and hydrolysis in *Shewanella woodyi*. Biochemistry *51*, 2087–2099.

Madigan, M., and Martinko, J. (2006). Brock Biology of Microorganisms (Pearson Education, Inc., San Francisco, CA, USA).

Maki, J., and Mitchell, R. (1985). Involvement of lectins in the settlement and metamorphosis of marine invertebrate larvae. Bull. Mar. Sci. *37*, 675–683.

Maki, J.S., and Mitchell, R. (2002). Biofouling in the marine environment. In Encyclopedia of Environmental Microbiology (John Wiley & Sons, New York, NY, USA).

Mansfeld, F., Tsai, R., Shih, H., Little, B., Ray, R., and Wagner, P. (1992). An electrochemical and surface analytical study of stainless steels and titanium exposed to natural seawater. Corr. Sci. 33, 445–456.

Marshall, K., Stout, R., and Mitchell, R. (1971). Mechanism of the initial events in the sorption of marine bacteria to surfaces. J. Gen. Microbiol. *68*, 337–348.

Maybruck, B.T., and Rogerson, A. (2004). Protozoan Epibionts on the Prop Roots of the Red Mangrove Tree, Rhizaphora mangle. Protistology 3, 265–272.

Mceldowney, S., and Fletcher, M. (1986). Effect of growth conditions and surface characteristics of aquatic bacteria on their attachment to solid surfaces. J. Gen. Microbiol. *132*, 513–523.

McNamara, C.J., Lee, K.B., Russell, M.A., Murphy, L.E., and Mitchell, R. (2009). Analysis of bacterial community composition in concretions formed on the USS Arizona, Pearl Harbor, HI. J. Cult. Herit. *10*, 232–236.

Miao, L., and Qian, P.-Y. (2005). Antagonistic antimicrobial activity of marine fungi and bacteria isolated from marine biofilm and seawaters of Hong Kong. Aquat. Microb. Ecol. *38*, 231–238.

Molino, P.J., and Wetherbee, R. (2008). The biology of biofouling diatoms and their role in the development of microbial slimes. Biofouling *24*, 365–379.

Moss, J.A., Nocker, A., Lepo, J.E., and Snyder, R.A. (2006). Stability and change in estuarine biofilm bacterial community diversity. Appl. Environ. Microbiol. *72*, 5679–5688.

Musker, A. (1980). Universal roughness functions for naturally-occurring surfaces. Trans. Can. Soc. Mech. Engin. *6*, 1–6.

Muthukrishnan, T., Abed, R.M.M., Dobretsov, S., Kidd, B., and Finnie, A.A. (2014). Long-term microfouling on commercial biocidal fouling control coatings. Biofouling *30*, 1155–1164.

Nadell, C.D., Xavier, J.B., and Foster, K.R. (2009). The sociobiology of biofilms. FEMS Microbiol. Rev. *33*, 206–224.

Natalio, F., André, R., Hartog, A.F., Stoll, B., Jochum, K.P., Wever, R., and Tremel, W. (2012). Vanadium pentoxide nanoparticles mimic vanadium haloperoxidases and thwart biofilm formation. Nat. Nanotechnol. *7*, 530–535.

Ninnemann, H., and Maier, J. (1996). Indications for the occurrence of nitric oxide synthases in fungi and plants and the involvement in photoconidiation of *Neurospora crassa*. Photochem. Photobiol. *64*, 393–398.

Nocker, A., Lepo, J.E., and Snyder, R.A. (2004). Influence of an oyster reef on development of the microbial heterotrophic community of an estuarine biofilm. Appl. Environ. Microbiol. *70*, 6834–6845.

Oh, H.-S., Yeon, K.-M., Yang, C.-S., Kim, S.-R., Lee, C.-H., Park, S.Y., Han, J.Y., and Lee, J.-K. (2012). Control of membrane biofouling in MBR for wastewater treatment by quorum quenching bacteria encapsulated in microporous membrane. Environ. Sci. Technol. *46*, 4877–4884.

O'Toole, G., Kaplan, H.B., and Kolter, R. (2000). Biofilm formation as microbial development. Annu. Rev. Microbiol. *54*, 49–79.

Park, S.-Y., Kang, H.-O., Jang, H.-S., Lee, J.-K., Koo, B.-T., and Yum, D.-Y. (2005). Identification of extracellular N-acylhomoserine lactone acylase from a *Streptomyces* sp. and its application to quorum quenching. Appl. Environ. Microbiol. *71*, 2632–2641.

Parsek, M.R., and Greenberg, E.P. (2000). Acyl-homoserine lactone quorum sensing in gram-negative bacteria: a signaling mechanism involved in associations with higher organisms. Proc. Natl. Acad. Sci. U.S.A. *97*, 8789–8793.

Pereira, M.O., Kuehn, M., Wuertz, S., Neu, T., and Melo, L.F. (2002). Effect of flow regime on the architecture of a *Pseudomonas fluorescens* biofilm. Biotechnol. Bioeng. *78*, 164–171.

Piehler, M.F., and Paerl, H.W. (1996). Enhanced biodegradation of diesel fuel through the addition of particulate organic carbon and inorganic nutrients in coastal marine waters. Biodegradation 7, 239–247.

Pitts, B., Hamilton, M.A., Zelver, N., and Stewart, P.S. (2003). A microtiter-plate screening method for biofilm disinfection and removal. J. Microbiol. Methods 54, 269–276.

Plate, L., and Marletta, M.A. (2012). Nitric oxide modulates bacterial biofilm formation through a multicomponent cyclic-di-GMP signaling network. Mol. Cell 46, 449–460.

Qian, P.-Y., Lau, S., Dahms, H.-U., Dobretsov, S., and Harder, T. (2007). Marine biofilms as mediators of colonization by marine macroorganisms: implications for antifouling and aquaculture. Mar. Biotechnol. 9, 399–410.

Railkin, A.I. (2003). Marine Biofouling: Colonization Processes and Defenses (CRC Press, Boca Raton, FL, USA).

Rajamani, S., Teplitski, M., Kumar, A., Krediet, C.J., Sayre, R.T., and Bauer, W.D. (2011). *N*-Acyl homoserine lactone lactonase, AiiA, inactivation of quorum-sensing agonists produced by *Chlamydomonas reinhardtii* (Chlorophyta) and characterization of *aiiA* transgenic algae. J. Phycol. 47, 1219–1227.

Rao, D., Webb, J.S., and Kjelleberg, S. (2005). Competitive interactions in mixed-species biofilms containing the marine bacterium *Pseudoalteromonas tunicata*. Appl. Environ. Microbiol. 71, 1729–1736.

Rao, D., Webb, J.S., Holmström, C., Case, R., Low, A., Steinberg, P., and Kjelleberg, S. (2007). Low densities of epiphytic bacteria from the marine alga *Ulva australis* inhibit settlement of fouling organisms. Appl. Environ. Microbiol. 73, 7844–7852.

Rasmussen, T.B., Skindersoe, M.E., Bjarnsholt, T., Phipps, R.K., Christensen, K.B., Jensen, P.O., Andersen, J.B., Koch, B., Larsen, T.O., and Hentzer, M. (2005). Identity and effects of quorum-sensing inhibitors produced by Penicillium species. Microbiology 151, 1325–1340.

Römling, U., and Amikam, D. (2006). Cyclic di-GMP as a second messenger. Curr. Opin. Microbiol. 9, 218–228.

Römling, U., Gomelsky, M., and Galperin, M.Y. (2005). C-di-GMP: the dawning of a novel bacterial signalling system. Mol. Microbiol. 57, 629–639.

Roux, A., Payne, S.M., and Gilmore, M.S. (2009). Microbial telesensing: probing the environment for friends, foes, and food. Cell Host Microbe 6, 115–124.

Ruby, E.G., and Lee, K.-H. (1998). The *Vibrio fischeri-Euprymna* scolopes light organ association: current ecological paradigms. Appl. Environ. Microbiol. 64, 805–812.

Salta, M., Wharton, J.A., Blache, Y., Stokes, K.R., and Briand, J.F. (2013). Marine biofilms on artificial surfaces: structure and dynamics. Environ. Microbiol. 15, 2879–2893.

Sauer, K., Camper, A.K., Ehrlich, G.D., Costerton, J.W., and Davies, D.G. (2002). *Pseudomonas aeruginosa* displays multiple phenotypes during development as a biofilm. J. Bacteriol. 184, 1140–1154.

Schlag, S., Nerz, C., Birkenstock, T.A., Altenberend, F., and Götz, F. (2007). Inhibition of staphylococcal biofilm formation by nitrite. J. Bacteriol. 189, 7911–7919.

Schultz, M., Bendick, J., Holm, E., and Hertel, W. (2011). Economic impact of biofouling on a naval surface ship. Biofouling 27, 87–98.

Sio, C.F., Otten, L.G., Cool, R.H., Diggle, S.P., Braun, P.G., Bos, R., Daykin, M., Cámara, M., Williams, P., and Quax, W.J. (2006). Quorum quenching by an *N*-acyl-homoserine lactone acylase from *Pseudomonas aeruginosa* PAO1. Infect. Immun. 74, 1673–1682.

Steinberg, P.D., and De Nys, R. (2002). Chemical mediation of colonization of seaweed surfaces. J. Phycol. 38, 621–629.

Stoodley, P., Sauer, K., Davies, D., and Costerton, J.W. (2002). Biofilms as complex differentiated communities. Annu. Rev. Microbiol. 56, 187–209.

Swain, G. (1998). Biofouling control: a critical component of drag reduction. Paper presented at: Proceedings of the International Symposium on Seawater Drag Reduction (Newport, RI, USA).

Tait, K., Joint, I., Daykin, M., Milton, D.L., Williams, P., and Camara, M. (2005). Disruption of quorum sensing in seawater abolishes attraction of zoospores of the green alga Ulva to bacterial biofilms. Environ. Microbiol. 7, 229–240.

Tao, Y., Misko, T., Howlett, A., and Klein, C. (1997). Nitric oxide, an endogenous regulator of *Dictyostelium discoideum* differentiation. Development 124, 3587–3595.

Taylor, M.W., Schupp, P.J., Baillie, H.J., Charlton, T.S., De Nys, R., Kjelleberg, S., and Steinberg, P.D. (2004). Evidence for acyl homoserine lactone signal production in bacteria associated with marine sponges. Appl. Environ. Microbiol. 70, 4387–4389.

Teplitski, M., Mathesius, U., and Rumbaugh, K.P. (2010). Perception and degradation of N-acyl homoserine lactone quorum sensing signals by mammalian and plant cells. Chem. Rev. 111, 100–116.

Thompson, R., Norton, T., and Hawkins, S. (2004). Physical stress and biological control regulate the producer-consumer balance in intertidal biofilms. Ecology 85, 1372–1382.

Thompson, R., Moschella, P., Jenkins, S., Norton, T., and Hawkins, S.J. (2005). Differences in photosynthetic marine biofilms between sheltered and moderately exposed rocky shores. Mar. Ecol. Prog. Ser. 296, 53–63.

Thornton, D.C., Dong, L.F., Underwood, G.J., and Nedwell, D.B. (2002). Factors affecting microphytobenthic biomass, species composition and production in the Colne Estuary (UK). Aquat. Microb. Ecol. 27, 285–300.

Underwood, G.J. (2005). Microalgal (Microphytobenthic) biofilms in shallow coastal waters: how important are species? Proc. Calif. Acad. Sci. 56, 162.

Underwood, G., Perkins, R.G., Consalvey, M.C., Hanlon, A., Oxborough, K., Baker, N., and Paterson, D. (2005). Patterns in microphytobenthic primary productivity: species-specific variation in migratory rhythms and photosynthetic efficiency in mixed-species biofilms. Limnol. Oceanogr. 50, 755–767.

Valle, J., Da Re, S., Henry, N., Fontaine, T., Balestrino, D., Latour-Lambert, P., and Ghigo, J.-M. (2006). Broad-spectrum biofilm inhibition by a secreted bacterial polysaccharide. Proc. Natl. Acad. Sci. U.S.A. 103, 12558–12563.

Van Alst, N.E., Picardo, K.F., Iglewski, B.H., and Haidaris, C.G. (2007). Nitrate sensing and metabolism modulate motility, biofilm formation, and virulence in *Pseudomonas aeruginosa*. Infect. Immun. 75, 3780–3790.

Vandevivere, P., and Kirchman, D.L. (1993). Attachment stimulates exopolysaccharide synthesis by a bacterium. Appl. Environ. Microbiol. 59, 3280–3286.

Videla, H.A., and Herrera, L.K. (2005). Microbiologically influenced corrosion: looking to the future. Int. Microbiol. *8*, 169.

Wahl, M. (1996). Living attached: fouling, epibiosis. In Fouling Organisms of the Indian Ocean: Biology and Control Technology, R. Nagabhushanam, and M.F. Thompson, eds. (Oxford & IBH Publishers, New Delhi, India), pp. 31–84.

Waters, C.M., and Bassler, B.L. (2005). Quorum sensing: cell-to-cell communication in bacteria. Annu. Rev. Cell Dev. Biol. *21*, 319–346.

Webster, N.S., Smith, L.D., Heyward, A.J., Watts, J.E., Webb, R.I., Blackall, L.L., and Negri, A.P. (2004). Metamorphosis of a scleractinian coral in response to microbial biofilms. Appl. Environ. Microbiol. *70*, 1213–1221.

Weinbauer, M.G. (2004). Ecology of prokaryotic viruses. FEMS Microbiol. Rev. *28*, 127–181.

Wieczorek, S.K., and Todd, C.D. (1998). Inhibition and facilitation of settlement of epifaunal marine invertebrate larvae by microbial biofilm cues. Biofouling *12*, 81–118.

Wiese, J., Thiel, V., Nagel, K., Staufenberger, T., and Imhoff, J.F. (2009). Diversity of antibiotic-active bacteria associated with the brown alga laminaria saccharina from the baltic sea. Mar. Biotechnol. *11*, 287–300.

Wigglesworth-Cooksey, B., and Cooksey, K.E. (2005). Use of fluorophore-conjugated lectins to study cell-cell interactions in model marine biofilms. Appl. Environ. Microbiol. *71*, 428–435.

Wilken, M., and Huchzermeyer, B. (1999). Suppression of mycelia formation by NO produced endogenously in *Candida tropicalis*. Eur. J. Cell Biol. *78*, 209–213.

Williams, P. (2007). Quorum sensing, communication and cross-kingdom signalling in the bacterial world. Microbiology *153*, 3923–3938.

Williams, P., Winzer, K., Chan, W.C., and Cámara, M. (2007). Look who's talking: communication and quorum sensing in the bacterial world. Philos. Trans. R. Soc. B Biol. Sci. *362*, 1119–1134.

Wong, C.S., Yin, W.F., Choo, Y.M., Sam, C.K., Koh, C.L., and Chan, K.G. (2012). Coexistence of quorum-quenching and quorum-sensing in tropical marine Pseudomonas aeruginosa strain MW3A. World J. Microbiol. Biotechnol. *28*, 453–461.

Yebra, D.M., Kiil, S., Weinell, C.E., and Dam-Johansen, K. (2006). Presence and effects of marine microbial biofilms on biocide-based antifouling paints. Biofouling *22*, 33–41.

Yin, W.F., Purmal, K., Chin, S., Chan, X.Y., Koh, C.L., Sam, C.K., and Chan, K.G. (2012). *N*-Acyl homoserine lactone production by *Klebsiella pneumoniae* isolated from human tongue surface. Sensors *12*, 3472–3483.

Zhang, L.H., and Dong, Y.H. (2004). Quorum sensing and signal interference: diverse implications. Mol. Microbiol. *53*, 1563–1571.

Zobell, C.E. (1943). The effect of solid surfaces upon bacterial activity. J. Bacteriol. *46*, 39–56.

Zumft, W.G. (1993). The biological role of nitric oxide in bacteria. Arch. Microbiol. *160*, 253–264.

Microbial Biodiversity in Natural Biofilms

4

Katharina Besemer

Abstract

Natural biofilms in aquatic ecosystems exist as complex and dynamic communities, which harbour a considerable microbial biodiversity, typically dominated by Alpha-, Beta- and Gamma-proteobacteria, bacteroidetes and cyanobacteria. A number of publications have highlighted the importance of biofilm biodiversity for ecosystem processes in a range of environments. The aim of this chapter is to give an overview of different aspects of microbial diversity in natural biofilms, including local diversity (alpha-diversity), among-patch diversity (beta-diversity), taxonomic and functional diversity. While its major focus lies on biofilms in the benthic and hyporheic zone of streams and rivers, this chapter also includes examples from lake, tidal and marine sediments, and biofilms associated with suspended aggregates. The local and regional processes that have been proposed to sustain or constrain biofilm biodiversity and community assembly, such as environmental heterogeneity, biotic interactions and dispersal dynamics are reviewed. Finally, this chapter considers the relationship between taxonomic diversity, functional diversity and ecosystem processes.

Introduction

Biofilms are an important form of microbial life in many aquatic ecosystems (Davey and O'Toole, 2000; Hall-Stoodley *et al.*, 2004; see also Chapter 1), where they contribute to major ecosystem processes and even to globally important biogeochemical fluxes (Battin *et al.*, 2003a, 2008). Biofilms develop in virtually every interfacial environment (Battin *et al.*, 2008; Hall-Stoodley *et al.*, 2004) and on a wide variety of organic and inorganic substrates, which makes them prominent players in organic matter remineralization (Das *et al.*, 2007; Golladay and Sinsabaugh, 1991) and in the stabilization of fine sediments against resuspension (Decho, 2000; Gobet *et al.*, 2012), for instance. In ecosystems with large surface-to-volume-ratios, such as streams, benthic (the interface between stream water and sediment exposed to light) and hyporheic (the streambed sediment where stream water and groundwater mix) biofilms typically dominate microbial life (Battin, 2000; Battin *et al.*, 2003ab; Boulton *et al.*, 1998; Marmonier *et al.*, 2012), while biofilms on submerged plants or attached to suspended aggregates ('mobile biofilms'; Battin *et al.*, 2008) are of major importance in larger rivers, estuaries and lakes (Luef *et al.*, 2007; Simon *et al.*, 2002; see Chapter 1). Furthermore, biofilms play a critical role in some marine ecosystems, such as intertidal mats, which are important locations of primary productivity and the main food resource for many benthic grazers (Decho, 2000; Russell *et al.*, 2007).

The taxonomic and functional diversity harboured by natural biofilms in aquatic ecosystems is huge. Extensive three-dimensional structures (see Chapter 2), a complex network of interactions and a surprising level of multi-cellular behaviour (see Chapter 3) act in concert to provide diverse micro-habitats favouring taxonomic and functional biodiversity (Flemming and Wingender, 2010; Stoodley *et al.*, 2002; Watnick and Kolter, 2000; Webb *et al.*, 2003). Bacteria, archaea, algae, fungi, protozoa and viruses contribute to the biodiversity of these dynamic communities (Jackson and Jackson, 2008; Lock *et al.*, 1984; Romaní, 2010; Stoodley *et al.*, 2002). The rapid development of novel technologies in the last decades has pushed the limits to which the diversity of microbial communities can currently be explored, and unveiled the vast diversity of the microbial world. Still, a comprehensive appraisal of microbial biodiversity is elusive for many ecosystems (Quince *et al.*, 2008; Zinger *et al.*, 2012) and aquatic biofilms are no exception to this. Nevertheless, the unprecedented accuracy with which microbial diversity can now be analysed has fostered new insights in the distribution of biofilm biodiversity and the underlying mechanisms (Besemer *et al.*, 2013; Wang *et al.*, 2012c, 2013). Novel statistical methods (Quince *et al.*, 2008) and meta-analyses of published data enable comparisons between studies and help to set findings in a broader ecological context (Lozupone and Knight, 2007).

This chapter reviews findings on the microbial diversity contained in natural biofilms in a range of aquatic habitats, the mechanisms generating and maintaining this biodiversity, and the implications of biodiversity for ecosystem functioning. While the major focus of this chapter lies on the prokaryotic diversity of benthic and hyporheic biofilms in streams and rivers, examples from other environments and from eukaryotic microbial groups are included to provide a more comprehensive picture of the diversity contained in natural biofilms.

Biofilm community assembly – stochastic versus deterministic processes

Biofilm formation is initiated when single cells or microbial aggregates from the overlying water column attach to a submerged surface (Costerton *et al.*, 1995; Hödl *et al.*, 2011), where some will start to grow and eventually form a biofilm (Stoodley *et al.*, 2002). This initial colonization may be largely stochastic, as it relies on dispersal from the source community suspended in the water (Battin *et al.*, 2007; Jackson, 2003). However, the aptitude to attach to surfaces or to other microbes differs between species (Rickard *et al.*, 2003; Yawata *et al.*, 2014), which may select for certain species already during this early phase of biofilm succession. Further, Hödl and colleagues (2011) found suggestive evidence that biotic interactions, such as cell-to-cell signalling or surface conditioning, play an important role already during primary colonization and early biofilm formation. As biofilm growth progresses, interactions between species and with the abiotic environment may gain further importance, denoting a shift from predominantly stochastic to deterministic processes governing biofilm formation (Battin *et al.*, 2007; Besemer *et al.*, 2007).

Succession of biofilms is driven by differential species performance in dispersal, survival and reproduction (Burns and Ryder, 2001). Based on a study on freshwater biofilms, Jackson and colleagues (2001) developed a conceptual model of diversity dynamics during biofilm succession. This model proposes that initial colonization leads to a rapid increase in species diversity, which subsequently declines as the biofilm becomes dominated by a limited number of strong competitors. Finally, as biofilms mature, their complex three-dimensional structure may facilitate greater diversity through habitat diversification and broader resource diversity. Trophic interactions within the biofilm are likely to become more important than competition during this late successional state, thereby fostering a switch from bottom-up to top-down control of biofilm diversity (Jackson, 2003).

Several empirical studies support the predictions of this conceptual model. For instance, epilithic river biofilms showed a rapid increase, followed by a decrease and, during late succession, again a slight increase in diversity during a biomass accrual phase (Lyautey *et al.*, 2005). Further, river biofilms developing on exposed glass slides (Boulêtreau *et al.*, 2014) and on decaying leaves (Fischer *et al.*, 2009) showed a decrease in bacterial richness after a peak during early succession, potentially caused by competitive exclusion. In contrast, Lear and colleagues (2008) found that epilithic biofilm diversity was still increasing after 7–9 weeks of growth, indicating that biofilm succession in these streams proceeded slowly. It is in fact plausible to assume that differences in resource supply, physical habitat conditions or disturbance regime in the various ecosystems modulate the successional patterns of aquatic biofilms. For instance, Fierer and colleagues (2010) predicted different successional trajectories for microbial communities dominated by autotrophic organisms (e.g. algae-dominated biofilms), heterotrophic microbial communities fuelled by external inputs of organic carbon (e.g. hyporheic biofilms), and communities whose major carbon sources are contained within the substrate itself (e.g. biofilms growing on decaying leaves). Changes in the quality and quantity of resources as they are modified by the microorganisms during biofilm growth are indeed likely to explain successional differences between epilithic (Lear *et al.*, 2008) and leaf litter (Fischer *et al.*, 2009) biofilms, for instance. Interestingly, successional patterns in experimental biofilms were also shown to be altered by shear stress. Biofilms grown at intermediate shear showed the pattern predicted by Jackson's model, whereas biofilms grown at high shear rates showed no second increase of diversity, and biofilms grown at low shear showed high diversity throughout the experiment (Rochex *et al.*, 2008).

Disturbances change the microbial landscape and set biofilm communities back to an earlier successional state by removing a part of the biodiversity, which then becomes available for colonization of new substrata (Battin *et al.*, 2007; Burns and Ryder, 2001). The temporal and spatial scale of such disturbances may influence diversity by altering the relative importance of biotic interactions and dispersal dynamics. When disturbance frequency is low, the extended time for species to recruit to habitat patches and for local interactions to take place is expected to favour competitively superior species, which are able to suppress or displace inferior competitors under the given suite of environmental conditions (Tilman, 1987). At high disturbance frequency, conditions may favour dominance by species that are either good dispersers and able to quickly colonize empty habitat patches, or are resistant against disturbances (Cadotte, 2007; Cardinale *et al.*, 2006a). In streams with a dynamic hydrologic regime, frequent disturbances might therefore preserve pioneering biofilm taxa by resetting the process of succession, while in more stable environments, like lakes for instance, biofilm communities are more likely to reach a mature state (Burns and Ryder, 2001). Several studies demonstrated the effects of disturbance frequency on biofilm diversity and succession. Cardinale and colleagues (2006a) showed that algal species diversity in stream biofilms could be explained by the joint effects of disturbance frequency, which created new niche opportunities for species, and by the rate at which biomass accrual lead to successional displacement of inferior by superior competitors. In an experimental study, Milferstedt and colleagues (2013) found that biofilm diversity was able to recover between weekly disturbances indicating community resilience, while daily disturbances led to a community dominated by a single, disturbance-resistant organism. Further, bacterial richness in stream biofilms during succession increased faster in a Mediterranean stream subjected to frequent drying and flooding events than in a more stable Central European stream, suggesting that the Mediterranean biofilm was dominated by taxa adapted to fast re-colonization after disturbances, whereas diversity in the Central European stream was likely controlled by biotic interactions (Artigas *et al.*, 2012). Given the spatial and temporal heterogeneity of disturbances and hydrodynamic conditions experienced by biota in aquatic environments (Lake, 2000) it may, therefore, be assumed that natural biofilms exist as assemblages of patches with differing successional trajectories and at different successional states (Besemer *et al.*, 2007, 2009a; Cardinale *et al.*, 2002).

Metacommunity ecology – the roles of dispersal and environmental conditions

At the regional scale, natural aquatic biofilms exist as part of a larger microbial network in which local communities are linked by dispersal of multiple interacting species to form a metacommunity (Battin *et al.*, 2007; Holyoak *et al.*, 2005). Numerous studies have assessed the relative importance of regional (dispersal dynamics, landscape patterns) and local processes (abiotic habitat conditions, deterministic biotic interactions) for microbial community composition and diversity. In a conceptual model for freshwater bacterioplankton, Logue and Lindström (2008) proposed that in aquatic ecosystems with long retention times such as lakes, community assembly is mainly driven by species sorting, where the local environment and biotic interactions select from the metacommunity. Systems with short residence times such as streams, on the other hand, are strongly affected by mass effects through the constant input of species, which maintain species in less favourable habitats and lessen the co-variance between environmental conditions and community composition (Logue and Lindström, 2008). However, the attachment of microbial cells to a stable surface increases their residence time relative to those in the water column, which potentially makes them more susceptible to local environmental conditions.

Indeed, environmental conditions, rather than geographic distance, were repeatedly found to explain biofilm community composition indicating that species sorting would be a candidate mechanism for biofilm community assembly. For instance, Fierer and colleagues (2007) found that pH could predict most of the variation between bacterial communities inhabiting fine benthic organic matter in streams, and, similarly, Horner-Devine and colleagues (2004) found that bacterial community composition in salt marsh sediments was primarily driven by water chemistry and plant cover, rather than geographic distance. However, combined environmental and dispersal dynamics were observed to explain community variation among marine sediment bacteria (Xiong *et al.*, 2014), and ammonia-oxidizing bacterial communities in salt-marsh sediments, which showed imprints of dispersal limitation at local, but not at regional scale (Martiny *et al.*, 2011). Furthermore, community composition and diversity of epilithic stream biofilms across New Zealand were related to both geographical location and environmental variables, though the relationship with the environmental properties was clearly stronger (Lear *et al.*, 2013). Lear and colleagues suggested that the structure of bacterial communities is largely determined by niche-based processes (that is, species sorting due to environmental factors) rather than by dispersal limitations. To evaluate species sorting as a possible mechanism of biofilm assembly, the biofilm communities from boreal (Besemer *et al.*, 2012) and glacier-fed streams (Wilhelm *et al.*, 2013) were compared with random subsamples of the respective stream water communities, as they might result from purely stochastic immigration to an empty habitat patch from a source community. It was found that such stochastic immigration was unlikely to explain the observed community composition of these biofilms. Collectively, these findings suggest that the local environment and biotic interactions select microorganisms from the stream water for biofilm formation (Fig. 4.1). In contrast, stream water communities can likely be explained by mass effects, determined by large cell influx rates and short residence times limiting the influence of environmental factors (Logue and Lindström, 2008). Streams collect microorganisms from various sources, such as the surrounding terrestrial landscape, groundwater and upstream tributaries, collectively forming the upstream metacommunity (Crump *et al.*, 2007, 2012; Wilhelm *et al.*, 2013). The notion of stream water as an important integrator of microbial diversity is indeed supported by the fact that significantly higher diversity was found in the boreal and glacier-fed stream water communities than in the biofilms (Besemer *et al.*, 2012; Wilhelm *et al.*, 2013). Similar results have been reported for a marsh of Lake Hallwil in Switzerland, were the microbial diversity of the lake water was slightly higher than that in sediments, and clearly higher than on plant litter and in epiphytic biofilms (Buesing *et al.*, 2009). In contrast, biofilm diversity was found to be higher than planktonic bacterial diversity in Lake Baikal (Parfenova *et al.*, 2013) and in mountain lakes in the Pyrenees (Bartrons *et al.*, 2012). According to the conceptual model of Logue and Lindström (2008), it can be expected that continuous inflow of microbes could support the microbial diversity in stream water and, to a lesser extent, in lakes with a short retention time (for instance, 3.8 years in lake Hallwil; Buesing and Gessner, 2006), while species sorting and competitive interactions might constrain the diversity in a lake with such a high retention time as lake Baikal (> 300 years; Jasechko *et al.*, 2013). However, the mountain lakes studied by Bartrons and colleagues (2010, 2012) showed higher phylogenetic diversity in biofilms than in the plankton despite the short water residence times of these lakes.

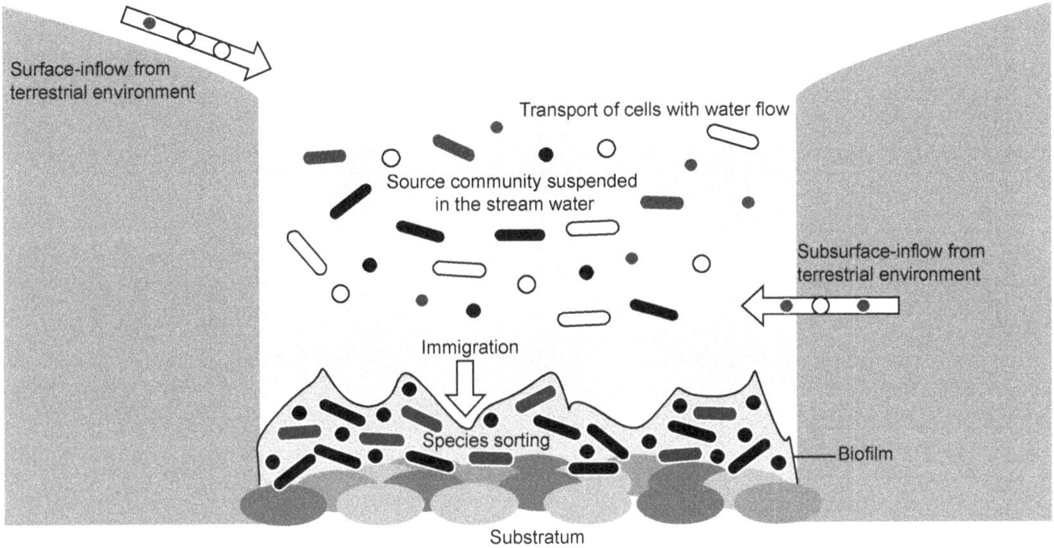

Figure 4.1 Theoretical scheme of stream biofilm assembly. The stream water collects and transports microorganisms from the surrounding landscape into a source community, which eventually seeds the benthic biofilm. The attachment to a stable surface and the encapsulation of microbial cells in the biofilm matrix increase their residence time, making them more susceptible to the ambient environmental conditions. Species sorting is therefore a candidate mechanism for the assembly of benthic biofilm communities. The microbial diversity in the stream water, on the other hand, may be supported by mass effects through the constant influx of species from the surrounding environment.

Landscape topography and water flow affect dispersal in aquatic systems, but also generate microhabitats that differ in turbulence and shear (Battin *et al.*, 2003b, 2007; Rickard *et al.*, 2004). Experimenting with stream mesocosms containing streambed landscapes, epilithic biofilm community composition was found to be related to the spatial variation of hydrodynamic conditions (Besemer *et al.*, 2009a). The resulting species turnover between habitat patches differing in hydrodynamic conditions generated a gradient of beta-diversity that increased with habitat heterogeneity at the landscape scale. It can be assumed that the flow landscape sorted species according to their ability to cope with the local hydrodynamic environment, which agrees with earlier findings showing that flow velocity and shear stress modify biofilm structure (Battin *et al.*, 2003b). Using the same experimental set-up, Woodcock and colleagues (2013) showed that the beta-diversity patterns among habitat patches could be sufficiently reproduced by a neutral model of flow induced dispersal when the hydrodynamic regime in the mesocosm was homogenous, that is, when the streambed was flat. This result suggests that dispersal has a role in structuring biofilm composition at small scales, and agrees with the findings from Martiny and colleagues (2011) in salt marsh sediments. However, this relatively simple model failed to capture the complexity of the heterogeneous flow landscapes. Furthermore, Woodcock and colleagues (2013) found evidence that the importance of immigration from the stream water varied between microhabitats differing in water turbulence and flow velocity. These findings point to flow heterogeneity as a factor likely influencing biofilm biodiversity by altering both dispersal dynamics and physical habitat conditions.

The importance of flow-induced dispersal for biofilm diversity at the scale of an entire stream network was assessed by sampling more than 100 streams in a pre-alpine catchment (Besemer *et al.*, 2013). Confluences, as conspicuous nodes in the fluvial network, were expected to accumulate species from multiple catchments and, thus, increase the size of the metacommunity from which the local biofilm communities assemble. If random invasion from the metacommunity has a role in structuring a local community, the diversity of this community will increase with the size of the metacommunity from which it is drawn (Curtis and Sloan, 2004). Alpha-diversity was therefore predicted to increase downstream of confluences, and, at the network scale, from headwaters to larger streams. However, biodiversity and evenness of microbial biofilms were, on average, lower downstream of confluences throughout the fluvial network, ultimately causing biodiversity to decrease downstream (Besemer *et al.*, 2013). These results suggest that the local environment and biotic interactions, such as competition, may modify the influence of metacommunity connectivity on local biofilm biodiversity throughout a stream network. The larger regional species pool downstream of confluences is more likely to include superior competitors adapted to the streambed environment. Cascading through the network, this accumulation of competitive taxa could ultimately decrease mean alpha-diversity.

Furthermore, beta-diversity among headwaters was higher than between larger streams, evoking dispersal limitation as a driver of beta-diversity among these systems (Besemer *et al.*, 2013). Headwaters also encompass a larger geographical area compared with downstream catchments, which results in a wider range of environmental conditions among these systems, potentially enhancing beta-diversity (Clarke *et al.*, 2008; Fierer *et al.*, 2007). Controlling for a possible distance effect, it was found that the higher beta-diversity among headwaters could not be explained exclusively by the larger geographical (and potentially environmental) distance among them. This suggests that the dendritic nature of fluvial networks constrains microbial dispersal and leads to elevated beta-diversity between headwaters. Pronounced variability of alpha-diversity among headwaters further indicated an imprint of dispersal limitation, in combination with higher habitat and resource variation among headwaters (Besemer *et al.*, 2013; Brown *et al.*, 2011).

A negative effect of high metacommunity connectivity, as indicated by this study, has indeed been predicted by a theoretical model (Mouquet and Loreau, 2003) and has also been shown for experimental metacommunities of marine algal biofilms (Eggers *et al.*, 2012; Matthiessen *et al.*, 2010). Matthiessen and colleagues (2010) found that habitat heterogeneity in terms of varying light intensities among local patches increased local and regional diversity, maintaining inferior competitors in less favourable patches. However, increasing dispersal rates among local communities homogenized the metacommunity by distributing the best competitor across the whole metacommunity, thereby decreasing diversity and evenness. Using a similar set-up to test for the combined effects of habitat connectivity and heat stress, Eggers and colleagues (2012) reported that high dispersal rates decreased local diversity and evenness, while, at the same time, generating a rescue effect for extinction-prone rare species. Taken together, these studies indicate that the complex network of interactions, which characterizes microbial biofilms (Stoodley *et al.*, 2002), is likely to play an important role in the assembly and diversity of natural biofilm communities.

Spatial and temporal species turnover – the role of beta-diversity

Beta-diversity across space

Regional diversity depends on both local diversity and community variability between habitat patches. The biodiversity contained in a region can therefore be large even when local diversity of individual habitat patches is low, given that the species turnover between habitat patches is high (Nemergut *et al.*, 2013). Spatial beta-diversity was found to be an important component of biofilm diversity at multiple spatial scales, comparing streams from multiple catchments (Wilhelm *et al.*, 2013), streams within a watershed (Besemer *et al.*, 2013; Fierer *et al.*, 2007), habitat patches within a stream (Fischer *et al.*, 2009; Wang *et al.*, 2012a) and even architectural features within a biofilm (Besemer *et al.*, 2009b). Strikingly, beta-diversity was higher among biofilms than among stream water samples in glacier-fed streams, contrasting the patterns observed for alpha-diversity (Wilhelm *et al.*, 2013). This highlights the potential of biofilms to harbour significant biodiversity at the regional scale, even if their alpha-diversity is lower than that of the water column in some environments.

The high alpha-diversity found in some habitats may actually reflect high beta-diversity between micro-habitats pooled in a single sample (Nemergut *et al.*, 2013). This might be especially true for biofilms developing on and in sediments, were strong environmental gradients and high structural complexity are assumed to enhance microbial diversity (Brunke and Gonser, 1997; Torsvik *et al.*, 2002). Conducting a comprehensive meta-analysis of environmental samples, Lozupone and Knight (2007) found that sediments harbour exceptionally high phylogenetic diversity, even higher than soils. Further, sediments were reported to feature high beta-diversity at the regional scale because of their heterogeneity in terms of redox gradients and nutrient availability (Nemergut *et al.*, 2013). Similarly, in a study comparing bacterial communities from different habitats, the highest spatial beta-diversity was found in subsurface soils and sediments, followed by mountain stream biofilms, surface sediments and lake water (Wang *et al.*, 2013). Collectively, these studies suggest aquatic sediments as one of the most important reservoirs for microbial diversity and underline the necessity to consider both alpha- and beta-diversity in the study of sediment-attached microbial communities.

Environmental heterogeneity is a key factor maintaining beta-diversity between habitat patches. Human-induced loss of geomorphological streambed heterogeneity was found to decrease beta-diversity and gamma-diversity in benthic algal biofilms, while alpha-diversity remained high in the altered streams (Passy and Blanchet, 2007) – similar to the patterns that were found for bacterial beta-diversity in stream biofilms (Besemer *et al.*, 2009a). On a larger spatial scale, headwaters, as the smallest but most abundant components of a fluvial network, harbour the highest degree of habitat heterogeneity (Brown *et al.*, 2011) and the highest level of microbial beta-diversity (Besemer *et al.*, 2013). Habitat loss (e.g. by stream burial, mountain-top mining) and homogenization (e.g. through stream regulation and streambed degradation) is, therefore, likely to have severe impacts for the microbial diversity in streams and rivers and for the ecosystem functions and services they provide.

Temporal beta-diversity

Another aspect of beta-diversity is the temporal turnover of communities, which can promote regional diversity through asynchronous environmental fluctuations in different habitat patches (Chesson, 2000). Microbial communities typically consist of a set of abundant taxa, which are assumed to perform most ecosystem functions, and a huge diversity of rare taxa, potentially constituting a seed bank (Lennon and Jones, 2011; Pedrós-Alió, 2006). This seed bank comprises populations that are potentially capable of being resuscitated following environmental change (Lennon and Jones, 2011). Dormancy is a common response to environmental stress among microorganisms, which can prolong their persistence and allows species to contend with temporal variability of environmental conditions (Jones and Lennon, 2010). The accumulation of dormant organisms in the seed bank promotes co-existence of competing species by constituting a temporal storage effect, which may be especially relevant in harsh and fluctuating environments (Chesson, 2000; Chesson and Huntly, 1997; Lennon and Jones, 2011).

Epilithic and epipsammic biofilms in glacier-fed streams, as an example of an environment experiencing strong seasonal and diurnal fluctuations of water chemistry and temperature, were found to harbour a large number of rare taxa, which were more abundant in the active community (based on rRNA analysis) than in the bulk community (based on analysis of the rRNA gene; Wilhelm *et al.*, 2014). This suggests that rare taxa contribute disproportionately to microbial community dynamics in glacier-fed streams. Transient dormancy and high community turnover, where taxa repeatedly enter and leave the seed bank, could therefore be candidate mechanisms that maintain high biofilm diversity in these systems. This agrees with findings from hyporheic sediments from glacier-fed streams (Freimann *et al.*, 2013a) where pronounced temporal turnover in community structure and function was observed, compared to groundwater-fed streams. Furthermore, community turnover in glacier-fed hyporheic communities was more closely related to changes in water chemistry than in the comparatively stable groundwater-fed streams (Freimann *et al.*, 2013b). Decreasing glacial runoff due to climate change could lead to reduced temporal beta-diversity and, thus, to a loss of diversity at the landscape scale (Freimann *et al.*, 2013b; Wilhelm *et al.*, 2013).

In a completely different environment, namely epiphytic biofilms on brown macro-algae in the Baltic Sea, Stratil and colleagues (2013) found indications that members of the rare biosphere could become abundant and potentially contribute to community functions when environmental conditions became appropriate, thus supporting the seed bank hypothesis. In contrast, temporal variation in biofilm communities attached to sand grains of coastal sediments could be explained by constant input of transient taxa from other locations by advection and physical mixing (Gobet *et al.*, 2012). Though significant temporal community turnover was observed in this system, most was due to changes in the rare biosphere, and rare taxa tended to stay rare, arguing against a dynamic seed bank. Obviously, various mechanisms may drive temporal – as spatial – beta-diversity patterns in various ecosystems.

Taxonomic composition of biofilms

Numerous studies have elucidated the taxonomic and phylogenetic structure of microbial biofilms across a broad range of habitats, of which only a few are mentioned here. The following paragraphs give a short overview of the most abundant microbial phyla within biofilms. The bacterial diversity of biofilms is typically dominated by Alpha-, Beta- and Gamma-proteobacteria, bacteroidetes and cyanobacteria (Romaní *et al.*, 2013). Beta-proteobacteria is often the most abundant bacterial phylum in freshwater habitats (Barberán and Casamayor, 2010; Glöckner *et al.*, 1999; Newton *et al.*, 2011), while Alpha- and Gamma-proteobacteria are commonly found to dominate marine systems (Glöckner *et al.*, 1999; Rusch *et al.*, 2007). Accordingly, Beta-proteobacteria has repeatedly been found to be the most important bacterial group in freshwater biofilms, including epilithic, epipsammic and hyporheic stream and river biofilms (Araya *et al.*, 2003; Battin *et al.*, 2001; Beier *et al.*, 2008; Besemer *et al.*, 2012; Bricheux *et al.*, 2013; Gao *et al.*, 2005; Marxsen *et al.*, 2010; Olapade and Leff, 2005) and aggregate-associated microbial communities in lakes (Schweitzer *et al.*, 2001; Tang *et al.*, 2009), while Alpha- and Gamma-proteobacteria are often the most abundant bacterial groups in marine biofilms (Gobet *et al.*, 2012; Lachnit *et al.*, 2011; McKew *et al.*, 2011). However, a surprisingly large number of studies found that Alpha-proteobacteria were also as abundant – or even more abundant – than Beta-proteobacteria in freshwater biofilms ranging from epilithic and hyporheic stream biofilms (Anderson-Glenna *et al.*, 2008; Hall *et al.*, 2012; Lear *et al.*, 2009a; Romaní *et al.*, 2014; Wang *et al.*, 2012b) to biofilms on living or decaying plants (Buesing *et al.*, 2009; McNamara and Leff, 2004), and to diatom-aggregates in lakes (Knoll *et al.*, 2001). While some studies (Knoll *et al.*, 2001; Schweitzer *et al.*, 2001) observed a switch from an Alpha-proteobacteria dominated to a Beta-proteobacteria dominated community during biofilm development, others (Manz *et al.*, 1999) found the opposite pattern – decreasing importance of Beta-proteobacteria, and increasing dominance of Alpha-proteobacteria. Further, a study on sediment bacteria along a river from freshwater to the sea showed that both Alpha- and Beta-proteobacteria, together with *Acidobacteria* and *Nitrospira*, were overrepresented in freshwater sediments, while Gamma-proteobacteria were overrepresented in marine sediments and Epsilon-proteobacteria in intertidal sediments (Wang *et al.*, 2012c). Deviating strategies to cope with grazing pressure or low substrate availability (Newton *et al.*, 2011) might favour Alpha- or Beta-proteobacteria depending on habitat type and environmental conditions (Romaní *et al.*, 2014).

Members of the bacteroidetes phylum (formerly the *Cytophaga/Flavobacterium/Bacteroides* group) often represent a substantial – or even the dominant – part of biofilm communities (e.g. Anderson-Glenna *et al.*, 2008; Bartrons *et al.*, 2012; Gobet *et al.*, 2012; Hall *et al.*, 2012; Parfenova *et al.*, 2013) and especially aggregate-associated communities (Knoll *et al.*, 2001; Manz *et al.*, 1999; Schweitzer *et al.*, 2001). Gliding motility and the ability to use complex macromolecules could qualify members of this group to be important players in the degradation of suspended particles (Crump *et al.*, 1999; Kirchman, 2002; Newton *et al.*, 2011). For instance, Crump and colleagues (1999) reported that *Cytophaga* (bacteroidetes) were the most abundant group on estuarine particles, where they might contribute significantly to the estuarine food web. Further, bacteroidetes have been found to be associated with cyanobacterial blooms (Newton *et al.*, 2011) and can constitute the most abundant group on cyanobacterial aggregates (Tang *et al.*, 2009). The relative abundance of bacteroidetes might increase with particle age, when labile organic compounds are depleted and the particle increasingly consists of refractory organic material (Knoll *et al.*, 2001; Schweitzer *et al.*, 2001).

Cyanobacteria are a common phototrophic compartment of microbial biofilms and are the dominant organisms in the complex laminated stromatolitic mats (Paerl *et al.*, 2000). Cyanobacteria were found to be the most abundant microbial group in lake biofilms (Parfenova *et al.*, 2013), epipsammic (Hullar *et al.*, 2006) and epilithic (Wang *et al.*, 2012b) stream biofilms, and are often more abundant in biofilms than in the water column (Parfenova *et al.*, 2013; Ylla *et al.*, 2009). *Acidobacteria* can dominate the microbial community inhabiting fine benthic organic matter, especially at low pH (Fierer *et al.*, 2007). Further taxonomic groups typically present in biofilms at lower abundances include *Actinobacteria*, firmicutes, *Gemmatimonadetes*, Delta- and Epsilon-proteobacteria, *Verrucomicrobia*, *Planctomycetes*, and *Deinococcus-Thermus* (e.g. Bartrons *et al.*, 2012; Besemer *et al.*, 2009b, 2012; Hullar *et al.*, 2006; Lachnit *et al.*, 2011; Romaní *et al.*, 2014; Wang *et al.*, 2012b).

Archaea have generally been reported to constitute minor, or even undetectable, proportions of benthic biofilms in streams and rivers (Bricheux *et al.*, 2013; Lawrence *et al.*, 2004; Manz *et al.*, 1999; Marxsen *et al.*, 2010; Romaní *et al.*, 2014; Wilhelm *et al.*, 2014); however, higher relative abundances of archaea (in the range of 1–10%) have been found in streams (Battin *et al.*, 2001, 2004; Olapade and Leff, 2005) and saltmarsh sediments (McKew *et al.*, 2011). Buriánková and colleagues (2012) showed that methanogenic archaea can be an important component (>10%) of the microbial community in hyporheic sediments, indicating the presence of anoxic micro-habitats within hyporheic biofilms. A meta-analysis of archaeal sequences revealed that freshwater sediment communities contain lower archaeal biodiversity than planktonic freshwater communities (Auguet *et al.*, 2010). Furthermore, the lack of indicator species among freshwater sediment archaea in the study by Auguet and colleagues might indicate late archaeal colonization of this environment.

Among the eukaryotic organisms, algae, most commonly *Bacillariophyta* and *Chlorophyta*, are a major component of biofilms exposed to light (Arnon *et al.*, 2007; Artigas *et al.*, 2012; Battin *et al.*, 2003b; Besemer *et al.*, 2007; Wilhelm *et al.*, 2013). Fungi, especially *Ascomycota*, are a prominent structuring element of biofilms developing on submerged organic matter and are important players in the degradation of leaf litter and wood (Das *et al.*, 2007; Gessner *et al.*, 2007; Golladay and Sinsabaugh, 1991). Protists, including flagellates, ciliates, and amoebae can control biofilm growth and alter biofilm structure and function (Böhme *et al.*, 2009; Bott and Kaplan, 1989). Lastly, biofilms can also be an important reservoir of viruses in aquatic ecosystems, which may affect biofilms by structuring community composition and diversity (Jackson and Jackson, 2008; Luef *et al.*, 2009).

Biofilm diversity in different habitats

Biofilm community composition and diversity can differ significantly between different habitat types (e.g. benthic and hyporheic zone, living or decaying plant materials, suspended particles), even within the same environment (Buesing *et al.*, 2009; Wilhelm *et al.*, 2014). The nature of the substratum (organic or inorganic, surface area, stability), hydrologic conditions (shear stress, current velocity and turbulence influencing mass transfer), temporal and spatial gradients (e.g. light, temperature, nutrients, oxygen) differentiate biofilm habitats and are all likely to influence biofilm structure and function (Battin *et al.*, 2003b; Brunke and Gonser, 1997; Romaní, 2010). Comparing biofilms from different streambed substrates, Wilhelm and colleagues (2014) found that epipsammic biofilms harboured a higher bacterial diversity than epilithic biofilms. The larger specific surface area of sandy sediments may foster the adhesion of microbial cells from the water flushing through it, thereby supporting microbial diversity. Further, biofilms attached to surface sediment particles are more prone to translocation than epilithic biofilms (Hullar *et al.*, 2006), which might result in a patchy distribution of biofilm biodiversity and high beta-diversity among microhabitats (Nemergut *et al.*, 2013). Interestingly, the diversity of bacterial morphotypes showed the opposite pattern with higher diversity in epilithic than epipsammic biofilms, which might be related to grazing pressure (Romaní and Sabater, 2001). Bacterial diversity was found to increase with sediment depth in coastal sands (Gobet *et al.*, 2012), potentially reflecting increasing physicochemical stability of the habitat. Decreasing shear stress with sediment depth could contribute to this pattern (Brunke and Gonser, 1997; Rickard *et al.*, 2004). Storage effects through dormancy mechanisms may also be prevalent in deep sediments and in the hyporheic zone (Crump *et al.*, 2012), which can constitute a refuge for microorganisms during unfavourable periods (Timoner *et al.*, 2012).

While benthic or epiphytic biofilms and suspended aggregates in the photic zone of freshwater or marine systems contain substantial phototrophic components (Battin *et al.*, 2001; Besemer *et al.*, 2007; Romaní, 2010), biofilms within sediments or in the aphotic zone rely on external carbon sources (Romaní *et al.*, 2004) or, especially in deep-sea biofilms, on chemoautotrophic primary production (López-García *et al.*, 2003). The close spatial proximity of primary producers and heterotrophic microorganisms in biofilms may promote internal cycling of organic carbon (Battin *et al.*, 2003b; Singer *et al.*, 2011), but may also have important implications for the transformation and decomposition of allochthonous organic matter through priming, for instance (Danger *et al.*, 2013). The presence of algae might enhance the diversity of heterotrophic microorganisms, as observed for bacteria in benthic biofilms (Romaní *et al.*, 2004) and fungi in mixed-species biofilms on leaf litter (Danger *et al.*, 2013), potentially by increasing the availability and diversity of high-quality organic carbon sources. In contrast, studies on experimental stream biofilms observed a slightly negative (Lear *et al.*, 2009b) or no consistent effect (Romaní *et al.*, 2014) of light availability on microbial diversity, though the bacterial community composition was significantly affected in both studies. Negative effects of primary producers on the diversity of heterotrophic microorganisms could be caused by allelopathic compounds, which are produced by many algae and cyanobacteria (Leflaive and Ten-Hage, 2007), or competition between algae and bacteria (Hulot *et al.*, 2001). Indeed, a large field study in pre-alpine streams indicated a negative relationship between the relative abundance of cyanobacteria and the overall diversity and evenness of epilithic biofilm communities (Besemer *et al.*, 2013). Furthermore, heterotrophic bacteria attached to organic particles exhibited lower richness in lakes that experienced cyanobacterial blooms compared to lakes without such blooms (Tang *et al.*, 2009).

Suspended aggregates formed by microbes, their extracellular polymeric substances and detrital particles are hotspots of microbial biomass and activity that structure an otherwise poorly structured water column (Battin *et al.*, 2008). Contradictory findings have been reported regarding the relative importance of aggregate-associated microbes to overall microbial diversity. Higher aggregate-associated microbial diversity has been found in a wide range of environments, including lakes (Bižić-Ionescu *et al.*, *in press*), river floodplains (Besemer *et al.*, 2005), coastal marine environments (Mohit *et al.*, 2014; Zhang *et al.*, 2007), the open sea (Crespo *et al.*, 2013), and the deep ocean (Eloe *et al.*, 2011). In contrast, several studies indicated that the microbial diversity on suspended aggregates and particles was lower than in the free-living microbial community in lakes (Riemann and Winding, 2001), temperate marine (Acinas *et al.*, 1999; Moeseneder *et al.*, 2001) and Arctic marine environments (Kellogg and Deming, 2009), suggesting that particles recruit a fraction of the bacterial community present in the surrounding waters. However, those among the above cited studies using high-throughput sequencing methods, which can explore microbial diversity at a higher resolution than classical molecular methods (Zinger *et al.*, 2012), consistently reported higher diversity in the particle-attached microbial community (Bižić-Ionescu *et al.*, *in press*; Crespo *et al.*, 2013; Mohit *et al.*, 2014). Suspended aggregates in aquatic environments are highly diverse by means of their origin, composition and age and likely constitute a heterogeneous micro-environment harbouring a high number of potential niches for microbial growth (Bižić-Ionescu *et al.*, *in press*). Indeed, the diversity of aggregate-associated microbial communities was found to increase with availability and diversity of suspended particles (Bižić-Ionescu *et al.*, *in press*; Crespo *et al.*, 2013; Mohit *et al.*, 2014). These studies suggest that aggregate-associated microbes, as 'mobile biofilms', contribute significantly to the overall diversity of a range of aquatic habitats. In fact, the enormous microbial diversity of the oceans has been proposed to result in part from a cob-web like structure of polymers, colloids, gel particles and large aggregates constituting micro-niches in a seemingly homogeneous environment (Azam and Malfatti, 2007).

Biofilms associated with littoral plants or macro-algae exhibit a wide array of interactions with their host, ranging from symbiotic to pathological, and can even be necessary to maintain normal macro-algal morphology (Goecke *et al.*, 2010; Miranda *et al.*, 2013; Morris and Monier, 2003). Epiphytic biofilm communities have been found to differ depending on the macrophyte's taxonomic affiliation, tissue type, age and state of disease. For a *Planctomycetes*-dominated biofilm on kelp surface (*Laminaria hyperborea*), Bengtsson and Øvreås (2010) showed that overall biofilm diversity increased, while the relative abundance of *Planctomycetes* decreased with kelp tissue age. Ageing of the kelp tissue could be associated with reduced chemical defence against microbial colonization, resulting in a loss of the competitive advantage of *Planctomycetes* towards other bacterial groups. Together with the potential appearance of new niches involved in degradation of different kelp constituents, this might enable higher biofilm diversity (Bengtsson and Øvreås, 2010; Bengtsson *et al.*, 2012). This agrees with findings for the red macroalga *Delisea pulchra*, where thalli with bleaching disease harboured a more diverse biofilm than healthy thalli, potentially due to decreased selective chemical defence of the macroalga (Fernandes *et al.*, 2012). Similarly, community composition and diversity of biofilms growing on senescing macrophytes differed according to plant species and to the plant's elemental composition (Mille-Lindblom *et al.*, 2006). In contrast, experimental hyporheic biofilms fed with leaf leachates that differed in chemical composition showed clear changes in bacterial community composition, but no consistent change in diversity (Hall *et al.*, 2012).

The simplest and the most complex biofilm communities occur in extreme environments, including hot, nutrient-deprived, hypersaline, calcareous, acidic and high-irradiance ecosystems (Paerl *et al.*, 2000). Biofilms forming streamers in acidic, metal-rich waters, for instance, were found to harbour a limited bacterial diversity, dominated by Beta-proteobacteria (Hallberg *et al.*, 2006). Similarly, bacterial biofilms on substrates exposed to metal-rich hydrothermal vent emissions at the Mid-Atlantic Ridge consisted almost exclusively of Epsilon-proteobacteria, though considerable diversity was present within this group (López-García *et al.*, 2003). However, high bacterial diversity was found in adjacent sediments, indicating that fast-developing Epsilon-proteobacteria are the first colonizers where hydrogen sulfide is supplied and that, as communities evolve, they diversify and allow the creation of new niches (López-García *et al.*, 2003). The most complex biofilms are laminated cyanobacterial-bacterial mats, some of which facilitate the trapping and binding and/or the precipitation of minerals into organo-sedimentary structures to form stromatolites (Paerl *et al.*, 2000). In Antarctic desert streams, for instance, perennial, desiccation- and freezing tolerant microbial mats are the most prominent form of life and hotspots of productivity in an otherwise inhospitable environment (Stanish *et al.*, 2013). The diversity of diatoms in these mats was found to exhibit an inverse relationship with both bacterial and cyanobacterial diversity, and the diversity of these three groups showed different relationships to hydrology and water chemistry (Stanish *et al.*, 2013). In a contrasting environment, stromatolitic mats from the Bahamas, increased salinity resulted in a decrease in the relative abundance of cyanobacteria and an increase in the relative abundance of eukaryotic chloroplasts, accompanied by a dramatic shift in the morphology of the mat (Ahrendt *et al.*, 2014). These findings indicate differential environmental controls on the diversity of algae, cyanobacteria and heterotrophic bacteria in these complex microbial consortia.

Patterns of biofilm diversity along spatial and environmental gradients

Spatial patterns of biofilm diversity

The large-scale distribution of biodiversity on earth varies along spatial gradients, such as latitude, elevation and depth (Gaston, 2000), likely based on mechanisms related to spatial (e.g. the mid-domain effect, size of ecoclimatic zones) and climatic factors (e.g. light, temperature, water supply). For macro-organisms, it is widely accepted that species diversity within a specified area increases with decreasing latitudes and peaks close to the equator (Gaston, 2000; Hillebrand, 2004), while it remains debated whether such a patterns exists for microorganisms. Bacterial richness was found to increase towards lower latitudes in stream biofilms across New Zealand (Lear *et al.*, 2013; Washington *et al.*, 2013), which is in agreement with findings for marine (Fuhrman *et al.*, 2008; Pommier *et al.*, 2007; Sul *et al.*, 2013), and lake (Schiaffino *et al.*, 2011) bacterioplankton, but contrasts findings from soils (Chu *et al.*, 2010; Fierer and Jackson, 2006). This finding that stream biofilm microbes show a similar latitudinal pattern as macro-organisms is notable because the strength of the latitudinal diversity gradient has been shown to be weaker for small organisms, and for freshwater compared to marine and terrestrial environments (Hillebrand, 2004). High dispersal rates of microorganisms (e.g. Fenchel and Finlay, 2004) may in fact be one of the reasons for the weaker latitudinal diversity gradient in freshwater ecosystems, whose body mass distribution is skewed towards small organisms (Hillebrand, 2004).

Similar to latitude, species diversity may change with altitude, frequently showing either a monotonic decrease with altitude or a hump-shaped pattern (Rahbek, 2005), or, as reported for microorganisms living in soils and on plant-surfaces, no patterns at all (Fierer *et al.*, 2011). For stream biofilms, microbial diversity was found to be reduced at higher elevations in streams across New Zealand (Lear *et al.*, 2013), which agrees with findings from streams sampled above, at and below the tree-line in the Austrian Alps (Wilhelm *et al.*, *in press*). A decrease of diversity with elevation was also observed for benthic biofilms in alpine lakes, particularly related to the location of the lake above or below the tree-line (Bartrons *et al.*, 2012). The higher complexity of the surrounding ecosystem and the increasing diversity of organic matter sources below the treeline are likely to contribute to this pattern (Bartrons *et al.*, 2012). Further, Wilhelm and colleagues (2013) found that biofilm biodiversity decreased with elevation among glacier-fed streams, though this pattern was weaker than for planktonic communities of the same streams. These authors suggested that, at lower altitudes, more sources of microorganisms upstream in the catchment (e.g. subglacial, englacial and supraglacial runoff, groundwater, adjacent soils) contribute to the microbial community suspended in the stream water, which eventually seeds the biofilm. Species sorting according to ambient environmental conditions might have attenuated this altitudinal pattern in stream biofilms (Besemer *et al.*, 2012; Wilhelm *et al.*, 2013).

Contrasting these findings, stream biofilm richness and evenness increased with altitude in a Chinese stream spanning an altitudinal gradient of more than 2000 metres (Wang *et al.*, 2011, 2012b). Furthermore, phylogenetic clustering of biofilm bacteria increased with altitude, indicating that the high taxonomic diversity at high altitude streams likely consists of many closely related species (Wang *et al.*, 2012b). Large diel temperature variations, increasing supply of dissolved organic matter and frequent disturbances might have contributed to the high bacterial diversity at high altitudes (Wang *et al.*, 2011, 2012b). Indeed, large temporal variation in physical habitat conditions was proposed to enhance diversity already in the river continuum concept (Vannote *et al.*, 1980), a mechanism, which was later conceptualized as the temporal storage effect (Chesson, 2000). Availability and diversity of dissolved organic matter have been proposed to be highest in the headwater region of stream networks due to their intimate linkage to the terrestrial environment, which translates into increased habitat diversity for bacteria (McArthur *et al.*, 1992). Further, disturbances might contribute to diversity because superior competitors do not have the time to become dominant (Lake, 2000; Wang *et al.*, 2012b). In contrast to bacteria, Wang and colleagues (2011) found that diatom richness decreased monotonically with elevation in the same biofilms. This finding is surprising, given that both groups presumably disperse readily through water flow (Crump *et al.*, 2007; Finlay, 2002), and indicates different mechanisms governing their diversity.

In light of these contradictory observations, some of the data from a study of biofilms in pre-alpine streams (Besemer *et al.*, 2013) were re-analysed with regard to elevation as a potential force driving biofilm diversity. Elevation (ranging from 500 to 1800 metres above sea level) showed no significant correlation with diversity when tested alone, however, when catchment size and elevation were included in multiple regression analysis, diversity (as richness, Shannon and Simpson diversity) decreased significantly ($P < 0.05$) with both increasing catchment size (that is, downstream) and with increasing elevation (that is, upstream). Obviously, different mechanisms are counteracting here to determine biofilm diversity. The resulting elevational pattern of biofilm biodiversity in any stream will likely depend on the complex interplay of the spatial patterns of environmental conditions, dispersal dynamics constrained by the fluvial landscape, and the size of the metacommunity seeding individual streams (Besemer *et al.*, 2013; Wilhelm *et al.*, 2013).

Beyond the local scale, biofilm beta-diversity within and among streams has been shown to exhibit elevational patterns. Community turnover of biofilm bacteria along an altitudinal gradient was in the same order of magnitude as for diatoms and macro-invertebrates (Wang *et al.*, 2012a) and resulted in a significant distance-decay of similarity between biofilm communities (Wang *et al.*, 2013). Furthermore, biofilms beta-diversity among glacier-fed streams was found to decrease with increasing stream water temperature (Wilhelm *et al.*, 2013). This study suggested that warming temperatures at high altitudes may contribute to a homogenization of biofilm communities among glacier-fed streams – similar to patterns observed for invertebrates (Jacobsen *et al.*, 2012). Collectively, these studies indicate that high altitude streams, in spite of apparent harsh environmental conditions, constitute unique habitats for stream microbes with important implications for regional diversity.

Biofilm diversity along environmental gradients

Microbial diversity and community composition vary along multiple environmental factors, some of which are discussed in the following section. One of the driving forces behind latitudinal and elevational patterns of biodiversity is temperature, which influences the diversity distribution of many organisms across a range of spatial scales (Gaston, 2000). Temperature was identified as the most important factor governing the diversity of microbial mats, sediments and soil in geothermal areas ranging from 7°C to 99°C (Sharp *et al.*, 2014). Diversity along this large temperature gradient followed a unimodal distribution with a peak at 24°C. Similarly, Stratil and colleagues (2013) found that the diversity of epiphytic biofilms on *Fucus vesiculosus* was highest at 15°C and decreased towards lower and higher temperatures. At the temporal scale, the influence of temperature on biodiversity receives increasing attention in the context of global climate change, often in combination with other factors related to human-induced environmental change. Díaz-Villanueva and colleagues (2011) found that a temperature difference of 3°C had a significant effect on biofilm community composition in young biofilms, while in mature biofilms nutrients were more important; effects on biofilm richness were not significant. This indicates that water temperature changes in the range that can be expected for streams and rivers due to climate change could have significant impacts on biofilm composition, while the taxonomic diversity might be less affected.

Another important environmental variable governing microbial diversity is pH, which has been shown to explain much of the alpha- and beta-diversity patterns of soil bacteria (Fierer and Jackson, 2006). Epilithic biofilm diversity decreased with increasing acidity in streams ranging from relatively pristine to highly impacted by acid mine drainage (Lear *et al.*, 2009a), which agrees with the findings from soil microbial communities (Fierer and Jackson, 2006). However, in another study by Lear and colleagues (2013), biofilm diversity appeared unrelated to the pH in stream water or upstream soils. In the context of global change, ocean acidification due to rising levels of atmospheric carbon dioxide could impair microbial communities, especially lithifying microbial mats. Intertidal benthic algal biofilms exhibited decreasing algal and cyanobacterial richness when subjected to increasing temperature and CO_2 levels (Russell *et al.*, 2007); however, exposing lithifying microbial mats to elevated CO_2 concentrations had no profound effect on their diversity, community structure or carbon precipitation, which might indicate that these mats are adapted to fluctuating CO_2-concentrations (Ahrendt *et al.*, 2014).

Salinity is a major determinant of microbial community composition – more important than the distinction between aquatic and terrestrial environments – reflecting the importance of osmotic adaption (Lozupone and Knight, 2007; Tamames *et al.*, 2010). Patterns of microbial diversity along salinity gradients, on the other hand, are less clear. Bacterial richness and evenness in freshwater, intertidal and marine sediments were shown to decrease with increasing salinity along a river-estuary continuum (Wang *et al.*, 2012c), which agrees with findings for bacterioplankton from lakes and marine systems (Barberán and Casamayor, 2010). Furthermore, a meta-analysis of published studies found higher microbial diversity in freshwater than in marine environments (Tamames *et al.*, 2010), potentially reflecting higher environmental heterogeneity among freshwater systems. In contrast, suspended aggregate-associated bacteria were reported to be more diverse in the coastal sea than in lakes (Bižić-Ionescu *et al.*, *in press*), and the microbial diversity of a freshwater-wetland receiving saltwater intrusions during storm events increased when exposed to higher salinity (Jackson and Vallaire, 2009). The higher diversity found on marine aggregates could be related to the diversity of the available particles (Bižić-Ionescu *et al.*, *in press*), while the observed pattern in the wetland sediments might be caused by lowered dominance of strong competitors following the saltwater intrusion (Jackson and Vallaire, 2009).

The hydrologic regime, including the frequency and intensity of hydrologic disturbances (Allan, 2004), influences the diversity, community structure and functioning of aquatic biota (Marmonier *et al.*, 2012; Romaní *et al.*, 2013). Desiccation and flooding events promote physiological stress conditions for microbes, and the way that biofilm microbes respond to these disturbances can modulate the ecology and biogeochemistry of the environment they inhabit (Romaní *et al.*, 2013; Timoner *et al.*, 2012). Biofilms in salt marsh sediments, as an example of an environment experiencing regular cycles of desiccation and rewetting, showed only a slight reduction in biodiversity following desiccation, but a more pronounced loss of diversity after rewetting, indication intolerance of many dried cells to sudden rewetting (McKew *et al.*, 2011). The steady increase in salinity during desiccation favoured haloversatile taxa, which became dominant after rewetting, presumably because of the founder effect coupled with their capacity to tolerate the sudden hypo-osmotic stress (McKew *et al.*, 2011). Community recovery after desiccation might depend on the presence of persistent humid refuges, as was shown for a Central European and a Mediterranean stream (Marxsen *et al.*, 2010; Romaní *et al.*, 2013). After rewetting of desiccated stream sediments, bacterial diversity was higher and microbial activity recovered faster in the Central European stream, suggesting better survival of the indigenous microbial stream community due to higher sediment humidity (Marxsen *et al.*, 2010). Timoner and colleagues (2012) showed that autotrophic and heterotrophic microbes responded differently to desiccation and rewetting, indicating that the current increase in flow intermittency extent is likely to increase the relative importance of heterotrophic processes in stream ecosystems.

The global changes in catchment land use from pristine to human-dominated landscapes can affect and deteriorate the biodiversity of aquatic systems through nutrient enrichment, contaminant pollution or hydrologic alteration (Allan, 2004). Numerous studies have assessed the effects of human-induced alterations of aquatic environments on the structure and diversity of microbial communities. In studies on epilithic stream biofilms in New Zealand, catchment land use had a significant impact on the community structure (Lear *et al.*, 2013; Washington *et al.*, 2013), and Lear and colleagues (2013) found indications for a relationship between taxon richness and catchment properties, including land use. However, no significant effect of human activities in the upstream landscape on biofilm diversity was found in either of these studies. This agrees with findings from tropical streams, where prokaryotic and eukaryotic biofilm diversity showed no consistent patterns with respect to land use (Burgos-Caraballo *et al.*, 2014). In this study, nitrate concentration was the best single predictor of bacterial diversity.

Increased inputs of nutrients accompanying intensified agricultural land use, riparian deforestation and urbanization (Allan, 2004) can have significant effects on biofilm community composition; however, the responses of biofilm diversity vary markedly between studies. In the study by Burgos-Caraballo and colleagues (2014), bacterial diversity increased with nitrate concentrations. In contrast, epilithic stream biofilm diversity was lower in a highly impacted, nutrient rich urban stream than in more pristine streams (Lear *et al.*, 2009c) and bacterial diversity in wetland sediments decreased when exposed to increased nitrogen levels (Jackson and Vallaire, 2009). Lyautey and colleagues (2003) found similar bacterial richness in epilithic river biofilms upstream and downstream of a large urban centre, which increased the nitrogen and phosphorus load of the river, while the highest diversity of marine sediment bacteria was found at intermediate levels of nitrogen pollution (Xiong *et al.*, 2014). The mechanisms behind these contradictory patterns might include reduced interspecific competition at higher resource availability, which could support diversity (Burgos-Caraballo *et al.*, 2014); on the other hand, some members of a community might respond strongly to an increase in a limiting resource, thereby outcompeting other species and leading to a decline in species richness (Mittelbach *et al.*, 2001). The resulting effect of changing nutrient conditions may depend on the number of limiting resources in the system, as proposed by the resource-competition theory (Interlandi and Kilham, 2001).

Runoff from urban areas and industrial release increases the concentrations of heavy metals in aquatic systems, which get accumulated in biofilms and are then transferred to higher trophic levels (Ancion *et al.*, 2010). Exposure to zinc, copper and lead had lasting effects on community structure and decreased diversity in biofilms grown from stream water (Ancion *et al.*, 2010). Similarly, diversity of river biofilms decreased when exposed to nickel, an effect, which was alleviated by the concomitant supply of nutrients (organic carbon, nitrogen and phosphorus) at high oxygen levels (Lawrence *et al.*, 2004). Lawrence and colleagues (2004) further observed a reduction of catabolic functions by nickel, reflecting a loss of a functional group or a loss of diversity of functional groups.

How many microbial taxa are there in biofilms?

Due to the enormous microbial diversity in most environments, it is often impossible to get an exhaustive census of the microbial diversity contained in an environment (Caporaso *et al.*, 2011; Logue *et al.*, 2012; Sogin *et al.*, 2006). To get an idea of the true richness of a community it is therefore necessary to estimate the number of unobserved – additionally to the observed – species. A number of statistical techniques, including parametric and nonparametric, frequentist and Bayesian methods have been developed and used to assess the richness of microbial communities (Bunge *et al.*, 2014). However, it remains difficult to compare the richness estimates from different studies because of the different laboratory and statistical methods used. As one example, some richness estimates for biofilms are compared here, which have been made using 454-pyrosequencing of the 16S rRNA and its gene. The selected studies have the advantage of having used the same parametric method, the Diversity Estimation software of Quince and colleagues (2008) based on a 97% sequence similarity cutoff to distinguish operational taxonomic units (OTUs). The studies differ, however, in the primers used for PCR [Hall and colleagues (2012) and Wilhelm and colleagues (2013, 2014) used the same primers, Besemer and colleagues (2012) and Bengtsson and colleagues (2012) used different primers each], so the numbers presented here can only give a first rough idea of the true richness contained in microbial biofilms. Wilhelm and colleagues (2013, 2014) estimated that the microbial richness of epilithic and epipsammic biofilms in glacial streams ranged from 200 to 2500 OTUs and from 500 to 2000, respectively. This is similar to the richness found in hyporheic biofilms in boreal streams (1000–1300 OTUs; Besemer *et al.*, 2012) and in experimental hyporheic biofilms (600–2300 OTUs; Hall *et al.*, 2012). The estimated richness of marine epiphytic biofilms on kelp was lower, ranging from <100 to 600 (Bengtsson *et al.*, 2012), and Bengtsson and colleagues suggested that these kelp biofilms can be regarded as low bacterial diversity habitats. However, Bengtsson and colleagues fitted the OTU's abundance distributions to the Inverse-Gaussian distribution, while the above-mentioned studies applied the Sichel distribution (Quince *et al.*, 2008), which makes the results less comparable. Two of these studies also assessed the richness of the active biofilm community, which was, on average, lower than that of the bulk community (700–900 OTUs in boreal hyporheic biofilms; Besemer *et al.*, 2012; < 100–2200 OTUs and 400–1400 in epilithic and epipsammic biofilms, respectively, in glacial streams; Wilhelm *et al.*, 2014). For comparison, the estimated true richness of the stream water microbial community was 6100–6500 and 400–6100 OTUs in boreal and glacial streams, respectively, and lower (2800–3900 OTUs) in the active than in the bulk community in boreal streams (Besemer *et al.*, 2012; Wilhelm *et al.*, 2013).

Functional diversity in biofilms

The relationship between taxonomic diversity, functional diversity and ecosystem processes is a central issue of ecology (Gamfeldt *et al.*, 2013; Loreau *et al.*, 2005) and of uttermost importance in the light of global biodiversity loss (Loreau *et al.*, 2001; Sala and Knowlton, 2006; Worm *et al.*, 2006). While it is widely accepted that microbial diversity exerts influence over ecosystem functioning, the shape of this relationship and the underlying causes remain debated (Mokany *et al.*, 2008; Nielsen *et al.*, 2011). Diversity can enhance functioning through niche differentiation or facilitation, such that the performance of the community increases above the level expected by the performance of the individual contributing species, often subsumed under the term complementarity effect; and through stochastic processes involved in community assembly, such that more diverse communities have a higher probability of containing highly productive species, usually referred to as sampling or selection effects (Cardinale *et al.*, 2006b; Loreau *et al.*, 2001). The close proximity of microbial cells and the architectural and compositional differentiation of biofilms may foster functional complementarity (Besemer *et al.*, 2009b; Singer *et al.*, 2010, 2011). Indeed, indications for complementarity have been found for epilithic biofilms using experimental stream mesocosms differing in spatial flow heterogeneity (Singer *et al.*, 2010). Increasing flow heterogeneity resulted in higher beta-diversity between hydrodynamic microhabitats along the mesocosms and in a broader range of dissolved organic carbon compounds removed from the stream water. These results suggested that biofilm communities differentiated into functionally non-redundant local communities in habitat patches characterized by different hydrology, whose diversified metabolic capabilities then induced complementarity at the regional scale (Singer *et al.*, 2010). Further, niche complementarity among species was found to enhance biomass and nitrogen uptake of algal biofilms grown in heterogeneous flow environments (Cardinale, 2011). This study provided evidence that communities with more species were able to take greater advantage of the niche opportunities provided by the environment, which allowed the system to capture a greater proportion of the available resources. When, on the other hand, these niche opportunities were experimentally removed by making the flow environment homogenous, increased biomass in more diverse communities resulted from a species-specific selection effect (Cardinale, 2011). This agrees with earlier findings showing that the relative importance of complementarity and selection effects can change with the degree of spatial and temporal environmental heterogeneity (Cardinale *et al.*, 2000).

Changes of the underlying mechanisms with environmental context suggest that there may be no single, generalizable relationship between species diversity and ecosystem functioning (Cardinale *et al.*, 2000). Strong selection effects could result in an idiosyncratic relationship between community diversity and function, where community composition and functional identity are more important than species diversity *per se;* further, functional redundancy within a community can lead to a diversity–function relationship which saturates at low levels of diversity (Nielsen *et al.*, 2011). Especially for microbes, assumed high levels of redundancy and metabolic plasticity have been suggested to preclude a clear relationship between microbial diversity and universal functions, such as productivity or respiration (Langenheder *et al.*, 2000; Nielsen *et al.*, 2011). For instance, Gobet and colleagues (2012) found that taxonomic diversity varied more than microbial functions in coastal sand biofilms, suggesting that a limited number of continuously abundant, resident taxa performed most of the functions and/or a certain level of functional redundancy among taxa. Still, microbial abundance and extra-cellular enzymatic activity were related to major changes in community composition in this study (Gobet *et al.*, 2012). In contrast, a significant positive relationship between bacterial evenness and production was found in epiphytic kelp biofilms (Bengtsson *et al.*, 2012), which in this system indicates increasing productivity with decreasing dominance of a few *Planctomycetes*-associated OTUs. These results suggest that a few slow-growing *Planctomycetes* dominate on living kelp surfaces while a more even bacterial community may be necessary for the degradation of fragmented kelp. While universal functions may often be buffered by a high number of similarly performing species (Nielsen *et al.*, 2011), specific functions could be prone to be lost as result of decreasing microbial diversity. Indeed, Peter and colleagues (2011) found that the likelihood of sustaining multi-functionality, measured as the activity of a range of extracellular enzymes, decreased with decreasing diversity in freshwater biofilms, indicating a limited level of functional redundancy. This pattern was more pronounced when biofilms were fed only recalcitrant organic matter than when a labile carbon source was added (Peter *et al.*, 2011).

Functional diversity of a community may often be a more suitable predictor of ecosystem processes than is taxonomic diversity (Lecerf and Richardson, 2010; Tilman *et al.*, 1997). For epiphytic biofilms, Burke and colleagues (2011) found that a core of functional genes associated with a surface-attached mode of life were represented in a variety of microbial taxa, indicating the presence of functional guilds formed by functionally redundant taxa. Their findings suggest that the specific species composition in a biofilm is driven by stochastic colonization of members of these guilds from the source community, which would result in a weak coupling between taxonomy and function. Therefore, although taxonomic and functional diversity appear to be correlated at some point at least (Fierer *et al.*, 2012), this relationship can be modified by the degree of functional redundancy in the community (Hooper *et al.*, 2005; Nielsen *et al.*, 2011).

Biofilm functional diversity, as taxonomic diversity, has been shown to respond to changing environmental conditions, ultimately affecting the biogeochemical cycling of organic matter and nutrients in aquatic systems (Romaní *et al.*, 2014; Ylla *et al.*, 2014). Warming of stream water by 3°C increased functional diversity in stream biofilms, mainly due to a wider use of carbohydrates and polymers (Ylla *et al.*, 2014). In contrast, experimental stream biofilms showed a decrease in functional diversity under elevated (2°C) water temperature towards a specialized use of a few carbohydrates when grown under light, while dark-grown biofilms showed a slight increase in functional diversity at higher temperature (Romaní *et al.*, 2014). Furthermore, young biofilms appeared to be less specialized in the use of organic carbon sources than mature biofilms, indicating that the capacity to use a wide range of organic compounds might be advantageous for pioneering species (Romaní *et al.*, 2014). The structural and functional diversity of biofilms, reflecting the spatial and temporal heterogeneity of their environment, could thus have broad consequences for globally relevant carbon fluxes (Singer *et al.*, 2010). This might be especially true in systems in which biofilms dominate microbial life, such as stream networks (Battin *et al.*, 2008). Global loss, deterioration and homogenization of aquatic habitats make it imperative to understand the links between taxonomic diversity, functional diversity and ecosystem processes of microbial biofilms in aquatic systems.

Acknowledgements

I thank Tom J. Battin, Brigitte Rosenwirth, Jürgen Besemer and Michael Forisch for their helpful comments on this manuscript. Katharina Besemer is supported by an Erwin Schrödinger fellowship (Austrian Science Fund: J3542–B22).

References

Acinas, S.G., Antón, J., and Rodríguez-Valera, F. (1999). Diversity of free-living and attached bacteria in offshore Western Mediterranean waters as depicted by analysis of genes encoding 16S rRNA. Appl. Environ. Microbiol. *65*, 514–522.

Ahrendt, S., Mobberley, J., Visscher, P., Koss, L., and Foster, J. (2014). Effects of elevated carbon dioxide and salinity on the microbial diversity in lithifying microbial mats. Minerals *4*, 145–169.

Allan, J.D. (2004). Landscapes and riverscapes: the influence of land use on stream ecosystems. Annu. Rev. Ecol. Evol. Syst. *35*, 257–284.

Ancion, P.-Y., Lear, G., and Lewis, G.D. (2010). Three common metal contaminants of urban runoff (Zn, Cu and Pb) accumulate in freshwater biofilm and modify embedded bacterial communities. Environ. Pollut. *158*, 2738–2745.

Anderson, M.J., Crist, T.O., Chase, J.M., Vellend, M., Inouye, B.D., Freestone, A.L., Sanders, N.J., Cornell, H.V., Comita, L.S., Davies, K.F., *et al.* (2011). Navigating the multiple meanings of β diversity: a roadmap for the practicing ecologist. Ecol. Lett. *14*, 19–28.

Anderson-Glenna, M.J., Bakkestuen, V., and Clipson, N.J.W. (2008). Spatial and temporal variability in epilithic biofilm bacterial communities along an upland river gradient. FEMS Microbiol. Ecol. *64*, 407–418.

Araya, R., Tani, K., Takagi, T., Yamaguchi, N., and Nasu, M. (2003). Bacterial activity and community composition in stream water and biofilm from an urban river determined by fluorescent *in situ* hybridization and DGGE analysis. FEMS Microbiol. Ecol. *43*, 111–119.

Arnon, S., Packman, A.I., Peterson, C.G., and Gray, K.A. (2007). Effects of overlying velocity on periphyton structure and denitrification. J. Geophys. Res. *112*, G01002.

Artigas, J., Fund, K., Kirchen, S., Morin, S., Obst, U., Romaní, A.M., Sabater, S., and Schwartz, T. (2012). Patterns of biofilm formation in two streams from different bioclimatic regions: analysis of microbial community structure and metabolism. Hydrobiologia *695*, 83–96.

Auguet, J.-C., Barberan, A., and Casamayor, E.O. (2010). Global ecological patterns in uncultured Archaea. ISME J. *4*, 182–190.

Azam, F., and Malfatti, F. (2007). Microbial structuring of marine ecosystems. Nat. Rev. Microbiol. *5*, 782–791.

Barberán, A., and Casamayor, E.O. (2010). Global phylogenetic community structure and β-diversity patterns in surface bacterioplankton metacommunities. Aquat. Microb. Ecol. *59*, 1–10.

Bartrons, M., Camarero, L., and Catalan, J. (2010). Isotopic composition of dissolved inorganic nitrogen in high mountain lakes: variation with altitude in the Pyrenees. Biogeosciences *7*, 1469–1479.

Bartrons, M., Catalan, J., and Casamayor, E.O. (2012). High bacterial diversity in epilithic biofilms of oligotrophic mountain lakes. Microb. Ecol. *64*, 860–869.

Battin, T.J. (2000). Hydrodynamics is a major determinant of streambed biofilm activity: from the sediment to the reach scale. Limnol. Oceanogr. *45*, 1308–1319.

Battin, T.J., Wille, A., Sattler, B., and Psenner, R. (2001). Phylogenetic and functional heterogeneity of sediment biofilms along environmental gradients in a glacial stream. Appl. Environ. Microbiol. *67*, 799–807.

Battin, T.J., Kaplan, L.A., Newbold, J.D., and Hansen, C.M.E. (2003a). Contributions of microbial biofilms to ecosystem processes in stream mesocosms. Nature *426*, 439–442.

Battin, T.J., Kaplan, L.A., Newbold, J.D., Cheng, X., and Hansen, C. (2003b). Effects of current velocity on the nascent architecture of stream microbial biofilms. Appl. Environ. Microbiol. *69*, 5443–5452.

Battin, T.J., Wille, A., Psenner, R., and Richter, A. (2004). Large-scale environmental controls on microbial biofilms in high-alpine streams. Biogeosciences *1*, 159–171.

Battin, T.J., Sloan, W.T., Kjelleberg, S., Daims, H., Head, I.M., Curtis, T.P., and Eberl, L. (2007). Microbial landscapes: new paths to biofilm research. Nat. Rev. Microbiol. *5*, 76–81.

Battin, T.J., Kaplan, L.A., Findlay, S., Hopkinson, C.S., Marti, E., Packman, A.I., Newbold, J.D., and Sabater, F. (2008). Biophysical controls on organic carbon fluxes in fluvial networks. Nat. Geosci. *1*, 95–100.

Beier, S., Witzel, K.-P., and Marxsen, J. (2008). Bacterial community composition in Central European running waters examined by temperature gradient gel electrophoresis and sequence analysis of 16S rRNA genes. Appl. Environ. Microbiol. *74*, 188–199.

Bengtsson, M.M., and Øvreås, L. (2010). *Planctomycetes* dominate biofilms on surfaces of the kelp *Laminaria hyperborea*. BMC Microbiol. *10*, 261.

Bengtsson, M.M., Sjøtun, K., Lanzén, A., and Ovreås, L. (2012). Bacterial diversity in relation to secondary production and succession on surfaces of the kelp *Laminaria hyperborea*. ISME J. *6*, 2188–2198.

Besemer, K., Moeseneder, M.M., Arrieta, J.M., Herndl, G.J., and Peduzzi, P. (2005). Complexity of bacterial communities in a river-floodplain system (Danube, Austria). Appl. Environ. Microbiol. *71*, 609–620.

Besemer, K., Singer, G., Limberger, R., Chlup, A.-K., Hochedlinger, G., Hödl, I., Baranyi, C., and Battin, T.J. (2007). Biophysical controls on community succession in stream biofilms. Appl. Environ. Microbiol. 73, 4966–4974.

Besemer, K., Singer, G., Hödl, I., and Battin, T.J. (2009a). Bacterial community composition of stream biofilms in spatially variable-flow environments. Appl. Environ. Microbiol. 75, 7189–7195.

Besemer, K., Hödl, I., Singer, G., and Battin, T.J. (2009b). Architectural differentiation reflects bacterial community structure in stream biofilms. ISME J. 3, 1318–1324.

Besemer, K., Peter, H., Logue, J.B., Langenheder, S., Lindström, E.S., Tranvik, L.J., and Battin, T.J. (2012). Unraveling assembly of stream biofilm communities. ISME J. 6, 1459–1468.

Besemer, K., Singer, G., Quince, C., Bertuzzo, E., Sloan, W., and Battin, T.J. (2013). Headwaters are critical reservoirs of microbial diversity for fluvial networks. Proc. R. Soc. Ser. B *280*, 20131760.

Bižić-Ionescu, M., Zeder, M., Ionescu, D., Orlić, S., Fuchs, B.M., Grossart, H.-P., and Amann, R. Comparison of bacterial communities on limnic versus coastal marine particles reveals profound differences in colonization. Environ. Microbiol. In press.

Böhme, A., Risse-Buhl, U., and Küsel, K. (2009). Protists with different feeding modes change biofilm morphology. FEMS Microbiol. Ecol. *69*, 158–169.

Bott, T.L., and Kaplan, L.A. (1989). Densities of benthic protozoa and nematodes in a Piedmont stream. J. North Am. Benthol. Soc. *8*, 187–196.

Boulêtreau, S., Lyautey, E., Dubois, S., Compin, A., Delattre, C., Touron-Bodilis, A., Mastrorillo, S., and Garabetian, F. (2014). Warming-induced changes in denitrifier community structure modulate the ability of phototrophic river biofilms to denitrify. Sci. Total Environ. *466–467*, 856–863.

Boulton, A.J., Findlay, S., Marmonier, P., Stanley, E.H., and Valett, H.M. (1998). The functional significance of the hyporheic zone in streams and rivers. Annu. Rev. Ecol. Syst. *29*, 59–81.

Bricheux, G., Morin, L., Le Moal, G., Coffe, G., Balestrino, D., Charbonnel, N., Bohatier, J., and Forestier, C. (2013). Pyrosequencing assessment of prokaryotic and eukaryotic diversity in biofilm communities from a French river. Microbiologyopen 2, 402–414.

Brown, B.L., Swan, C.M., Auerbach, D.A., Campbell Grant, E.H., Hitt, N.P., Maloney, K.O., and Patrick, C. (2011). Metacommunity theory as a multispecies, multiscale framework for studying the influence of river network structure on riverine communities and ecosystems. J. North Am. Benthol. Soc. *30*, 310–327.

Brunke, M., and Gonser, T. (1997). The ecological significance of exchange processes between rivers and groundwater. Freshw. Biol. *37*, 1–33.

Buesing, N., and Gessner, M.O. (2006). Benthic bacterial and fungal productivity and carbon turnover in a freshwater marsh. Appl. Environ. Microbiol. 72, 596–605.

Buesing, N., Filippini, M., Bürgmann, H., and Gessner, M.O. (2009). Microbial communities in contrasting freshwater marsh microhabitats. FEMS Microbiol. Ecol. *69*, 84–97.

Bunge, J., Willis, A., and Walsh, F. (2014). Estimating the number of species in microbial diversity studies. Annu. Rev. Stat. Its Appl. *1*, 427–445.

Burgos-Caraballo, S., Cantrell, S.A., and Ramírez, A. (2014). Diversity of benthic biofilms along a land use gradient in tropical headwater streams, Puerto Rico. Microb. Ecol. *68*, 47–59.

Buriánková, I., Brablcová, L., Mach, V., Hýblová, A., Badurová, P., Cupalová, J., Čáp, L., and Rulík, M. (2012). Methanogens and methanotrophs distribution in the hyporheic sediments of a small lowland stream. Fundam. Appl. Limnol. *181*, 87–102.

Burke, C., Steinberg, P., Rusch, D., Kjelleberg, S., and Thomas, T. (2011). Bacterial community assembly based on functional genes rather than species. Proc. Natl. Acad. Sci. U.S.A. *108*, 14288–14293.

Burns, A., and Ryder, D.S. (2001). Potential for biofilms as biological indicators in Australian riverine systems. Ecol. Manag. Restor. *2*, 53–64.

Cadotte, M.W. (2007). Competition-colonization trade-offs and disturbance effects at multiple scales. Ecology *88*, 823–829.

Caporaso, J.G., Lauber, C.L., Walters, W.A., Berg-Lyons, D., Lozupone, C.A., Turnbaugh, P.J., Fierer, N., and Knight, R. (2011). Global patterns of 16S rRNA diversity at a depth of millions of sequences per sample. Proc. Natl. Acad. Sci. U.S.A. *108*, 4516–4522.

Cardinale, B.J. (2011). Biodiversity improves water quality through niche partitioning. Nature 472, 86–89.

Cardinale, B.J., Nelson, K., and Palmer, M.A. (2000). Linking species diversity to the functioning of ecosystems: on the importance of environmental context. OIKOS *91*, 175–183.

Cardinale, B.J., Palmer, M.A., Swan, C.M., Brooks, S., and Poff, N.L. (2002). The influence of substrate heterogeneity on biofilm metabolism in a stream ecosystem. Ecology 83, 412–422.

Cardinale, B.J., Hillebrand, H., and Charles, D.F. (2006a). Geographic patterns of diversity in streams are predicted by a multivariate model of disturbance and productivity. J. Ecol. *94*, 609–618.

Cardinale, B.J., Srivastava, D.S., Duffy, J.E., Wright, J.P., Downing, A.L., Sankaran, M., and Jouseau, C. (2006b). Effects of biodiversity on the functioning of trophic groups and ecosystems. Nature *443*, 989–992.

Chao, A., and Jost, L. (2012). Coverage-based rarefaction and extrapolation: standardizing samples by completeness rather than size. Ecology 93, 2533–2547.

Chao, A., Chiu, C.-H., and Jost, L. (2010). Phylogenetic diversity measures based on Hill numbers. Phil. Trans. R. Soc. Lond. B. Biol. Sci. *365*, 3599–3609.

Chesson, P. (2000). Mechanisms of maintenance of species diversity. Annu. Rev. Ecol. Syst. *31*, 343–358.

Chesson, P., and Huntly, N. (1997). The roles of harsh and fluctuating conditions in the dynamics of ecological communities. Am. Nat. *150*, 519–553.

Chu, H., Fierer, N., Lauber, C.L., Caporaso, J.G., Knight, R., and Grogan, P. (2010). Soil bacterial diversity in the Arctic is not fundamentally different from that found in other biomes. Environ. Microbiol. *12*, 2998–3006.

Clarke, A., Mac Nally, R., Bond, N., and Lake, P.S. (2008). Macroinvertebrate diversity in headwater streams: a review. Freshw. Biol. *53*, 1707–1721.

Costerton, J.W., Lewandowski, Z., Caldwell, D.E., Korber, D.R., and Lappin-Scott, H.M. (1995). Microbial biofilms. Annu. Rev. Microbiol. *49*, 711–745.

Crespo, B.G., Pommier, T., Fernández-Gómez, B., and Pedrós-Alió, C. (2013). Taxonomic composition of the particle-attached and free-living bacterial assemblages in the Northwest Mediterranean Sea analyzed by pyrosequencing of the 16S rRNA. Microbiologyopen *2*, 541–552.

Crump, B.C., Armbrust, E.V., and Baross, J.A. (1999). Phylogenetic analysis of particle-attached and free-living bacterial communities in the Columbia river, its estuary, and the adjacent coastal ocean. Appl. Environ. Microbiol. *65*, 3192–3204.

Crump, B.C., Adams, H.E., Hobbie, J.E., and Kling, G.W. (2007). Biogeography of bacterioplankton in lakes and streams of an Arctic tundra catchment. Ecology *88*, 1365–1378.

Crump, B.C., Amaral-Zettler, L.A., and Kling, G.W. (2012). Microbial diversity in arctic freshwaters is structured by inoculation of microbes from soils. ISME J. *6*, 1629–1639.

Curtis, T.P., and Sloan, W.T. (2004). Prokaryotic diversity and its limits: microbial community structure in nature and implications for microbial ecology. Curr. Opin. Microbiol. *7*, 221–226.

Danger, M., Cornut, J., Chauvet, E., Chavez, P., Elger, A., and Lecerf, A. (2013). Benthic algae stimulate leaf litter decomposition in detritus-based headwater streams: a case of aquatic priming effect? Ecology *94*, 1604–1613.

Das, M., Royer, T.V., and Leff, L.G. (2007). Diversity of fungi, bacteria, and actinomycetes on leaves decomposing in a stream. Appl. Environ. Microbiol. *73*, 756–767.

Davey, M.E., and O'Toole, G.A. (2000). Microbial biofilms: from ecology to molecular genetics. Microbiol. Mol. Biol. Rev. *64*, 847–867.

Decho, A.W. (2000). Microbial biofilms in intertidal systems: an overview. Cont. Shelf Res. *20*, 1257–1273.

Devictor, V., Mouillot, D., Meynard, C., Jiguet, F., Thuiller, W., and Mouquet, N. (2010). Spatial mismatch and congruence between taxonomic, phylogenetic and functional diversity: the need for integrative conservation strategies in a changing world. Ecol. Lett. *13*, 1030–1040.

Díaz Villanueva, V., Font, J., Schwartz, T., and Romani, A.M. (2011). Biofilm formation at warming temperature: acceleration of microbial colonization and microbial interactive effects. Biofouling *27*, 59–71.

Dirzo, R., and Raven, P.H. (2003). Global state of biodiversity and loss. Annu. Rev. Environ. Resour. *28*, 137–167.

Eggers, S.L., Eriksson, B.K., and Matthiessen, B. (2012). A heat wave and dispersal cause dominance shift and decrease biomass in experimental metacommunities. Oikos *121*, 721–733.

Eloe, E.A., Shulse, C.N., Fadrosh, D.W., Williamson, S.J., Allen, E.E., and Bartlett, D.H. (2011). Compositional differences in particle-associated and free-living microbial assemblages from an extreme deep-ocean environment. Environ. Microbiol. Rep. *3*, 449–458.

Fenchel, T., and Finlay, B.J. (2004). The ubiquity of small species: patterns of local and global diversity. BioScience *54*, 777–784.

Fernandes, N., Steinberg, P., Rusch, D., Kjelleberg, S., and Thomas, T. (2012). Community structure and functional gene profile of bacteria on healthy and diseased thalli of the red seaweed *Delisea pulchra*. PLoS One *7*, e50854.

Fierer, N., and Jackson, R.B. (2006). The diversity and biogeography of soil bacterial communities. Proc. Natl. Acad. Sci. U.S.A. *103*, 626–631.

Fierer, N., Morse, J.L., Berthrong, S.T., Bernhardt, E.S., and Jackson, R.B. (2007). Environmental controls on the landscape-scale biogeography of stream bacterial communities. Ecology *88*, 2162–2173.

Fierer, N., Nemergut, D., Knight, R., and Craine, J.M. (2010). Changes through time: integrating microorganisms into the study of succession. Res. Microbiol. *161*, 635–642.

Fierer, N., McCain, C.M., Meir, P., Zimmermann, M., Rapp, J.M., Silman, M.R., and Knight, R. (2011). Microbes do not follow the elevational diversity patterns of plants and animals. Ecology *92*, 797–804.

Fierer, N., Leff, J.W., Adams, B.J., Nielsen, U.N., Bates, T.S., Lauber, C.L., Owens, S., Gilbert, J.A., Wall, D.H., and Caporaso, J.G. (2012). Cross-biome metagenomic analyses of soil microbial communities and their functional attributes. Proc. Natl. Acad. Sci. U.S.A. *109*, 21390–21395.

Finlay, B.J. (2002). Global dispersal of free-living microbial eukaryote species. Science *296*, 1061–1063.

Fischer, H., Bergfur, J., Goedkoop, W., and Tranvik, L. (2009). Microbial leaf degraders in boreal streams: bringing together stochastic and deterministic regulators of community composition. Freshw. Biol. *54*, 2276–2289.

Flemming, H.-C., and Wingender, J. (2010). The biofilm matrix. Nat. Rev. Microbiol. *8*, 623–633.

Freimann, R., Bürgmann, H., Findlay, S.E.G., and Robinson, C.T. (2013a). Response of lotic microbial communities to altered water source and nutritional state in a glaciated alpine floodplain. Limnol. Oceanogr. *58*, 951–965.

Freimann, R., Bürgmann, H., Findlay, S.E.G., and Robinson, C.T. (2013b). Bacterial structures and ecosystem functions in glaciated floodplains: contemporary states and potential future shifts. ISME J. *7*, 2361–2373.

Fuhrman, J.A., Steele, J.A., Hewson, I., Schwalbach, M.S., Brown, M.V., Green, J.L., and Brown, J.H. (2008). A latitudinal diversity gradient in planktonic marine bacteria. Proc. Natl. Acad. Sci. U.S.A. *7*, 7774–7778.

Gamfeldt, L., Snäll, T., Bagchi, R., Jonsson, M., Gustafsson, L., Kjellander, P., Ruiz-Jaen, M.C., Fröberg, M., Stendahl, J., Philipson, C.D., *et al.* (2013). Higher levels of multiple ecosystem services are found in forests with more tree species. Nat. Commun. *4*, 1340.

Gao, X., Olapade, O.A., and Leff, L.G. (2005). Comparison of benthic bacterial community composition in nine streams. Aquat. Microb. Ecol. *40*, 51–60.

Gaston, K.J. (2000). Global patterns in biodiversity. Nature *405*, 220–227.

Gessner, M.O., Gulis, V., Kuehn, K.A., Chauvet, E., and Suberkropp, K. (2007). Fungal decomposers of plant litter in aquatic ecosystems. In Environmental and Microbial Relationships, C.P. Kubice and I.S. Druzhinina, eds. (Springer-Verlag, Berlin/Heidelberg:), pp. 301–324.

Glöckner, F.O., Fuchs, B.M., and Amann, R. (1999). Bacterioplankton compositions of lakes and oceans: a first comparison based on fluorescence *in situ* hybridization. Appl. Environ. Microbiol. *65*, 3721–3726.

Gobet, A., Böer, S.I., Huse, S.M., van Beusekom, J.E.E., Quince, C., Sogin, M.L., Boetius, A., and Ramette, A. (2012). Diversity and dynamics of rare and of resident bacterial populations in coastal sands. ISME J. *6*, 542–553.

Goecke, F., Labes, A., Wiese, J., and Imhoff, J.F. (2010). Chemical interactions between marine macroalgae and bacteria. Mar. Ecol. Prog. Ser. *409*, 267–299.

Golladay, S.W., and Sinsabaugh, R.L. (1991). Biofilm development on leaf and wood surfaces in a boreal river. Freshw. Biol. *25*, 437–450.

Haegeman, B., Hamelin, J., Moriarty, J., Neal, P., Dushoff, J., and Weitz, J.S. (2013). Robust estimation of microbial diversity in theory and in practice. ISME J. *7*, 1092–1101.

Hall, E.K., Besemer, K., Kohl, L., Preiler, C., Riedel, K., Schneider, T., Wanek, W., and Battin, T.J. (2012). Effects of resource chemistry on the composition and function of stream hyporheic biofilms. Front. Microbiol. *3*, 35.

Hallberg, K.B., Coupland, K., Kimura, S., and Johnson, D.B. (2006). Macroscopic streamer growths in acidic, metal-rich mine waters in north Wales consist of novel and remarkably simple bacterial communities. Appl. Environ. Microbiol. *72*, 2022–2030.

Hall-Stoodley, L., Costerton, J.W., and Stoodley, P. (2004). Bacterial biofilms: from the natural environment to infectious diseases. Nat. Rev. Microbiol. *2*, 95–108.

Hill, M.O. (1973). Diversity and evenness: a unifying notation and its consequences. Ecology *54*, 427–432.

Hillebrand, H. (2004). On the generality of the latitudinal diversity gradient. Am. Nat. *163*, 192–211.

Hödl, I., Hödl, J., Wörman, A., Singer, G., Besemer, K., and Battin, T.J. (2011). Voronoi tessellation captures very early clustering of single primary cells as induced by interactions in nascent biofilms. PLoS One *6*, e26368.

Holyoak, M., Leibold, M.A., Mouquet, N., Holt, R.D., and Hoopes, M.F. (2005). Metacommunities: a framework for large-scale community ecology. In Metacommunities: Spatial Dynamics and Ecological Communities, Holyoak, M., Leibold, M.A., and Holt, R.D., eds. (University of Chicago Press, Chicago, IL, USA), pp. 1–31.

Hooper, D.U., Chapin III, F.S., Ewel, J.J., Hector, A., Inchausti, P., Lavorel, S., Lawton, J.H., Lodge, D.M., Loreau, M., Naeem, S., *et al.* (2005). Effects of biodiversity on ecosystem functioning: A consensus of current knowledge. Ecol. Monogr. *75*, 3–35.

Horner-Devine, M.C., Lage, M., Hughes, J.B., and Bohannan, B.J.M. (2004). A taxa–area relationship for bacteria. Nature *432*, 750–753.

Hullar, M.A.J., Kaplan, L.A., and Stahl, D.A. (2006). Recurring seasonal dynamics of microbial communities in stream habitats. Appl. Environ. Microbiol. *72*, 713–722.

Hulot, F.D., Morin, P.J., and Loreau, M. (2001). Interactions between algae and the microbial loop in experimental microcosms. Oikos *95*, 231–238.

Interlandi, S.J., and Kilham, S.S. (2001). Limiting resources and the regultation of diversity in phytoplankton communities. Ecology *82*, 1270–1282.

Jackson, C.R. (2003). Changes in community properties during microbial succession. Oikos *101*, 444–448.

Jackson, C.R., and Vallaire, S.C. (2009). Effects of salinity and nutrients on microbial assemblages in Louisiana wetland sediments. Wetlands *29*, 277–287.

Jackson, E.F., and Jackson, C.R. (2008). Viruses in wetland ecosystems. Freshw. Biol. *53*, 1214–1227.

Jackson, C.R., Churchill, P.F., and Roden, E.E. (2001). Successional changes in bacterial assemblage structure during epilithic biofilm development. Ecology *82*, 555–566.

Jacobsen, D., Milner, A.M., Brown, L.E., and Dangles, O. (2012). Biodiversity under threat in glacier-fed river systems. Nat. Clim. Chang. *2*, 361–364.

Jasechko, S., Sharp, Z.D., Gibson, J.J., Birks, S.J., Yi, Y., and Fawcett, P.J. (2013). Terrestrial water fluxes dominated by transpiration. Nature *496*, 347–350.

Jones, S.E., and Lennon, J.T. (2010). Dormancy contributes to the maintenance of microbial diversity. Proc. Natl. Acad. Sci. U.S.A. *107*, 5881–5886.

Jost, L. (2006). Entropy and diversity. Oikos *113*, 363–375.

Kellogg, C.T.E., and Deming, J.W. (2009). Comparison of free-living, suspended particle, and aggregate-associated bacterial and archaeal communities in the Laptev Sea. Aquat. Microb. Ecol. *57*, 1–18.

Kirchman, D.L. (2002). The ecology of *Cytophaga-Flavobacteria* in aquatic environments. FEMS Microbiol. Ecol. *39*, 91–100.

Knoll, S., Zwisler, W., and Simon, M. (2001). Bacterial colonization of early stages of limnetic diatom microaggregates. Aquat. Microb. Ecol. *25*, 141–150.

Lachnit, T., Meske, D., Wahl, M., Harder, T., and Schmitz, R. (2011). Epibacterial community patterns on marine macroalgae are host-specific but temporally variable. Environ. Microbiol. *13*, 655–665.

Lake, P.S. (2000). Disturbance, patchiness, and diversity in streams. J. North Am. Benthol. Soc. *19*, 573–592.

Langenheder, S., Lindström, E.S., and Tranvik, L.J. (2000). Weak coupling between community composition and functioning of aquatic bacteria. Limnol. Oceanogr. *50*, 957–967.

Lawrence, J.R., Chenier, M.R., Roy, R., Beaumier, D., Fortin, N., Swerhone, G.D.W., Neu, T.R., and Greer, C.W. (2004). Microscale and molecular assessment of impacts of nickel, nutrients, and oxygen level on structure and function of river biofilm communities. Appl. Environ. Microbiol. *70*, 4326–4339.

Lear, G., Anderson, M.J., Smith, J.P., Boxen, K., and Lewis, G.D. (2008). Spatial and temporal heterogeneity of the bacterial communities in stream epilithic biofilms. FEMS Microbiol. Ecol. *65*, 463–473.

Lear, G., Niyogi, D., Harding, J., Dong, Y., and Lewis, G. (2009a). Biofilm bacterial community structure in streams affected by acid mine drainage. Appl. Environ. Microbiol. *75*, 3455–3460.

Lear, G., Turner, S.J., and Lewis, G.D. (2009b). Effect of light regimes on the utilisation of an exogenous carbon source by freshwater biofilm bacterial communities. Aquat. Ecol. *43*, 207–220.

Lear, G., Boothroyd, I.K.G., Turner, S.J., Roberts, K., and Lewis, G.D. (2009c). A comparison of bacteria and benthic invertebrates as indicators of ecological health in streams. Freshw. Biol. *54*, 1532–1543.

Lear, G., Washington, V., Neale, M., Case, B., Buckley, H., and Lewis, G. (2013). The biogeography of stream bacteria. Glob. Ecol. Biogeogr. *22*, 544–554.

Lecerf, A., and Richardson, J.S. (2010). Biodiversity-ecosystem function research: insights gained from streams. River Res. Appl. *26*, 45–54.

Leflaive, J., and Ten-Hage, L. (2007). Algal and cyanobacterial secondary metabolites in freshwaters: a comparison of allelopathic compounds and toxins. Freshw. Biol. *52*, 199–214.

Leibold, M.A., Holyoak, M., Mouquet, N., Amarasekare, P., Chase, J.M., Hoopes, M.F., Holt, R.D., Shurin, J.B., Law, R., Tilman, D., *et al.* (2004). The metacommunity concept: a framework for multi-scale community ecology. Ecol. Lett. *7*, 601–613.

Leinster, T., and Cobbold, C.A. (2012). Measuring diversity: the importance of species similarity. Ecology *93*, 477–489.

Lennon, J.T., and Jones, S.E. (2011). Microbial seed banks: the ecological and evolutionary implications of dormancy. Nat. Rev. Microbiol. *9*, 119–130.

Lock, M.A., Wallace, R.R., Costerton, J.W., Ventullo, R.M., and Charlton, S.E. (1984). River epilithon: a structural-functional toward model. Oikos *42*, 10–22.

Logue, J.B., and Lindström, E.S. (2008). Biogeography of bacterioplankton in inland waters. Freshw. Rev. *1*, 99–114.

Logue, J.B., Langenheder, S., Andersson, A.F., Bertilsson, S., Drakare, S., Lanzén, A., and Lindström, E.S. (2012). Freshwater bacterioplankton richness in oligotrophic lakes depends on nutrient availability rather than on species–area relationships. ISME J. *6*, 1127–1136.

López-García, P., Duperron, S., Philippot, P., Foriel, J., Susini, J., and Moreira, D. (2003). Bacterial diversity in hydrothermal sediment and epsilonproteobacterial dominance in experimental microcolonizers at the Mid-Atlantic Ridge. Environ. Microbiol. *5*, 961–976.

Loreau, M., Naeem, S., Inchausti, P., Bengtsson, J., Grime, J.P., Hector, A., Hooper, D.U., Huston, M.A., Raffaelli, D., Schmid, B., *et al.* (2001). Biodiversity and ecosystem functioning: current knowledge and future challenges. Science (80-.). *294*, 804–808.

Loreau, M., Mouquet, N.M., and Holt, R.D. (2005). From metacommunities to metaecosystems. In Metacommunities: Spatial Dynamics and Ecological Communities, Holyoak, M., Leibold, M.A., and Holt, R.D., eds. (University of Chicago Press, Chicago, IL, USA), pp. 418–438.

Lozupone, C.A., and Knight, R. (2007). Global patterns in bacterial diversity. Proc. Natl. Acad. Sci. U.S.A. *104*, 11436–11440.

Luef, B., Aspetsberger, F., Hein, T., Huber, F., and Peduzzi, P. (2007). Impact of hydrology on free-living and particle-associated microorganisms in a river floodplain system (Danube, Austria). Freshw. Biol. *52*, 1043–1057.

Luef, B., Neu, T.R., and Peduzzi, P. (2009). Imaging and quantifying virus fluorescence signals on aquatic aggregates: a new method and its implication for aquatic microbial ecology. FEMS Microbiol. Ecol. *68*, 372–380.

Lyautey, E., Teissier, S., Charcosset, J., Rols, J., and Garabétian, F. (2003). Bacterial diversity of epilithic biofilm assemblages of an anthropised river section, assessed by DGGE analysis of a 16S rDNA fragment. Aquat. Microb. Ecol. 33, 217–224.

Lyautey, E., Jackson, C.R., Cayrou, J., Rols, J.-L., and Garabétian, F. (2005). Bacterial community succession in natural river biofilm assemblages. Microb. Ecol. *50*, 589–601.

Magurran, A.E. (2004). Measuring biological diversity (Blackwell Publishing Ltd, Oxford, UK).

Manz, W., Wendt-Potthoff, K., Neu, T.R., Szewzyk, U., and Lawrence, J.R. (1999). Phylogenetic composition, spatial structure, and dynamics of lotic bacterial biofilms investigated by fluorescent *in situ* hybridization and confocal laser scanning microscopy. Microb. Ecol. 37, 225–237.

Marmonier, P., Archambaud, G., Belaidi, N., Bougon, N., Breil, P., Chauvet, E., Claret, C., Cornut, J., Datry, T., Dole-Olivier, M.-J., *et al.* (2012). The role of organisms in hyporheic processes: gaps in current knowledge, needs for future research and applications. Int. J. Limnol. *48*, 253–266.

Martiny, J.B.H., Eisen, J.A., Penn, K., Allison, S.D., and Horner-Devine, M.C. (2011). Drivers of bacterial β-diversity depend on spatial scale. Proc. Natl. Acad. Sci. U.S.A. *108*, 7850–7854.

Marxsen, J., Zoppini, A., and Wilczek, S. (2010). Microbial communities in streambed sediments recovering from desiccation. FEMS Microbiol. Ecol. *71*, 374–386.

Matthiessen, B., Mielke, E., and Sommer, U. (2010). Dispersal decreases diversity in heterogeneous metacommunities by enhancing regional competition. Ecology *91*, 2022–2033.

McArthur, J.V., Leff, L.G., and Smith, M.H. (1992). Genetic diversity of bacteria along a stream continuum. J. North Am. Benthol. Soc. *11*, 269–277.

McKew, B.A., Taylor, J.D., McGenity, T.J., and Underwood, G.J.C. (2011). Resistance and resilience of benthic biofilm communities from a temperate saltmarsh to desiccation and rewetting. ISME J. *5*, 30–41.

McNamara, C.J., and Leff, L.G. (2004). Bacterial community composition in biofilms on leaves in a northeastern Ohio stream. J. North Am. Benthol. Soc. *23*, 677–685.

Milferstedt, K., Santa-Catalina, G., Godon, J.-J., Escudié, R., and Bernet, N. (2013). Disturbance frequency determines morphology and community development in multi-species biofilm at the landscape scale. PLoS One *8*, e80692.

Mille-Lindblom, C., Fischer, H., and Tranvik, L.J. (2006). Litter-associated bacteria and fungi – a comparison of biomass and communities across lakes and plant species. Freshw. Biol. *51*, 730–741.

Miranda, L.N., Hutchison, K., Grossman, A.R., and Brawley, S.H. (2013). Diversity and abundance of the bacterial community of the red Macroalga *Porphyra umbilicalis*: Did bacterial farmers produce macroalgae? PLoS One *8*, e58269.

Mittelbach, G.G., Steiner, C.F., Scheiner, S.M., Gross, K.L., Reynolds, H.L., Waide, R.B., Willig, M.R., Dodson, S.I., and Gough, L. (2001). What is the observed relationship between species richness and productivity? Ecology *82*, 2381–2396.

Moeseneder, M.M., Winter, C., and Herndl, G.J. (2001). Horizontal and vertical complexity of attached and free-living bacteria of the eastern Mediterranean Sea, determined by 16S rDNA and 16S rRNA fingerprints. Limnol. Oceanogr. *46*, 95–107.

Mohit, V., Archambault, P., Toupoint, N., and Lovejoy, C. (2014). Phylogenetic differences in attached and free-living bacterial communities in a temperate coastal lagoon during summer, revealed via high-throughput 16S rRNA gene sequencing. Appl. Environ. Microbiol. *80*, 2071–2083.

Mokany, K., Ash, J., and Roxburgh, S. (2008). Functional identity is more important than diversity in influencing ecosystem processes in a temperate native grassland. J. Ecol. *96*, 884–893.

Morris, C.E., and Monier, J.-M. (2003). The ecological significance of biofilm formation by plant-associated bacteria. Annu. Rev. Phytopathol. *41*, 429–453.

Mouquet, N., and Loreau, M. (2003). Community patterns in source-sink metacommunities. Am. Nat. *162*, 544–557.

Nemergut, D.R., Schmidt, S.K., Fukami, T., O'Neill, S.P., Bilinski, T.M., Stanish, L.F., Knelman, J.E., Darcy, J.L., Lynch, R.C., Wickey, P., *et al.* (2013). Patterns and processes of microbial community assembly. Microbiol. Mol. Biol. Rev. 77, 342–356.

Newton, R.J., Jones, S.E., Eiler, A., McMahon, K.D., and Bertilsson, S. (2011). A guide to the natural history of freshwater lake bacteria. Microbiol. Mol. Biol. Rev. 75, 14–49.

Nielsen, U.N., Ayres, E., Wall, D.H., and Bardgett, R.D. (2011). Soil biodiversity and carbon cycling: a review and synthesis of studies examining diversity–function relationships. Eur. J. Soil Sci. *62*, 105–116.

Noss, R.F. (1990). Indicators for monitoring biodiversity: a hierarchical approach. Conserv. Biol. *4*, 355–364.

Olapade, O.A., and Leff, L.G. (2005). Seasonal response of stream biofilm communities to dissolved organic matter and nutrient enrichments. Appl. Environ. Microbiol. *71*, 2278–2287.

Paerl, H.W., Pinckney, J.L., and Steppe, T.F. (2000). Cyanobacterial-bacterial mat consortia: examining the functional unit of microbial survival and growth in extreme environments. Environ. Microbiol. *2*, 11–26.

Parfenova, V.V., Gladkikh, A.S., and Belykh, O.I. (2013). Comparative analysis of biodiversity in the planktonic and biofilm bacterial communities in Lake Baikal. Microbiology *82*, 91–101.

Passy, S.I., and Blanchet, F.G. (2007). Algal communities in human-impacted stream ecosystems suffer beta-diversity decline. Divers. Distrib. *13*, 670–679.

Pedrós-Alió, C. (2006). Marine microbial diversity: can it be determined? Trends Microbiol. *14*, 257–263.

Peter, H., Ylla, I., Gudasz, C., Romanı, A.M., Sabater, S., and Tranvik, L.J. (2011). Multifunctionality and diversity in bacterial biofilms. PLoS One *6*, e23225.

Pommier, T., Canbäck, B., Riemann, L., Boström, K.H., Simu, K., Lundberg, P., Tunlid, A., and Hagstrlöm, A. (2007). Global patterns of diversity and community structure in marine bacterioplankton. Mol. Ecol. *16*, 867–880.

Quince, C., Curtis, T.P., and Sloan, W.T. (2008). The rational exploration of microbial diversity. ISME J. *2*, 997–1006.

Rahbek, C. (2005). The role of spatial scale and the perception of large-scale species-richness patterns. Ecol. Lett. *8*, 224–239.

Rickard, A.H., Mcbain, A.J., Ledder, R.G., Handley, P.S., and Gilbert, P. (2003). Coaggregation between freshwater bacteria within biofilm and planktonic communities. FEMS Microbiol. Lett. *220*, 133–140.

Rickard, A.H., Mcbain, A.J., Stead, A.T., and Gilbert, P. (2004). Shear rate moderates community diversity in freshwater biofilms. Appl. Environ. Microbiol. *70*, 7426–7435.

Riemann, L., and Winding, A. (2001). Community dynamics of free-living and particle-associated bacterial assemblages during a freshwater phytoplankton bloom. Microb. Ecol. *42*, 274–285.

Rochex, A., Godon, J., Bernet, N., and Escudié, R. (2008). Role of shear stress on composition, diversity and dynamics of biofilm bacterial communities. Water Res. *42*, 4915–4922.

Romaní, A.M. (2010). Freshwater biofilms. In Biofouling, S. Dürr and J.C. Thomason, eds. (Blackwell Publishing Ltd, Oxford, UK), pp. 137–153.

Romaní, A.M., and Sabater, S. (2001). Structure and activity of rock and sand biofilms in a Mediterranean stream. Ecology *82*, 3232–3245.

Romaní, A.M., Guasch, H., Muñoz, I., Ruana, J., Vilalta, E., Schwartz, T., Emtiazi, F., and Sabater, S. (2004). Biofilm structure and function and possible implications for riverine DOC dynamics. Microb. Ecol. *47*, 316–328.

Romaní, A.M., Amalfitano, S., Artigas, J., Fazi, S., Sabater, S., Timoner, X., Ylla, I., and Zoppini, A. (2013). Microbial biofilm structure and organic matter use in mediterranean streams. Hydrobiologia *719*, 43–58.

Romaní, A.M., Borrego, C.M., Díaz-Villanueva, V., Freixa, A., Gich, F., and Ylla, I. (2014). Shifts in microbial community structure and function in light- and dark-grown biofilms driven by warming. Environ. Microbiol. *16*, 2550–2567.

Rusch, D.B., Halpern, A.L., Sutton, G., Heidelberg, K.B., Williamson, S., Yooseph, S., Wu, D., Eisen, J.A., Hoffman, J.M., Remington, K., *et al.* (2007). The Sorcerer II Global Ocean Sampling expedition: northwest Atlantic through eastern tropical Pacific. PLoS Biol. *5*, e77.

Russell, B.D., Connell, S.D., Findlay, H.S., Tait, K., Widdicombe, S., and Mieszkowska, N. (2007). Ocean acidification and rising temperatures may increase biofilm primary productivity but decrease grazer consumption. Philos. Trans. R. Soc. Lond. B. Biol. Sci. *368*, 20120438.

Sala, E., and Knowlton, N. (2006). Global marine biodiversity trends. Annu. Rev. Environ. Resour. *31*, 93–122.

Schiaffino, R.M., Unrein, F., Gasol, J.M., Massana, R., Balagué, V., and Izaguirre, I. (2011). Bacterial community structure in a latitudinal gradient of lakes: the roles of spatial versus environmental factors. Freshw. Biol. *56*, 1973–1991.

Schweitzer, B., Huber, I., Amann, R., Ludwig, W., and Simon, M. (2001). a- and b-*Proteobacteria* control the consumption and release of amino acids on lake snow aggregates. Appl. Environ. Microbiol. *67*, 632–645.

Sharp, C.E., Brady, A.L., Sharp, G.H., Grasby, S.E., Stott, M.B., and Dunfield, P.F. (2014). Humboldt's spa: microbial diversity is controlled by temperature in geothermal environments. ISME J. *8*, 1166–1174.

Simon, M., Grossart, H.-P., Schweitzer, B., and Ploug, H. (2002). Microbial ecology of organic aggregates in aquatic ecosystems. Aquat. Microb. Ecol. *28*, 175–211.

Singer, G., Besemer, K., Schmitt-Kopplin, P., Hödl, I., and Battin, T.J. (2010). Physical heterogeneity increases biofilm resource use and its molecular diversity in stream mesocosms. PLoS One *5*, e9988.

Singer, G., Besemer, K., Hochedlinger, G., Chlup, A.-K., and Battin, T.J. (2011). Monomeric carbohydrate uptake and structure–function coupling in stream biofilms. Aquat. Microb. Ecol. *62*, 71–83.

Sogin, M.L., Morrison, H.G., Huber, J.A., Welch, D.M., Huse, S.M., Neal, P.R., Arrieta, J.M., and Herndl, G.J. (2006). Microbial diversity in the deep sea and the underexplored 'rare biosphere.' Proc. Natl. Acad. Sci. U.S.A. *103*, 12115–12120.

Stanish, L.F., O'Neill, S.P., Gonzalez, A., Legg, T.M., Knelman, J., McKnight, D.M., Spaulding, S., and Nemergut, D.R. (2013). Bacteria and diatom co-occurrence patterns in microbial mats from polar desert streams. Environ. Microbiol. *15*, 1115–1131.

Stoodley, P., Sauer, K., Davies, D.G., and Costerton, J.W. (2002). Biofilms as complex differentiated communities. Annu. Rev. Microbiol. *56*, 187–209.

Stratil, S.B., Neulinger, S.C., Knecht, H., Friedrichs, A.K., and Wahl, M. (2013). Temperature-driven shifts in the epibiotic bacterial community composition of the brown macroalga *Fucus vesiculosus*. Microbiologyopen *2*, 338–349.

Sul, J.W., Oliver, T.A., Ducklow, H.W., Amaral-Zettler, L.A., and Sogin, M.L. (2013). Marine bacteria exhibit a bipolar distribution. Proc. Natl. Acad. Sci. U.S.A. *110*, 2342–2347.

Tamames, J., Abellán, J.J., Pignatelli, M., Camacho, A., and Moya, A. (2010). Environmental distribution of prokaryotic taxa. BMC Microbiol. *10*, 85.

Tang, X., Gao, G., Qin, B., Zhu, L., Chao, J., Wang, J., and Yang, G. (2009). Characterization of bacterial communities associated with organic aggregates in a large, shallow, eutrophic freshwater lake (Lake Taihu, China). Microb. Ecol. *58*, 307–322.

Tilman, D. (1987). On the meaning of competition and the mechanisms of competitive superiority. Funct. Ecol. *1*, 304–315.

Tilman, D., Knops, J., Wedin, D., Reich, P., Ritchie, M., and Siemann, E. (1997). The influence of functional diversity and composition on ecosystem processes. Science (80-.). *277*, 1300–1302.

Timoner, X., Acuña, V., Von Schiller, D., and Sabater, S. (2012). Functional responses of stream biofilms to flow cessation, desiccation and rewetting. Freshw. Biol. *57*, 1565–1578.

Torsvik, V., Øvreås, L., and Thingstad, T.F. (2002). Prokaryotic diversity – magnitude, dynamics, and controlling factors. Science (80-.). *296*, 1064–1066.

Vannote, R.L., Minshall, G.W., Cummins, K.W., Sedell, J.R., and Cushing, C.E. (1980). The river continuum concept. Can. J. Fish. Aquat. Sci. *37*, 130–137.

Wang, J., Soininen, J., Zhang, Y., Wang, B., Yang, X., and Shen, J. (2011). Contrasting patterns in elevational diversity between microorganisms and macroorganisms. J. Biogeogr. *38*, 595–603.

Wang, J., Soininen, J., Zhang, Y., Wang, B., Yang, X., and Shen, J. (2012a). Patterns of elevational beta diversity in micro- and macroorganisms. Glob. Ecol. Biogeogr. *21*, 743–750.

Wang, J., Soininen, J., He, J., and Shen, J. (2012b). Phylogenetic clustering increases with elevation for microbes. Environ. Microbiol. Rep. *4*, 217–226.

Wang, J., Shen, J., Wu, Y., Tu, C., Soininen, J., Stegen, J.C., He, J., Liu, X., Zhang, L., and Zhang, E. (2013). Phylogenetic beta diversity in bacterial assemblages across ecosystems: deterministic versus stochastic processes. ISME J. *7*, 1310–1321.

Wang, Y., Sheng, H.-F., He, Y., Wu, J.-Y., Jiang, Y.-X., Tam, N.F.-Y., and Zhou, H.-W. (2012c). Comparison of the levels of bacterial diversity in freshwater, intertidal wetland, and marine sediments by using millions of illumina tags. Appl. Environ. Microbiol. *78*, 8264–8271.

Washington, V.J., Lear, G., Neale, M.W., and Lewis, G.D. (2013). Environmental effects on biofilm bacterial communities: a comparison of natural and anthropogenic factors in New Zealand streams. Freshw. Biol. *58*, 2277–2286.

Watnick, P., and Kolter, R. (2000). Biofilm, city of microbes. J. Bacteriol. *182*, 2675–2679.

Webb, J.S., Givskov, M., and Kjelleberg, S. (2003). Bacterial biofilms: prokaryotic adventures in multicellularity. Curr. Opin. Microbiol. *6*, 578–585.

Whittaker, R.H. (1972). Evolution and measurement of species diversity. Taxon *21*, 213–251.

Whittaker, R.J., Willis, K.J., and Field, R. (2001). Scale and species richness: towards a general, hierarchical theory of species diversity. J. Biogeogr. *28*, 453–470.

Wilhelm, L., Singer, G.A., Fasching, C., Battin, T.J., and Besemer, K. (2013). Microbial biodiversity in glacier-fed streams. ISME J. *7*, 1651–1660.

Wilhelm, L., Besemer, K., Fasching, C., Urich, T., Singer, G.A., Quince, C., and Battin, T.J. (2014). Rare but active taxa contribute to community dynamics of benthic biofilms in glacier-fed streams. Environ. Microbiol. *16*, 2514–2524.

Wilhelm, L., Besemer, K., Fragner, L., Peter, H., Weckwerth, W., and Battin, T.J. Altitudinal patterns of diversity and functional traits of metabolically active microorganisms in stream biofilms. ISME J. [Epub ahead of print].

Woodcock, S., Besemer, K., Battin, T.J., Curtis, T.P., and Sloan, W.T. (2013). Modelling the effects of dispersal mechanisms and hydrodynamic regimes upon the structure of microbial communities within fluvial biofilms. Environ. Microbiol. *15*, 1216–1225.

Worm, B., Barbier, E.B., Beaumont, N., Duffy, J.E., Folke, C., Halpern, B.S., Jackson, J.B.C., Lotze, H.K., Micheli, F., Palumbi, S.R., *et al.* (2006). Impacts of biodiversity loss on ocean ecosystem services. Science (80-.). *314*, 787–790.

Xiong, J., Ye, X., Wang, K., Chen, H., Hu, C., Zhu, J., and Zhang, D. (2014). Biogeography of the sediment bacterial community responds to a nitrogen pollution gradient in the East China Sea. Appl. Environ. Microbiol. *80*, 1919–1925.

Yawata, Y., Cordero, O.X., Menolascina, F., Hehemann, J.-H., Polz, M.F., and Stocker, R. (2014). Competition-dispersal tradeoff ecologically differentiates recently speciated marine bacterioplankton populations. Proc. Natl. Acad. Sci. U.S.A. *111*, 5622–5627.

Ylla, I., Borrego, C., Romaní, A.M., and Sabater, S. (2009). Availability of glucose and light modulates the structure and function of a microbial biofilm. FEMS Microbiol. Ecol. *69*, 27–42.

Ylla, I., Canhoto, C., and Romaní, A.M. (2014). Effects of warming on stream biofilm organic matter use capabilities. Microb. Ecol. *68*, 132–145.

Zhang, R., Liu, B., Lau, S.C.K., Ki, J.-S., and Qian, P.-Y. (2007). Particle-attached and free-living bacterial communities in a contrasting marine environment: Victoria Harbor, Hong Kong. FEMS Microbiol. Ecol. *61*, 496–508.

Zinger, L., Gobet, A., and Pommier, T. (2012). Two decades of describing the unseen majority of aquatic microbial diversity. Mol. Ecol. *21*, 1878–1896.

Aquatic Biofilms and Biogeochemical Processes

5

Laura Leff, Jonathon B. Van Gray, Eugènia Martí, Stephanie N. Merbt and Anna M. Romaní

Abstract

In this chapter, we discuss the biogeochemistry of biofilms with an emphasis on the unique features of biofilms that impact oxidation-reduction reactions, such as retention of materials (nutrients, organic molecules, including enzymes), shortened uptake lengths, intercellular interactions due to proximity of the organisms, and redox gradients over short spatial distances. Our focus is on carbon, nitrogen and phosphorus with some inclusion of micronutrients. The role of biofilm structure and microbial community composition (including algal–bacterial relationships) in determining the spatiotemporal variations in biogeochemistry is discussed as well as the spatial scales of biogeochemical reactions within biofilms. Methods for measuring biogeochemical processes and environmental conditions in biofilms are reviewed, from molecular tools and fine scale measurements, to large scale biogeochemical measurements in flumes and mesocosms or natural systems. The relevance of the spatial scale is highlighted as well as the challenge for scaling-up of methods.

Introduction

Biofilms are ubiquitous when surfaces are exposed to water and microorganisms, and have unique physical and chemical properties (e.g. reviewed by Costerton *et al.*, 1987; see Chapter 1). The close physical spacing of cells and properties of the extracellular matrix provide opportunities for significant biogeochemical interactions allowing biofilms to serve as 'hotspots' of biogeochemical cycling. These properties change over time as the biofilm develops and vary with environmental conditions of sites were they develop, including the nature of the substrate to which the biofilm is attached. Ultimately, the characteristics of biofilms and their biogeochemical reactivity have implications for ecosystem-scale processes, such as nutrient uptake, in aquatic ecosystems.

As spatiotemporal refugia from the dynamic flow of the bulk fluid, biofilms provide a unique immobilized microenvironment in which the biological, chemical, and physical properties of the biofilm matrix facilitate abiotic–biotic interactions necessary to drive biogeochemical reactions. This respite from the transience of the bulk fluid permits the aggregate microbial populations to persist and diversify in close proximity (Hansen *et al.*, 2007; Poltak and Cooper, 2010), allowing for greater exchange of genetic materials and metabolites within and between populations (Battin *et al.*, 2003). These interactions result in a high uptake efficiency and increased retention time for bioreactive solutes (i.e. nutrients, organic compounds, and extracellular enzymes) within the biofilm and provide a dynamic resource pool for internal use by those organisms within the sphere of the diffusive capabilities of the associated matrix (Battin *et al.*, 2003; Flemming and Wingender, 2010; Stewart, 2003).

The extracellular polymeric substances (EPS) matrix serves to capture nutrients from water, retain extracellular enzymes, and provide protection from disturbance and predation (Sutherland, 2001). Biofilms accumulate and concentrate nutrients (Lock *et al.*, 1984) which are available for assimilatory or dissimilatory processes. The three dimensional structure of the biofilm includes passageways (channels and pores) by which nutrients can be transported (DeBeer and

Schramm, 1999). In addition, biofilms may alter adsorption. Collectively, these features allow biofilms to play a major role in ecosystem function. The importance of this growth form has led to a large number of studies in freshwater ecosystems, such as those focused on production of extracellular enzymes and nitrogen cycling (e.g. Romaní *et al.*, 2013; von Shiller *et al.*, 2009). The nature of the role of biofilms and the extent of our knowledge varies among ecosystems; arguably there is more known about biofilm function in streams and rivers than in lakes or wetlands. This is likely attributable to the more widely acknowledged and studied role that benthos play in streams.

One factor that contributes to differences in biofilms among aquatic ecosystems is the nature of the benthic substrate (see Chapter 1). Substrates for biofilm development can be generally considered organic (such as leaves or wood) or inorganic (such as rocks). The surface properties of the substrate, such as charge, roughness, etc., influence the biofilm (Palmer *et al.*, 2007) as does leaching of materials from the underlying substrate. Research on biogeochemistry has been received special attention for biofilms on rocks and sand, as well as on leaves.

Physical proximity of cells in biofilms provides the opportunity for inter-cellular communication, sharing of resources, modification of the physical environment that enhances adhesion, and alteration of redox conditions at a small spatial scale. For example, Reidel *et al.* (2001) discovered that cross-species intercellular communication occurred in mixed species biofilms. Studies such as this one, as well as those on co-aggregation, have often focused on interactions among organisms in laboratory communities and many of these phenomena have not been investigated in aquatic ecosystems and are more widely known from oral biofilms (Kolenbrander, 2000). However, these types of interactions are potentially important in biogeochemistry of aquatic biofilms. Unlike some other aspects of biofilms, this physical positioning and structure is inherently connected to community composition, particularly in light of the specific inter-species interactions that have been demonstrated in laboratory studies (James *et al.*, 1995).

In this chapter, we discuss the biogeochemistry of aquatic biofilms with an emphasis on the unique features of biofilms that impact oxidation-reduction reactions and retention of materials. We focus on carbon, nitrogen and phosphorus with some inclusion of other elements (e.g. S, Fe, etc.). The role of both extrinsic (environmental conditions, inorganic and organic nutrient availability), and intrinsic (biofilm structure) factors in determining the spatiotemporal variations in biogeochemistry and the spatial scales of biogeochemical reactions within biofilms is discussed. Methods for measuring biogeochemical processes and environmental conditions in biofilms are introduced, such as use of molecular tools, assessment of microbial function linked to biogeochemical processes (extracellular enzymes, respiration), fine scale measurements (microelectrodes, confocal microscopy) and the use of larger-scale methods (flumes, mesocosms, use of stable isotopes).

Biofilm biogeochemical environment

While the biofilm environment greatly enhances the efficacy of the biogeochemical processes occurring therein (Battin *et al.*, 2003; Poltak and Cooper, 2011; Burmølle *et al.*, 2014), capabilities of individual biofilms vary based on the specific extrinsic and intrinsic factors as they affect the biological, chemical, and physical components of the biofilm matrix. Extrinsic factors can be generalized as the external physico-chemical drivers affecting biofilm formation and stability (Lyautey *et al.*, 2005), while intrinsic properties can be broadly characterized as factors associated with the internal heterogeneity of the microbial composition of the biofilm – each of which will be explored in greater detail in the following sections (Fig. 5.1).

Extrinsic biofilm properties affecting biogeochemical processes

The ubiquity of biofilms within aquatic systems suggests that biofilms form regardless of environmental conditions as long as adequate resources are provided (e.g. nutrients and space to colonize). Yet extrinsic physical and chemical drivers play a

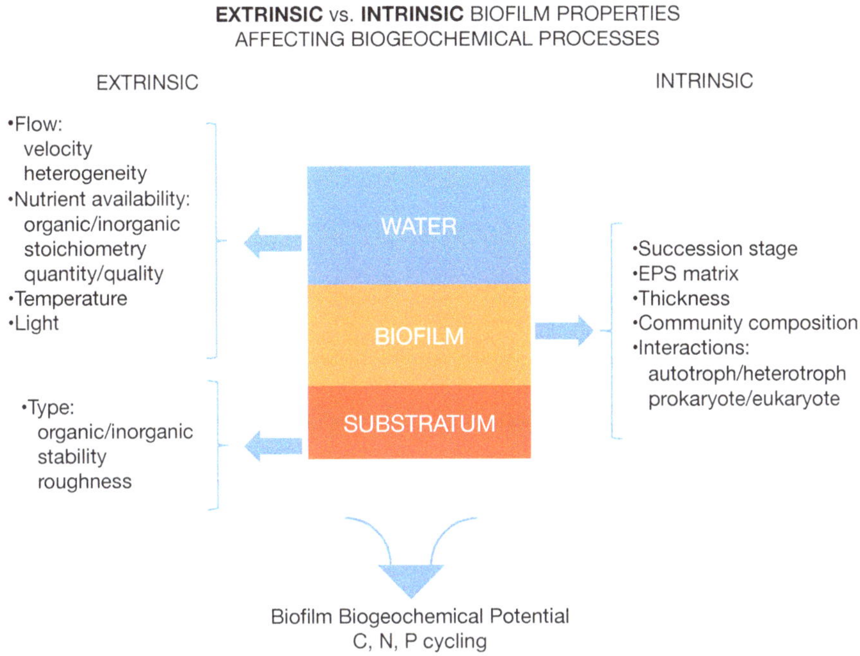

Figure 5.1 Diagram of the extrinsic and intrinsic factors affecting biofilm biogeochemical processes, and finally determining the biofilm potential for nutrient cycling. Extrinsic factors mainly include those related to physical and chemical drivers, while intrinsic ones are those derived from biofilm structure and composition.

significant role in affecting biofilm biogeochemical processes, as the course of biofilm development and the subsequent functional capabilities of the community are contingent upon these factors. For our purposes here, we will focus on abiotic factors that affect biofilm biogeochemistry, including the substrata type, hydrological variability, temperature, and nutrient availability.

The effects of substrata on biofilm biogeochemistry are closely linked to the nature of the substrata (e.g. organic versus inorganic) and the physical and chemical characteristics of substrata surfaces. Organic (such as leaves and woody debris) and inorganic (such as rocks or sand grains, clay tiles and glass slides or beads) substrata differ in their reactivity with the biofilm (Whitehead and Verran, 2009). Organic surfaces provide a pool of supplemental resources (such as reduced carbon, organic nitrogen, etc.) for heterotrophic use and may promote higher colonization rates of aquatic fungi (Golladay and Sinsabaugh, 1991), thereby increasing resource availability via substrate degradation. Inorganic substrata tend to be less reactive or completely inert as a resource pool for microbial organisms in biofilms (Whitehead and Verran, 2008). Comparisons among substrata demonstrate that surface properties impact rates of biofilm formation, density, diversity, and resource retention (Golladay and Sinsabaugh, 1991; Sinsabaugh *et al.*, 1991; Barbiero, 2000; Khatoon *et al.*, 2007; Bergey *et al.*, 2010). Physical characteristics of the substrata are significant to biofilm biogeochemistry (Whitehead and Verran, 2008). Substratum stability and topography have been shown to positively affect benthic algae colonization, which act as a major source of endogenous oxygen production and carbon fixation (see Burkholder, 1996). Similarly, in a field study comparing the effects of nutrient supply on biofilms forming on rock or sand substrata, Romani *et al.* (2004a) identified substratum stability and roughness as key determinants of biofilm structure and heterotrophic metabolism. In a larger scale, in a comparison of physical heterogeneity of stream substrate on biological activity, Cardinale *et al.* (2002) found significant increases in algal productivity and biofilm respiration as a function of greater substrate heterogeneity.

By their very nature, aquatic biofilms are subjected to stresses associated with the flow of the bulk fluid. Examination of physical response to flow velocities in a mixed bioreactor community (Beyenal and Lewandowski, 2002) found that the internal structure of the biofilm matrix is the result of an intricate balance between two contrasting properties: the preservation of structural integrity through increased density so as to withstand shear stress and increasing diffusivity of nutrients deeper into the biofilm matrix (i.e. reduced density). For instance, increased flow velocity has been implicated in reducing biofilm density and spatial heterogeneity (Battin *et al.*, 2003b; Coundoul *et al.*, 2014), while simultaneously increasing the mass transfer of nutrients from the bulk fluid into the biofilm matrix (Donlan, 2002; Taherzadeh *et al.*, 2012). In the same study, Battin *et al.* (2003b) found biofilms developed under slow velocity conditions to accrue more biomass, but were also more reliant on internal resource cycling than those developed under elevated velocities. While these results provide great insight into the potential effects of flow on the biogeochemistry of biofilms, they are based on manipulative studies that compare contrasting scenarios of low versus high flows rather than the heterogeneity found in natural systems, failing to capture details into the effects of flow heterogeneity on community development. In a pair of studies, Besemer *et al.* examined the effects of flow velocity and heterogeneity (laminar, transitional, or turbulent flows) on bacterial community composition (Besemer *et al.*, 2009) and biofilm community succession (Besemer *et al.*, 2007; see also Chapter 4). These studies found increasing bacterial diversity with flow heterogeneity (Besemer *et al.*, 2009), as well as that flow velocity and heterogeneity influenced early microbial successional trajectories. However, community composition converged upon biofilm maturation as algal populations became dominant and were believed to exert greater control over the internal biofilm environment (Besemer *et al.*, 2007). Taken together, these results suggest that during early succession, flow plays a significant role in determining the phylogenetic composition of the biofilm community and that, as maturation continues, the effects of flow are reduced as structural stability with the biofilm is increased due to algal interactions.

Temperature plays a significant role in mediating biogeochemical activity in aquatic habitats due to the inherent ties with the metabolic activity, chemical kinetics, and compound stability. Due to taxon-specific tolerances, temperature often dictates what organisms can persist in the environment and contribute to biofilm formation (Goller and Romeo, 2008; Hostacka *et al.*, 2010) and the biogeochemical potential of the resulting community. Increased temperatures have been linked to increases in algal growth rates (Butterwick *et al.*, 2005), bacterial productivity (Fischer *et al.*, 2002; Adams *et al.*, 2010), resource diffusivity (Bissett *et al.*, 2008), enzymatic activity (Bouletreau *et al.*, 2012; Ylla *et al.*, 2014), as well as to modifications in community composition (Rubin and Leff, 2007) and enzyme capabilities (Ylla *et al.*, 2012). Additionally, changes in temperature can alter the overall metabolic state of the biofilm, as increased temperatures have been shown to shift biofilms towards heterotrophy (Hancke and Glud, 2004; Degerman *et al.*, 2013).

Aquatic resource dynamics have an immeasurable effect on biofilm activity. In aquatic systems the bulk fluid represents the largest pool of dissolved nutrients available to the biofilm community, especially to those developing on inorganic substrata. Consisting of a diverse mixture of allochthonous and autochthonous C, N, and P compounds, the bulk fluid resource pool is highly heterogeneous, and as such, specific resource availability varies based on the concentrations of resources in the bulk fluid and the ability of the biofilm community to satisfy their stoichiometric requirements through resource assimilation. Thus, the bulk fluid resource pool can effectively contribute to resource limitation or saturation within the biofilm matrix, thereby altering the structural and functional dynamics of the community (Tank and Dodds, 2003; Hoellein *et al.*, 2011). For instance, as described in a subsequent section, a number of studies found changes in the aqueous dissolved organic carbon (DOC) pool to significantly affect biofilm community composition (Olapade and Leff, 2006), P cycling (Ardón and Pringle, 2007) and N dynamics (Johnson *et al.*, 2012). Studies looking solely into the effects of nutrient (e.g. N or P) additions saw similar results; increased fungal and bacterial activity with N and P additions (Gulis *et al.*, 2008), organic and inorganic N treatments resulted in increased hyporheic biofilm nitrification (Findlay and Sinsabaugh, 2003), N and P enrichment stimulated algal biomass accumulation (Artigas *et al.*, 2013), and so forth. These results highlight not only the complexity of the effects of aquatic resources on biofilm biogeochemistry, but also the depth of potential alterations that can occur as the result of modifications to the aquatic resource pool.

Obviously light availability determines development of photoautotrophic community and many adaptations to either high/low light intensity have been described for algae and cyanobacteria and thus influencing community composition (e.g. Corcoll *et al.*, 2012). For the purpose of this chapter, however, our discussion of the effect of light is mainly focused on the consequence of photosynthetic organisms in the biofilm for biogeochemical processes as well as their significance for autotrophic–heterotrophic relationships, as described below.

Intrinsic biofilm properties affecting biogeochemical processes

The biogeochemical potential of the biofilm matrix is inextricably linked to its intrinsic properties. Autogenic changes associated with community succession and biofilm maturation result in compositional shifts in microbial populations and with them, the metabolic diversity and biogeochemical potential of the biofilm environment (Jackson *et al.*, 2001; Lyautey *et al.*, 2005; Burmølle *et al.*, 2014). During the early stages of succession, the biogeochemical environment of the biofilm is constrained by a lack of physical complexity, as early colonizers in most aquatic systems are likely dominated by metabolically generalist taxa reliant on resources present within the overlying bulk fluid (Sobczak, 1996; Romaní *et al.*, 2014a). Biofilm maturation occurs as competitively adept populations expand and additional planktonic taxa are integrated into the matrix via coadhesion/coaggregation mechanisms (Rickard *et al.*, 2003), forming a stratified biological architecture. EPS excretion by active cells creates a spatially heterogeneous biofilm architecture that includes numerous interstitial voids as well as particulates captured from the bulk fluid. Successional stratification of microbial cells and populations results in a high degree of spatial heterogeneity at the micro-scale, with community depth varying based on cell density, ranging from 1–2 microns in monolayered communities, to several millimetres in low-disturbance, highly stratified biofilms (Wimpenny *et al.*, 2000; Battin *et al.*, 2003b).

Concurrently, as biofilm biomass and physical complexity increase, the chemical concentrations of endogenous and exogenous solutes are affected as alterations in resource production/consumption and diffusion rates lead to the development of extensive physico-chemical micro-gradients (Stewart, 2003; Stewart and Franklin, 2008; Renslow *et al.*, 2010). Stewart and Franklin (2008) identified three distinct gradient patterns that develop as a function of metabolite concentration: decreasing metabolic substrate concentrations with biofilm depth, metabolic products increase with biofilm depth, and metabolic intermediaries originating within the biofilm will diffuse to areas of lower concentrations, regardless of depth. For instance, in many aquatic biofilms oxygen gradients are formed as a function of autotrophy occurring within the surface layers of the biofilm. Autotrophic oxygenation near the surface of the biofilm results in the formation of oxygen gradients, which can extend several hundred microns into the biofilm matrix, resulting in distinct vertical oxygen profiles in which O_2 concentrations become increasingly depleted with depth as aerobic respiration gives way to anaerobic metabolic processes (Stewart, 2008). Similar gradients have been identified in aquatic biofilms for a host of compounds including nitrite, nitrate, ammonium, sulfate, methane, and pH (Stewart, 2008).

In addition to the structural and physical and chemical changes, maturation affects the functional capabilities of the biofilm as niche differentiation and the localized set of symbiotic interactions and communications necessary to stimulate increased biogeochemical activity develop (Hansen *et al.*, 2007; Wintermute and Silver, 2010). Cross feeding and syntrophic interactions, the use of one organism's metabolic intermediates and by-products by another organism as a metabolic substrate, serve to increase internal nutrient cycling, reduce energy expenditures on resource acquisition from water column, and can create favourable environmental conditions for novel niche development (Pfeiffer and Bonheoffer, 2004; Okabe *et al.*, 2005; Bull and Harcombe, 2009; Wintermute and Silver, 2010; Pande *et al.*, 2013). Quorum sensing, a mechanism for intercellular communication within and between prokaryotic and eukaryotic populations, can promote synergistic or antagonistic interactions that affect the structural and functional attributes of the community (Valle *et al.*, 2004; Irie and Parsek, 2008; Atkinson and Williams, 2009; see Chapter 3).

Often-overlooked, the interactions between the heterotrophic and photoautotrophic components of the biofilm community are an integral part of biofilm biogeochemistry. During early stages of biofilm formation, the biofilm community is predominantly heterotrophic in nature and reliant on the resources available in the bulk fluid (Jackson, 2003). As the biofilm matures and autotrophic taxa are integrated, the heterotrophic community shifts to one dominated by autotrophy, with the heterotrophic community becoming increasingly reliant on the labile carbon provided by the photoautotrophic populations (Sobczak, 1996; Jackson, 2003). This metabolic shift greatly enhances the internalized cycling of resources within the biofilm as the complementary aspects of heterotrophic and autotrophic metabolism converge. For instance, Rier *et al.* (2007) found evidence to suggest that photosynthesis-mediated alterations to the physico-chemical state of the biofilm may enhance heterotrophic extracellular enzyme activity by altering biofilm pH and redox gradients. Similarly, Romani and Sabater (2000) saw increased enzyme activity in autotrophic biofilms and attributed this to the algal exudate and metabolite availability to heterotrophic taxa. Conversely, heterotrophic taxa also affect the autotrophic populations. Under nutrient-poor conditions, algal populations are often dependent upon endogenous bacterial metabolites to support productivity (Scott *et al.*, 2008). However, in the same study, Scott *et al.* (2008) found that increased nutrient loading may de-couple bacterial–algal interactions, suggesting that while the interactions of the autotrophic and heterotrophic populations within the biofilm may be highly influential on internal biogeochemical cycling, external modifications to the system can ultimately disrupt these processes.

Biogeochemical cycling

In this section, we highlight selected biogeochemical cycles (C, N and P cycling) in light of the context of extrinsic and intrinsic properties that influence biofilm function discussed above. Thus, for each element discussed consideration is given to factors that influence biogeochemical spatiotemporal changes. Although the three main elements are considered separately, there are many connections among them since biogeochemical processes. As shown in many of the examples included below for C, N and P cycling, in many occasions their interactions and stoichiometry matters and thus can be relevant for the understanding of a single nutrient cycling. As an example, using a whole stream N and C addition, Oviedo-Vargas *et al.* (2013) found that P cycling appeared to be N limited and that the form of N (ammonia versus nitrate) was critical. Moreover they conclude that C and P cycling are biologically coupled and that this coupling is mediated by N form.

Carbon cycling

Here we emphasize the forms and biogeochemical fate of organic matter (OM) with particular emphasis on differences among broad categories of OM: dissolved versus particulate, labile versus recalcitrant, allochthonous versus autochthonous. Because the OM pool is complex and undefined reliance on these coarse categories has been common (Münster and Chróst, 1990), although over time there have been advances in our ability to chemically characterize the OM pool.

Coarse particulate organic matter (CPOM) plays host to a well-developed biofilm community and decomposition of plant litter has been widely studied in aquatic ecosystems (e.g. Boyero *et al.*, 2011). However, only a portion of the studies on leaf litter decomposition include an explicitly stated focus on biofilms (e.g. Artigas *et al.*, 2011). Often in these studies, the biofilm per se is not the focus, but rather the researchers measure some aspect of microbial function which is presumed to be attributed to biofilm dwellers. Among them, those that examine extracellular enzymes are the most common while relatively fewer studies examined both microbial community structure and function. Given that in this particular section our focus is on biogeochemical cycling, generalizations about function based on these studies are presented here and microbial community composition is not extensively discussed. In addition, extracellular enzymes for N and P acquisition are described subsequently but, of course, although we discuss each element separately their biogeochemistry is very much intertwined. In fact, Sinsabaugh *et al.* (2010) demonstrated that the stoichiometric ratio of extracellular enzymes of biofilms tracked the C:N or C:P relative to planktonic communities; and thus, they conclude that biofilm and plankton communities have similar ecoenzymatic stoichiometric relationships with underlying microbiological community parameters.

Extracellular enzymes for degradation of plant polymers to monomers are among the most well studied aspects of aquatic biofilm biogeochemistry regardless of habitat and are widely studied in a range of ecosystems (Arnosti *et al.*, 2014). Leaves are the most commonly examined with fewer studies on wood; and the most widely assayed enzymes include beta-glucosidase, N-acetylglucosaminidase, beta-xylosidase, cellobiohydrolase and phenol oxidase (e.g. Artigas *et al.*, 2011; Frossard *et al.*, 2013). Enzymes are produced to varying extents by both fungi and bacteria; commonly, fungal biomass is thought to peak after bacterial biomass on leaves decomposing in streams. Also typically fungi are presumed to contribute more to the degradation of lignocellulose and some bacteria in biofilms may be solely relying on monomers produced by the degradation efforts of other microorganisms, and thus significant synergistic and antagonistic interactions between fungi and bacteria modulate plant litter decomposition (Romaní *et al.*, 2006). Factors that appear to control extracellular enzyme in leaf biofilms include concentrations of metals (such as Al), nutrient availability, algal biomass, and physical conditions, such as temperature (e.g. Jones and Lock, 1989; Tank and Webster, 1998; Rier *et al.*, 2007). These factors contribute to the spatial and temporal variation of the extracellular enzyme activities, together with the seasonality of litter inputs to the aquatic systems.

Impacts of dissolved organic matter (DOM) on biofilms have been examined in the laboratory and in the field; the latter has been accomplished via whole stream additions and nutrient diffusing artificial substrates among other approaches. Collectively, this range of approaches provides insight into biofilm biogeochemistry with each approach having inherent pluses and minuses. Regardless of the approach, there is an underlying pattern of spatiotemporal change linked to variability in both extrinsic and intrinsic factors. For example, in Mediterranean streams, biofilms change in OM use as seasonally OM sources transition from allochthonous to autochthonous (Romaní *et al.*, 2013). In these systems, seasonal drying reduces bacterial production and chlorophyll biomass, but the rapid recovery of both autotrophs and heterotrophs upon rewetting indicates their responsiveness to such seasonal changes. In the laboratory, biofilms on various surfaces including glass (slides or beads), ceramic tiles, leaf disks, etc. have been exposed to various DOM treatments. Response variables include those related to abundance or biomass, function (such as enzyme activity or productivity), and community structure. For example, Ghosh and Leff (2013), by using glass beads as the substrate, found that OM treatments induced changes in bacterial community composition and rates of N utilization. Ylla *et al.* (2009) used glass tiles and found that both glucose concentrations and light levels impacted biofilms, including altering beta-glucosidase and peptidase activity. In contrast, Freeman *et al.* (1990) found that high molecular weight DOM had an inhibitory effect on extracellular enzyme activity. Olapade and Leff (2005), who used nutrient diffusing artificial substrates to look at DOM and nutrient impacts on biofilms, found that the greatest impacts on biofilm were attributable to labile, low-molecular-weight DOM amendments at times of year when chlorophyll *a* concentrations were low. These four example studies highlight the relevance of both DOM quantity and quality on determining changes in the biofilm metabolism. In addition, it has been further shown that responses to specific properties of DOM vary among bacterial taxa (McNamara and Leff, 2004). At a larger scale, Singer *et al.* (2010), using stream flumes, demonstrated that heterogeneity in water flow altered DOC uptake and that such responses were likely linked to community composition. In addition, significant DOC, especially biodegradable DOC, uptake have been measured for light and dark grown biofilms in a fluvial system (Romaní *et al.*, 2004b).

Biofilms provide an opportunity for direct interaction between bacteria and algae, which can influence the dynamics of both organic and inorganic carbon. These interactions are perhaps most pronounced in microbial mats, especially those between heterotrophic bacteria and cyanobacteria in which production of biological available DOM excreted by the photoautotrophs can be used by heterotrophic bacteria (Marshall, 1989). Among the approaches used to examine bacterial-algal coupling in biofilms are those in which light and nutrients are manipulated experimentally. Both light and nutrients have been found to be limiting for periphyton (Hepinstall and Fuller, 1994). It has been shown that photosynthesis enhances peptidase activity and that this activity typically increases with higher chlorophyll content within the biofilm, likely due to the use of available labile peptides released by algae, but this tight link between algae and bacteria is weaken when glucose is added (Ylla *et al.*, 2009). In addition, photosynthesis can cause changes in pH that impact biofilm function. For example, Espeland and Wetzel (2001) found that such changes in pH reduced alpha and beta glucosidase activity and increased alkaline phosphatase.

Nitrogen cycling

The increased amount of biological available nitrogen derived from human activities has led to many studies on nitrogen in biofilms. Biofilms are highly metabolic active; and thus, assimilation of N from the water column is relevant to meet the biotic demand of N (von Schiller *et al.*, 2009). In fact, the relationship between the ^{15}N natural abundances of dissolved inorganic nitrogen and biofilms found across several stream locations within a catchment provides evidence for stream transfer into biofilms (Pastor *et al.*, 2013). The use of ^{15}N tracer additions in streams over the past decade has allowed quantifying *in situ* rates of assimilatory N uptake rates by biofilms developed on either inorganic and organic substrata. Overall, results from these studies highlight the active role of biofilms in N uptake and their contribution to in-stream N cycling (Mulholland *et al.*, 2000; Tank *et al.*, 2000; Dodds *et al.*, 2000; Hamilton *et al.*, 2001; Merriam *et al.*, 2002; Ashkenas *et al.*, 2004; Simon *et al.*, 2005; von Schiller *et al.*, 2009; Sobota *et al.*, 2012). In addition, the close interaction among organisms and the biogeochemical gradients within biofilms facilitate the coupling between dissimilatory oxidation and reduction reactions, such as nitrification and denitrification. Lastly, depending on environmental conditions, nitrogen fixation may be also important (Grimm and Petrone, 1997).

Biofilms, especially epilithic ones, tend to rely on nitrogen from the water column. Nutrient diffusing artificial substrates have been used (as shown for some studies of DOM) to examine the effect of nutrient availability mostly for biofilm growth and nutrient uptake. For instance, Tank and Dodds (2003) found that responses to nutrient amendment differed among biofilms dominated by heterotrophs and autotrophs, and also differed between organic and inorganic substrates. Moreover, increased biofilm growth and nitrogen uptake after nitrogen enrichment appears to be more relevant under light conditions (von Schiller *et al.*, 2007), suggesting a positive interaction between light and nutrients for biofilm primary production (Ylla *et al.*, 2007). Scott *et al.* (2009) also using nutrient diffusing artificial substrates found measurable nitrogen fixation and suggested that in the system studiednitrogen was limiting for primary producers. Likewise, Romaní *et al.* (2004) found that nutrient addition altered extracellular enzyme activity and that the nature of the response varied among biofilms developed in different substrates (rock versus sand). Other studies have investigated the response of biofilms in terms of N uptake to gradual increases in DIN concentration and found that while the response follows a Michaelis–Menten model for ammonium increases, the response for nitrate is absent or even negative (Kemp and Dodds, 2002;

O'Brien and Dodds, 2007; Ribot *et al.*, 2013). In the case of leaf decomposition, the effect of nitrate increases followed a Michaelis–Menten model response to breakdown rate (Ferreira *et al.*, 2006), and at the same time, leaf litter can be relevant as nitrate retention (Duan *et al.*, 2014). In addition, nutrients have been found to be limiting to biofilm communities on wood and leaves in studies involving whole stream nutrient additions (Stelzer *et al.*, 2003).

Various studies have examined dissimilatory processing of nitrogen in biofilms. Teissier *et al.* (2007) found that nitrifiers were present and able to compete with algae in river biofilms. In fact, nitrifier bacteria and archaea are already present at early stages of biofilm succession and their abundance seems to be favoured by ammonium concentration in stream water (Merbt *et al.*, 2011). Nitrification activity of biofilms have been measured using recirculating chambers and results show that environmental conditions of oxygen and ammonium concentrations favours nitrification rates (Kemp and Dodds, 2002). Other studies show that denitrification in biofilms can be relevant, thought its variability is related to biomass and oxygen concentrations (Teissier *et al.*, 2007) and nitrate concentration in the water column (Kemp and Doods, 2002). When the nitrogen cycle is viewed as a whole, there is evidence that nitrogen biogeochemistry associated with biofilms changes over succession. Teissier *et al.* (2007) indicated that at early stages, assimilation by epilithic biofilms exceeds mineralization and storage, whereas in more mature biofilms nitrogen mineralization is greater than uptake. In addition, results from a translocation experiment using ^{15}N natural abundances in mature biofilms, indicated that while biofilm N uptake rates seem to be controlled by environmental conditions (such as light and nutrient concentrations), N turnover rates are mostly determined by biofilm intrinsic characteristics (Peipoch *et al.*, 2014).

Phosphorus cycling

Because of its importance as a limiting nutrient, phosphorus has been extensively studied in freshwater ecosystems. Unlike nitrogen, the number of chemical forms available and transformations is restricted and thus much effort has been focused on assimilation of phosphate and production of alkaline phosphatase as well as on internal loading of phosphorus from sediments. Price and Carrick (2014) found that there were rapid and high levels of phosphorus exchange between biofilms and the overlying water and that seasonal changes strongly influenced phosphorus flux.

Nutrient diffusing artificial substrates have also been used to examine responses to phosphorus additions. For example, Rier *et al.* (2014) found that both light and phosphorus had a positive impact on biofilm algae and this, in turn, influenced extracellular enzyme activity, potentially as a result of priming by labile OM. Similarly, Ylla *et al.* (2007) found inorganic phosphorus uptake by the epilithic biofilm was enhanced at high light and high nutrient conditions. Other studies have also examined nitrogen and phosphorus together; for example, Sekar *et al.* (2002) found that light levels impacted algal communities as well as the N:P ratio of biofilms. As described for planktonic habitats, phosphatase activity in biofilms typically decreases when increasing P availability and increases when P is limiting, although this depends on N:P balance (Sala *et al.*, 2001). In a long-term nutrient enrichment experiment, phosphatase activity decreased in river biofilms and, at the same time, biofilms were enriched in P content after nine months of nutrient enrichment while biofilm nitrogen enrichment was only significant after 2 years of experiment (Sabater *et al.*, 2011).

Phosphorus cycling is especially relevant in sediment biofilms. In aquatic sediment, dissolved inorganic phosphorus constitutes a small percentage of the total phosphorus which is often limiting for microorganisms growth. Thus, hydrolysis of organic phosphorus compounds by means of hydrolytic phosphatases becomes the main mechanism for microorganisms to obtain phosphorus (e.g. Scholz and Marxsen, 1996). Phosphatase activity measurement in lake sediments indicate that phosphorus cycling is enhanced by higher organic matter availability (Wang *et al.*, 2012).

Cycling of sulfur and micronutrients

Many fewer studies have examined sulfur, metals, and micronutrients in general, cycling in biofilms relative to studies on carbon, nitrogen and phosphorus. Environments where these micronutrients have been studied, include those impacted by acid mine drainage in which forms and concentrations of micronutrients such as sulfur, iron and aluminium are altered. In addition, a number of studies have focused on wastewater treatment facilities in which biofilms develop. Generally, there is little information that is specific to freshwater natural biofilms regarding sulfur and micronutrients. However, presumably there are commonalities in the biotransformations of these elements that transcend the conditions of the specific ecosystem.

Sulfur is much more widely studied in marine ecosystems, including estuaries, than in freshwater ecosystems (Odum, 1998). In freshwater systems, sulfate-reducing bacteria which produce hydrogen sulfide are present and play a potential role in acid mine bioremediation.

Oxidation coupled to reduction of manganese or iron occurs in sediments and these might be linked to microbial processes since bacteria that large use manganese or iron as the sole terminal electron acceptor have been cultivated (Thamdrup, 2000). Iron is known to be taken up by biofilms and microbes play an important role in iron speciation through oxidation-reduction reactions. Also organic compounds released by bacteria influence iron dynamics in biofilms (Duckworth *et al.*, 2009). Julien *et al.* (2014) experimentally demonstrated that iron is not homogenous in biofilms, rather it appears in 'hotspots'. Iron bacteria that oxidize ferrous salts are rather restricted in distribution based on specific environmental conditions (Wetzel, 2001). Some of these species can also oxidize manganese salts.

Measurements of biogeochemical processing in biofilms

Biogeochemical cycling in aquatic biofilms typically plays a key role in organic and inorganic nutrient cycling in the aquatic habitat where it develops. The contribution of biofilms to emergent biogeochemical ecosystem functions depends on the characteristics of the aquatic ecosystem, as well as on the relative abundance of these compartments at whole-ecosystem scale. The influence of biofilm activity on water chemistry has been demonstrated in previous studies (Woodruff *et al.*, 1999; Mulholland *et al.*, 2000; Tank *et al.*, 2000). However, how biofilm activity contributes to whole-ecosystem processes is still an ongoing question in aquatic ecosystem ecology. Scaling-up activity of biofilms developed in different habitats to whole-ecosystem scale processes is a first to empirically tackle this question, but for this it is necessary to have good characterization of the biogeochemical activity of biofilms. In this section, we review some of the techniques that are currently used to characterize the biogeochemical activity of biofilms at different scales of organization, which could be merged to increase our knowledge on the relevance of biofilms at whole-ecosystem scale (Table 5.1).

Molecular tools and functional genomics

Biogeochemical processes in biofilms are mostly driven by microorganisms; although we can learn a lot from community and ecosystem scale experiments, molecular phylogenetics and functional genomics provide very powerful tools to understand which are the key microbial players driving biogeochemical processes. Functional genes encode for enzymes, which drive the biogeochemical process. All microorganisms with the same functional gene are put together in a functional group, which is linked to a biogeochemical process. However, the gene may be present in phylogenetic similar (monophyletic, e.g. ammonia-oxidizing bacteria) or in distinct microorganisms (polyphyletic, e.g. nitrogen-fixing bacteria or denitrifying bacteria), due to evolutionary processes or horizontal gene transfer. Therefore, members of the functional group may differ significantly in transformation rates and response to physical and chemical environmental conditions. Consequently, characterization of the microbial community composition as well as abundance and activity of the particular functional group is important for prediction of activity at the whole-biofilm scale (Fernàndez-Guerra and Casamayor, 2012).

Functional genes can be tracked in the environment. For instance, both the ammonia

Table 5.1 List of methods included in this chapter to study biofilm biogeochemical processes, including their primary application and approximate spatial scale

Method	Primary applications	Approximate spatial scale
Examine biofilm spatial heterogeneity		
Microsensors: optodes	Obtain gradients in biofilm depth and spatial distribution (e.g. for O_2 and N inorganic forms)	10^{-6}–10^{-3} m
Confocal microscopy combined with functional measurements	Localize specific metabolic activity within the biofilm structure (e.g. use of fluorescent substrata to detect extracellular enzyme activity, or fluorescent *in situ* hybridization (FISH) to identify microbial groups)	10^{-6}–10^{-3} m
Consider the biofilm as a blackbox		
Molecular tools (DNA, RNA)	Examine diversity of specific functional genes; single-cell analysis (e.g. FISH techniques)	10^{-6} m
Extracellular enzyme activities	Capacity of biofilm to degrade specific organic compounds	10^{-3}–10^{-2} m
Electron transport system (ETS) activity, O_2 and CO_2 mass balances, use of hydrometabolic tracers (e.g. Resazurin)	Obtain metabolic rates (respiration and gross primary production) of biofilms	10^{-3}–10^{-1} m
Spikes of inorganic nutrients or dissolved organic carbon (DOC) in recirculating chambers or flumes.	Obtain uptake/release rates and uptake kinetics for inorganic nutrients and DOC	10^{-2}–10^{-1} m
Addition of stable isotopes of inorganic nutrients and carbon	Follows fate of nutrients and organic carbon from the water column into biofilms	10^{-1}–10^{2} m

oxidizing bacteria and archaea encode for the enzyme ammonia monooxygenase and its alpha subunit (*amoA*) can be traced in environmental samples using specific primers [e.g. for amoA of bacteria (Rotthauwe *et al.*, 1997) and archaea (Francis *et al.*, 2005)] giving insights into composition, abundance, activity of functional genes and the organisms in which this gene occurs (Zak *et al.*, 2006).

The phylogenetic composition of the functional group can be characterized via molecular fingerprinting methods. After the amplification of the functional gene, use of denaturating gradient gel electrophoresis (DGGE) or terminal restriction fragment length polymorphism (T-RFLP) allows comparing the community composition of different samples among each other, but the measurement of diversity is not possible. In contrast, the sequencing of clone libraries allows determining diversity and evenness in a sample (Muyzer *et al.*, 1993; Muyzer and Wawer., 1998). Abundance of a functional group can be assessed using quantitative polymerase chain reaction (qPCR) where a fluorescent dye attaches to the double bound DNA subsequent of each elongation step and the emitted irradiance is measured. The quantification takes place by comparing with a standard DNA from genome or clone DNA (Fierer *et al.*, 2005). The activity of a functional group is measured via quantification of the messenger RNA of the functional gene by reverse-transcriptase qPCR. Here, the messenger RNA is extracted from the environmental sample, transcribed to copy DNA using the reverse transcriptase and subsequently quantified via qPCR (e.g. Nicol *et al.*, 2008).

All these methods are useful as long as the functional group is familiar and a specific primer available. However, if this is not the case, stable isotope probing (SIP) is an approach that can uncover active microorganisms in the biofilm assemblage avoiding the need of knowledge about the community composition. In this method, biofilm is incubated with isotopic enriched substrate (e.g. ^{13}C or ^{15}N). Active organisms in the assemblage incorporate stable isotopes in their DNA. Genomic DNA of the sample is extracted and the DNA enriched in heavy isotopes is separated from the not enriched via ultra centrifugation. The part of the DNA representing the 'active' microorganisms can be analysed via the amplification of the 16S ribosomal RNA and sequencing. This approach leads to sequences that are further compared with existing data bases (e.g. GenBank) to unveil the active part of the community composition of the sample.

These DNA based approaches are highly sensitive, but also prone to overestimation because gene copy numbers are not always adaptable to cell numbers. The target gene copy number per microbial genome is sometimes unknown and the number of genome copies per microbial cell can vary in different growth phases (Ludwig and Schleifer, 2000). In contrast, cells of a functional group can be visualized within a sample with fluorescent markers (fluorescent *in situ* hybridization, FISH) and further quantified by fluorescence microscopy (Daims and Wagner, 2007). Although FISH is a time-consuming and less sensitive method than DNA approaches, it allows direct visual assessment of the cells within their structure and hence minimization of biases. FISH is of high interest when investigating biofilm architecture or interactions between functional groups. In one analysis up to seven different populations can be assessed. The FISH method has been adapted to a great variety of questions to phylogenetic assignment (Clone-FISH), activity (FISH-MAR), quantification (SPIKE-FISH) and detection in complex samples when combined with confocal laser scanning microscopy (CLSM) (CARD-FISH, DOPE-FISH) (Wagner *et al.*, 2003).

Extracellular enzymes

Biogeochemical processes in biofilms have also been approached by the measurement of extracellular enzyme activities. The microorganisms within biofilms have the capacity to decompose organic matter by producing extracellular enzymes either located outside the cells or in the periplasmatic space (in Gram-negative bacteria, Chróst, 1990), or released into the biofilm EPS matrix (Freeman and Lock, 1995; Thomson and Sinsabaugh, 2000; Romaní *et al.*, 2008). Thanks to the enzyme activity, microbes can uptake simple compounds from high molecular weight compounds usually found in the environment. Extracellular enzymes are those involved in the hydrolysis of common polysaccharides from allochthonous/autochthonous origin (alpha- and beta-glucosidase, beta-xylosidase, cellobiohydrolase), peptides (peptidase), lipids (lipases), but also include those oxidative enzymes involved in the degradation of complex compounds such as lignin (phenol oxidase, peroxidase). Biofilm enzyme activities can be further involved in the degradation of chitin (from eukaryotes, for example; chitinases), acquisition of inorganic nutrients (phosphatase), and adhesion processes (certain proteases). Extracellular enzyme activity measured in biofilms informs us about internal nutrient and organic matter cycling and/or about use and transformation of materials from the flowing water, this may depend on the biofilm thickness and the stages of biofilm formation (Romaní *et al.*, 2012).

However, classical enzyme activity measurements consider the biofilm as a black box while enzymes and enzyme activity might be heterogeneously distributed within the biofilm.

Investigations of biofilm structure with confocal laser scanning microscopy (CLSM), have shown the patchiness structure of aquatic biofilms in three dimensions as well as allow detecting structural changes with nutrient enrichments, toxic substances and light intensity (Neu *et al.*, 2005; Lawrence *et al.*, 2005; Zippel and Neu, 2005; see Chapter 2). More recently, the use of ELF (enzyme labelled fluorescence)-substrata together with CLSM has allowed measuring extracellular enzyme activity and locating where it takes place within the biofilms. The measurement of cell-enzyme activities was first developed for planktonic environments by the use of ELF-substrata that crystallize at the site of enzyme activity. This methodology has been mainly developed for the measurement of phosphatase activity of attached bacteria (Van Ommen Kloeke and Geesey, 1999) and in planktonic algae (Nedoma *et al.*, 2003; Díaz-de- Quijano *et al.*, 2014). The first studies of cell-enzyme activity in attached communities (cyanobacterial mat, activated sludge) were performed by disrupting the biofilm structure to obtain a biofilm suspension (Sirová *et al.*, 2006; Schade and Lemmer, 2006). However, the use of the undisturbed biofilm is crucial to understand its functioning (Barranguet *et al.*, 2004). More recently, intact biofilms have been used to localize extracellular enzyme activity. Tielen *et al.* (2013) incubated *Pseudomonas aeruginosa* biofilms with ELF 97 palmitate and further CLSM visualization to localize lipase activity in the biofilm and showed its interaction with alginate matrix (Tielen *et al.*, 2013). In Nielsen *et al.* (2010), a combination of methods to describe cell activity in biofilms is described, including extracellular enzyme activity by a range of ELF-substrata and BODIPY-substrata, and fluorescence *in situ* hibridzation, which has been mainly applied in activated sludge communities (Kragelund *et al.*, 2011; Xia *et al.*, 2007, 2008). Recently, a combination of these techniques has been also used to further develop a model to determine fluid dynamic processes within the biofilm (Adav *et al.*, 2010). However, most of these new techniques are mainly used in engineered systems (activated sludge, biofilms in water treatment plants), and application to natural aquatic biofilms is still under experimental phase.

Biofilm metabolism at the fine scale

Fine-scale measurements of biogeochemical processes within biofilms can be inferred from vertical profiles by using microsensors. For instance, vertical profiles of oxygen and nitrous oxide within biofilms have been used to measure denitrification in biofilms (Revsbech *et al.*, 1988; Nielsen *et al.*, 1990). Since these early studies, microelectrodes have been used in many fields and are effective and promising research tools that are able to capture *in situ* bioreactivity of natural biofilms. For instance, the monitoring of biofilm development with microsensors was able to show that nitrification assemblages developed earlier than denitrification assemblages in biofilms (Li and Bishop, 2004). The use of microsensors has also been combined with *in situ* hydridization to link the presence of nitrifying bacteria to the nitrogen metabolism detected by the microsensors (Okabe *et al.*, 1999). The results from microelectrodes have been also used to model biofilm oxygen uptake kinetics, as described in Zhou *et al.* (2009). Microprofiles of oxygen concentration in biofilms have been used to investigate the effect of different toxicants on biofilms (e.g. Zou *et al.*, 2011). In this context, Hou *et al.* (2014), showed that the surface layers of a biofilm were more affected by ZnO nanoparticles, as measured by a reduction of respiration activity, than the deeper layers.

In parallel, development of planar optodes provides a tool for mapping the spatial distribution of oxygen concentration in two dimensions (Glud *et al.*, 1996; Kühl and Polerecky, 2008; Staal *et al.*, 2011a). Planar optodes use luminescent oxygen indicators immobilized in a polymeric matrix, which is permeable to oxygen and can be fixed on foils or glass surfaces (Glud *et al.*, 1996; Oguri *et al.*, 2006). Kühl *et al.* (2007) also presented an interesting combined method by applying confocal microscopy and imaging based on sensors for microscopic oxygen measurements to show oxygen decay in a biofilm flow chamber mounted on a microscope. Staal *et al.* (2011b) linked the biofilm biomass distribution to the oxygen spatial distribution and showed oxygen microenvironments linked to changes in flow conditions.

Influence of biofilm metabolism on water column biogeochemistry

In the paragraphs above, we described methods to examine the metabolism within the biofilms and the resulting gradients. To analyse how these internal processes impact water column biogeochemistry, biofilms can be examined as a whole after being placed into either recirculating chambers or artificial flumes. This laboratory experimental setting provides high reproducibility and therefore is highly valuable for study biofilm function in response to multiple treatments (Singer *et al.*, 2006). Overarching parameters to estimate biofilm metabolic activity are biofilm respiration and primary production. Respiration has been measured by incubating natural biofilms in the experimental setting, recording oxygen consumption and inferring it from an oxygen mass balance (e.g. Guasch and Sabater, 1995). These measurements have provided insights on the effects of environmental factors on biofilm respiration (e.g. temperature; Bouletreau *et al.*, 2012). Another method to measure respiration activity of biofilms is the estimation of the electron transport system activity (ETS) by the reduction of the electron transport acceptor INT (2–3 tetrazolium chloride) to INT-formazan (iodonitrotetrazolium formazan) (Blenkinsopp and Lock, 1990).

More recently, respiration in biofilms has been estimated by using a phenoxazine compound, resazurin (Raz) as 'smart tracer' (Haggerty *et al.*, 2008, 2009). In the presence of aerobic bacteria Raz reduces irreversibly to resorufin (Rru) and, at the same time, changes maximum excitation wavelength from 602 to 570 nm, respectively. The change from Raz to Rru has been shown to be proportional to aerobic respiration and can be measured using a spectrofluorometer (McNicholl *et al.*, 2007).

Recirculating chamber and flumes also allow measurement of biofilm nutrient and organic matter demand from the adjacent water column by estimating the uptake rates of these bioreactive solutes. Basically, these measurements are based on spikes of target bioreactive solutes into the mesocosms and the further trace of solute concentrations over time in the water column (e.g. O`Brien and Dodds, 2008; Kemp and Dodds, 2002; Ribot *et al.*, 2012). In these experiments, changes over time in recirculating chambers or along space in artificial flumes are used to estimate processing metrics associated with biofilm activity based on standard approaches for nutrient uptake in streams (Webster and Valett, 2006). For instance, the decrease of added NH_4^+ over time in recirculating chambers with biofilms is assumed to follow a first-order kinetic rate; and thus, the slope provides an estimate of the rate constant of a target process associated to biofilm activity. This approach is also useful to estimate kinetics of a biofilm in response to bioreactive solutes (Kemp and Dodds, 2002; Ribot *et al.*, 2012).

The measurement of dissimilatory uptake processes associated with the cycling of N by biofilms (e.g. nitrification or denitrification) can also be approached by the combined use of key substrate for the process and specific inhibitors of the particular process. For instant, nitrapyrine, acetylene and dicyandiamid are inhibitors of nitrification which have been used in mesocosm studies to estimate the nitrification rates associated with biofilm activity by comparing rates between inhibited and control treatments (Powell *et al.*, 1986; Teissier *et al.*, 2007; Smith and Schallenberg, 2013).

In recent years, use of stable isotopes has significantly increased our ability to assess the contribution of biofilms to nutrient, especially nitrogen, cycling at whole-reach scale. Patterns of natural abundance between dissolved inorganic nitrogen in the water column and in biofilms have allowed to examine not only the transfer of dissolved inorganic nitrogen into biofilms, but also rates at which nitrogen is incorporated into biofilms (Pastor *et al.*, 2014; Peipoch *et al.*, 2014). This technique is based on the fact that the lighter and heavier isotope vary in chemical and physical behaviour and that they undergo fractionation due to a preferential use of lighter in front of heavier isotopes in several biogeochemical processes. The predominantly enzymatic driven fractionation leads to a distinct composition of isotopes in source and product and hence natural abundance of stable isotopes can be used as tracers of metabolic activities (Kendall and Caldwell, 1998; Sulzman, 2007). In addition, stable isotopes have been used as tracers, which have allowed tracing the biogeochemical fate of bioreactive solutes in the water column and how different biotic compartments use them at ambient conditions at whole-ecosystem scale (Mulholland *et al.*, 2000). Addition of isotopically enriched nutrient tracers can be used in mesocosm experiments focusing on biofilms to identify rates of different processes or identifying the assimilatory responses of biofilm uptake in front of changes in nutrient availability (Ribot *et al.*, 2013).

In addition, contribution of biofilms to whole-reach rates of both metabolism and nutrient processing can be estimated by scaling the data derived from mesocosms studies if we assume that biofilms operate similarly under experimental and *in situ* conditions (Kemp and Dodds, 2002). Alternatively, in the case of nitrogen cycling, the use of stable isotopes can be done at the whole-stream reach scale and thus particular contribution of biofilms to whole.-reach uptake rates can be evaluated. Here the water column of streams and rivers can be enriched by stable isotopes via additions of isotopes without altering the background concentrations as well as the habitat conditions for particular biotic compartments. This approach has been used to provide uptake data not only at the whole-reach scale (Peterson *et al.*, 2001; Johnson *et al.*, 2012), but also to estimate the relative contribution of different biotic compartments occurring at whole-reach scale, including biofilms (Mulholland *et al.*, 2000). Similarly, phosphate oxygen isotope ratios have been used to link phosphorus cycling and microbial activity (phosphatase activity, Stout *et al.*, 2014). Yet, a more extensive data set is needed within this field and laboratory experimental context to reach consistent conclusions about the biogeochemical relevance of biofilms at the whole reach scale.

References

Adams, H.E., Crump, B.C., and Kling, G.W. (2010). Temperature controls on aquatic bacterial production and community dynamics in arctic lakes and streams. Environ. Microbiology, *12*, 1319–1333.

Ashkenas, L.R., Johnson, S.L., Gregory, S.V., Tank, J.L., and Wollheim, W.M. (2004). A stable isotope tracer study of nitrogen uptake and transformation in an old-growth forest stream. Ecology *85*, 1725–1739.

Adav, S.S., Lin, J.C.T., Yang, Z., Whiteley, C.G., Lee, D.J., Peng, X.F., and Zhang, Z.P. (2010). Stereological assessment of extracellular polymeric substances, exo-enzymes, and specific bacterial strains in bioaggregates using fluorescence experiments. Biotechnol. Adv. *28*, 255–280.

Ardon, M., and Pringle, C.M. (2007). The quality of organic matter mediates the response of heterotrophic biofilms to phosphorus enrichment of the water column and substratum. Freshwat. Biol. *52*, 1762–1772.

Arnosti, C., Bell, C., Moorhead, D.L., Sinsabaugh, R.L., Steen, A.D., Stromberger, M., Wallenstein, M., and Weintraub, M.N. (2014). Extracellular enzymes in terrestrial, freshwater, and marine environments: perspectives on system variability and common research needs. Biogeochem. *117*, 5–21.

Artigas, J., Gaudes, A., Muñoz, I., Romaní, A.M., and Sabater, S. (2011). Fungal and bacterial colonization of submerged leaf litter in a Mediterranean stream. Int. Rev. Hydrobiol. *96*, 221- 234.

Artigas, J., García-Berthou, E., Bauer, D.E., Castro, M.I., Cochero, J., Colautti, D.C., Cortelezzi, A., Donato, J.C., Elosegi, A., Feijoó, C., *et al.* (2013). Global pressures, specific responses: effects of nutrient enrichment in streams from different biomes. Environ. Res. Let. *8*, 014002.

Atkinson, S., and Williams, P. (2009). Quorum sensing and social networking in the microbial world. J. Roy. Soc. Interface *6*, 959–978.

Barbiero, R.P. (2000). A multi-lake comparison of epilithic diatom communities on natural and artificial substrates. Hydrobiol. *438*, 157–170.

Battin, T.J., Kaplan, L.A., Newbold, J.D., and Hansen, C.M. (2003). Contributions of microbial biofilms to ecosystem processes in stream mesocosms. Nature *426*, 439–442.

Battin, T.J., Kaplan, L.A., Newbold, J.D., Cheng, X., and Hansen, C. (2003). Effects of current velocity on the nascent architecture of stream microbial biofilms. Appl. Environ. Microbiol. *69*, 5443–5452.

Battin, T.J., Sloan, W.T., Kjelleberg, S., Daims, H., Head, I.M., Curtis, T.P., and Eberl, L. (2007). Microbial landscapes: new paths to biofilm research. Nature Rev. Microbiol. *5*, 76–81.

Bergey, E.A., Cooper, J.T., and Phillips, B.C. (2010). Substrate characteristics affect colonization by the bloom-forming diatom *Didymosphenia geminata*. Aquat. Ecol. *44*, 33–40.

Besemer, K., Singer, G., Limberger, R., Chlup, A.K., Hochedlinger, G., Hödl, I., Baranyi, C., and Battin, T.J. (2007). Biophysical controls on community succession in stream biofilms. Appl. Environ. Microbiol. *73*, 4966–4974.

Besemer, K., Singer, G., Hödl, I., and Battin, T.J. (2009). Bacterial community composition of stream biofilms in spatially variable-flow environments. Appl. Environ. Microbiol. *75*, 7189–7195.

Beyenal, H., and Lewandowski, Z. (2002). Internal and external mass transfer in biofilms grown at various flow velocities. Biotechnol. Prog. *18*, 55–61.

Bissett, A., Reimer, A., de Beer, D., Shiraishi, F., and Arp, G. (2008). Metabolic microenvironmental control by photosynthetic biofilms under changing macroenvironmental temperature and pH conditions. Appl. Environ. Microbiol. *74*, 6306–6312.

Blenkinsopp, S.A., and Lock, M.A. (1990). The measurement of Electron Transport System activity in river biofilms. Wat. Res. *24*, 441–445.

Boulêtreau, S., Salvo, E., Lyautey, E., Mastrorillo, S., and Garabetian, F. (2012). Temperature dependence of denitrification in phototrophic river biofilms. Sci. Tot. Environ. *416*, 323–328.

Boyero, L., Pearson, R.G., Gessner, M.O., Barmuta, L.A., Ferreira, V., Graça, M.A.S., Dudgeon, D., Boulton, A.J., Callisto, M., Chauvet, E., *et al.* (2011). A global experiment suggests climate warming will not accelerate litter decomposition in streams but might reduce carbon sequestration. Ecology Let. *14*, 289–294.

Bull, J.J., and Harcombe, W.R. (2009). Population dynamics constrain the cooperative evolution of cross-feeding. PLoS one *4*, e4115.

Burkholder, J.M. (1996). Interactions of benthic algae with their substrata. In Algal Ecology: Freshwater Benthic Ecosystems, Lowe, R.L., Stevenson, R.J., and Bothwell, M.L., eds. (Academic Press, New York, NY, USA), pp. 253–298.

Burmølle, M., Ren, D., Bjarnsholt, T., and Sørensen, S.J. (2014). Interactions in multispecies biofilms: do they actually matter? Trends Microbiol. *22*, 84–91.

Butterwick, C., Heaney, S.I., and Talling, J.F. (2005). Diversity in the influence of temperature on the growth rates of freshwater algae, and its ecological relevance. Freshwat. Biol. *50*, 291–300.

Cardinale, B.J., Palmer, M.A., Swan, C.M., Brooks, S., and Poff, N.L. (2002). The influence of substrate heterogeneity on biofilm metabolism in a stream ecosystem. Ecology *83*, 412–422.

Chróst, R.J. (1990). Microbial Ectoenzymes in Aquatic Environments. In Aquatic Microb. Ecol.: Biochemical and Molecular Approaches, J. Overbeck and R.J. Chróst, eds. (Springer-Verlag, New York), pp. 47–78.

Corcoll, N., Bonet, B., Leira, M., Montuelle, B., Tlili, A., and Guasch, H. (2012), Light History Influences the Response of Fluvial Biofilms to Zn Exposure. J. Phycol. *48*, 1411–1423.

Costerton, J.W., Cheng, K.J., Geesey, G.G., Ladd, T.I., Nickel, J.C., Dasgupta, M., and Marrie, T.J. (1987). Bacterial biofilms in nature and disease. Ann. Rev. Microbiol. *41*, 435–464.

Coundoul, F., Bonometti, T., Graba, M., Sauvage, S., Sanchez Pérez, J.M., and Moulin, F.Y. (2014). Role of local flow conditions in river biofilm colonization and early growth. River Res. Applic. doi: 10.1002/rra.2746.

de Beer, D., and Schramm, A. (1999). Micro-environments and mass transfer phenomena in biofilms studied with microsensors. Wat. Sci. Technol. *39*, 173–178.

Degerman, R., Dinasquet, J.V., Riemann, L., Sjöstedt de Luna, S., and Andersson, A. (2013). Effect of resource availability on bacterial community responses to increased temperature. Aquat. Microb. Ecol. *68*, 131–142.

Daims, H., and Wagner, M. (2007). Quantification of uncultured microorganisms by fluorescence microscopy and digital image analysis. App. Microbiol. Biotechnol. *75*, 237–248.

Diaz-de-Quijano, D., Palacios, P., Horňák, K., and Felip, M. (2014), 3D restoration microscopy improves quantification of enzyme-labeled fluorescence-based single-cell phosphatase activity in plankton. Cytometry. doi: 10.1002/cyto.a.22486

Dodds, W.K., Evans-White, M.A., Gerlanc, N.M., Gray, L., Gudder, D.A., Kemp, M.J., López, A.L., Stagliano, D., Strauss, E.A., Tank, J.L., *et al.* (2000). Quantification of the nitrogen cycle in a prairie stream. Ecosystems *3*, 574–589.

Donlan, R.M. (2002). Biofilms: microbial life on surfaces. Emerg. Infect. Dis. *8*, 881–890.

Duan, S., Delaney-Newcomb, K., Kaushal, S.S., Findlay, S.E.G., and Belt, K.T. (2014). Potential effects of leaf litter on water quality in urban watersheds. Biogeochemistry. *121*, 61–80.

Duckworth, O.W., Holmström, S.J., Peña, J., and Sposito, G. (2009). Biogeochemistry of iron oxidation in a circumneutral freshwater habitat. Chem. Geol. *260*, 149–158.

Ensign, S.H., and Doyle, M.W. (2006). Nutrient spiraling in streams and river networks. J. Geophys. Res. Biogeosciences *111*, G04009.

Espeland, E.M., and Wetzel, R.G. (2001). Effects of photosynthesis on bacterial phosphatase production in biofilms. Microb. Ecol. *42*, 328–337.

Fernàndez-Guerra, A., and Casamayor, E.O. (2012). Habitat-associated phylogenetic community patterns of microbial ammonia oxidizers. PLoS ONE *7*, e47330.

Ferreira, V., Gulis, V., Graça, M.-A.S. (2006). Whole-stream nitrate addition affects litter decomposition and associated fungi but not invertebrates. Oecologia *149*, 718–729.

Fierer, N., Jackson, J.A., Vilgalys, R., and Jackson, R.B. (2005). Assessment of Soil Microbial Community Structure by Use of Taxon-Specific Quantitative PCR Assays. App. Environ. Microbiol. *71*, 4117–4120.

Findlay, S., and Sinsabaugh, R.L. (2003). Response of hyporheic biofilm metabolism and community structure to nitrogen amendments. Aquat. Microb. Ecol. *33*, 127–136.

Fischer, H., Sachse, A., Steinberg, C.E., and Pusch, M. (2002). Differential retention and utilization of dissolved organic carbon by bacteria in river sediments. Limnol. Oceanogr. *47*, 1702–1711.

Flemming, H.C., and Wingender, J. (2010). The biofilm matrix. Nature Rev. Microbiol. *8*, 623–633.

Francis, C.A., Roberts, K.J., Beman, J.M., Santoro, A.E., and Oakley, B.B. (2005). Ubiquity and diversity of ammonia-oxidizing archaea in water columns and sediments of the ocean. P. Nat. Acad. Sci. USA *102*, 14683–14688.

Freeman, C., Lock, M., Marxsen, J., and Jones, S.E. (1990). Inhibitory effects of high molecular weight dissolved organic matter upon metabolic processes in biofilms from contrasting rivers and streams. Freshwat. Biol. *24*, 159–166.

Freeman, C., and Lock, M.A. (1995). The biofilm polysaccharide matrix: A buffer against changing organic substrate supply? Limnol. Oceanogr. *40*, 273–278.

Frossard, A., Gerull, L., Mutz, M., and Gessner, M.O. (2013). Litter supply as a driver of microbial activity and community structure on decomposing leaves: a test in experimental streams. Appl. Environ. Microbiol. *79*, 4965–4973.

Gaudes, A., Artigas, J., Romaní, A.M., Sabater, S., and Muñoz, I. (2009). Contribution of microbial and invertebrate communities to leaf litter colonization in a Mediterranean stream. J.N. Am. Benthol. Soc. *28*, 34–43.

Ghosh, S., and Leff, L.G. (2013). Impacts of labile organic carbon concentration on organic and inorganic nitrogen utilization by a stream biofilm bacterial community. Appl. Environ. Microbiol. *79*, 7130–7141.

Golladay, S.W., and Sinsabaugh, R.L. (1991). Biofilm development on leaf and wood surfaces in a boreal river. Freshwat. Biol. *25*, 437–450.

Goller, C.C., and Romeo, T. (2008). Environmental influences on biofilm development. In Bacterial Biofilms, Romeo, T., ed. (Springer, Berlin/Heidelberg, Germany), pp. 37–66.

Grimm, N.B., and Petrone, K.C. (1997). Nitrogen fixation in a desert stream ecosystem. Biogeochem. *37*, 33–61.

Guasch, H., and Subater, S. (1995). Seasonal variations in photosynthesis-irradiance responses by biofilms in Mediterranean streams. J. Phycol. *31*, 727–735.

Gulis, V., Suberkropp, K., and Rosemond, A.D. (2008). Comparison of fungal activities on wood and leaf litter in unaltered and nutrient-enriched headwater streams. Appl. Environ. Microbiol. *74*, 1094–1101.

Hamilton, S.K., Tank, J.L., Raikow, D.F., Wollheim, W.M., Peterson, B.J., and Webster, J.R. (2001). Nitrogen uptake and transformation in a midwestern US stream: a stable isotope enrichment study. Biogeochem. *54*, 297–340.

Hancke, K., and Glud, R.N. (2004). Temperature effects on respiration and photosynthesis in three diatom-dominated benthic communities. Aquat. Microb. Ecol. *37*, 265–281.

Hansen, S.K., Rainey, P.B., Haagensen, J.A., and Molin, S. (2007). Evolution of species interactions in a biofilm community. Nature *445*, 533–536.

Hepinstall, J.A., and Fuller, R.L. (1994). Periphyton reactions to different light and nutrient levels and the response of bacteria to these manipulations. Arch. Hydrobiol. *131*, 161–173.

Hoellein, T.J., Arango, C.P., and Zak, Y. (2011). Spatial variability in nutrient concentration and biofilm nutrient limitation in an urban watershed. Biogeochem. *106*, 265–280.

Hoštacká, A., Čižnár, I., and Štefkovičová, M. (2010). Temperature and pH affect the production of bacterial biofilm. Fol. Microbiol. *55*, 75–78.

Hou, J., Miao, L., Wang, C., Wang, P., Ao, Y., Qian, J., and Dai, S. (2014). Inhibitory effects of ZnO nanoparticles on aerobic wastewater biofilmsfrom oxygen concentration profiles determined by microelectrodes. 2014. J. Haz. Mat. *276*, 164–170.

Irie, Y., and Parsek, M.R. (2008). Quorum sensing and microbial biofilms. In Bacterial Biofilms, Romeo, T., ed. (Springer, Berlin/Heidelberg, Germany), pp. 67–84.

Jackson, C.R., Churchill, P.F., and Roden, E.E. (2001). Successional changes in bacterial assemblage structure during epilithic biofilm development. Ecology *82*, 555–566.

Jackson, C.R. (2003). Changes in community properties during microbial succession. Oikos *101*, 444–448.

James, G.A., Beaudette, L., and Costerton, J.W. (1995). Interspecies bacterial interactions in biofilms. J. Indust. Microbiol. *15*, 257–262.

Johnson, L.T., Royer, T.V., Edgerton, J.M., and Leff, L.G. (2012). Manipulation of the dissolved organic carbon pool in an agricultural stream: responses in microbial community structure, denitrification, and assimilatory nitrogen uptake. Ecosystems *15*, 1027–1038.

Jones, S.E., and Lock, M. (1989). Hydrolytic extracellular enzyme activity in heterotrophic biofilms from two contrasting streams. Freshwat. Biol. *22*, 289–296.

Julien, C., Laurent, E., Legube, B., Thomassin, J.H., Mondamert, L., and Labanowski, J. (2014). Investigation on the iron-uptake by natural biofilms. Wat. Res. *50*, 212–220.

Kemp, M.J., and Dodds, W.K. (2002). The influence of ammonium, nitrate, and dissolved oxygen concentrations on uptake, nitrification, and denitrification rates associated with prairie stream substrata. Limnol. Oceanogr. *47*, 1380–1393.

Khatoon, H., Yusoff, F., Banerjee, S., Shariff, M., and Bujang, J.S. (2007). Formation of periphyton biofilm and subsequent biofouling on different substrates in nutrient enriched brackishwater shrimp ponds. Aquaculture *273*, 470–477.

Kolenbrander, P.E. (2000). Oral microbial communities: biofilms, interactions, and genetic systems 1. Ann. Rev. Microbiol. *54*, 413–437.

Kragelund, C., Thomsen, T.R., Mielczarek, A.T., and Nielsen, P.H. (2011). Eikelboom's morphotype 0803 inactivated sludge belongs to the genus Caldilinea in the phylum Chloroflexi. FEMS Microbiol. Ecol. *76*, 451–462.

Kühl, M., Rickelt, L.F., and Thar, R. (2007). Combined Imaging of Bacteria and Oxygen in Biofilm. Appl. Environ. Microbiol. *73*, 6289–6295.

Lawrence, J.R., Swerhone, G.D.W., Wassenaar, L.I., and Neu, T.R. (2005). Effects of selected pharmaceuticals on riverine biofilm communities. Can. J. Microbiol. *51*, 655–669.

Li, J., and Bishop, P.L. (2004). Time course observations of nitrifying biofilm development using microelectrodes. J. Environ. Eng. Sci. *3*, 523–528.

Lock, M.A., Wallace, R.R., Costerton, J.W., Ventullo, R.M., and Charlton, S.E. (1984). River epilithon: toward a structural-functional model. Oikos *42*, 10–22.

Ludwig, W., and Schleifer, K.-H. (2000). How quantitative is quantitative PCR with respect to cell counts? System. Appl. Microbiol. *23*, 556–562.

Lyautey, E., Jackson, C.R., Cayrou, J., Rols, J.L., and Garabétian, F. (2005). Bacterial community succession in natural river biofilm assemblages. Microb. Ecol. *50*, 589–601.

Marshall, K.C. (1989). Cyanobacterial-heterotrophic bacterial interactions. In Microbial Mats: Physiological Ecology of Benthic Microbial Communities, Cohen, Y., and Rosenberg, E., eds. (ASM Press, Washington, DC, USA), pp. 239–245.

McNamara, C.J., and Leff, L.G. (2004). Response of biofilm bacteria to dissolved organic matter from decomposing maple leaves. Microb. Ecology *48*, 324–330.

Merbt, S.N., Auguet, J.-C., Casamayor, E.O., and Martı, E. (2011). Biofilm recovery in a wastewater treatment plant-influenced stream and spatial segregation of ammonia-oxidizing microbial populations. Limnol. Oceanogr. *56*, 1054–1064.

Merriam, J.L., McDowell, W.H., Tank, J.L., Wollheim, W.M., Crenshaw, C.L., and Johnson, S.L. (2002). Characterizing nitrogen dynamics, retention and transport in a tropical rainforest stream using an *in situ* ^{15}N addition. Freshwat. Biol. *47*, 143–160.

Meyer-Reil, L. (1994). Microbial life in sedimentary biofilms – The challenge to microbial ecologists. Mar. Ecol. Prog. Ser. Oldendorf, *112*, 303–311.

Münster, U., and Chróst, R.J. (1990). Origin, composition, and microbial utilization of dissolved organic matter. In Aquatic Microbial Ecology, Overbeck, J., and Chröst, R.J., eds. (Springer, New York, NY, USA), pp. 8–46.

Mulholland, P.J., Tank, J.L., Sanzone, D.M., Wollheim, W.M., Peterson, B.J., Webster, J.R., and Meyer, J.L. (2000). Nitrogen cycling in a forest stream determined by a ^{15}N tracer addition. Ecol. Mon. *70*, 471–493.

Muyzer, G., De Waal, E.C., and Uitterlinden, A.G. (1993). Profiling of complex microbial populations by denaturing gradient gel electrophoresis analysis of polymerase chain reaction-amplified genes coding for 16S rRNA. Appl. Environ. Microbiol. *59*, 695–700.

Muyzer, G., Brinkhoff, T., Nübel, U., Santegoeds, C., Schäfer, H., and Wawer, C. (1998). Denaturing gradient gel electrophoresis (DGGE) in microbial ecology. In Molecular Microbial Ecology Manual, Akkermans, A.D.L., van Elsas, J.D., and de Bruijn, F.J., eds. (Kluwer Academic Publishers, Dordrecht), pp. 1–27.

Nedoma, J., Strojsova, A., Vrba, J., Komárková, J., and Simek, K. (2003). Extracellular phosphatase activity of natural plankton studied with ELF97 phosphate: fluorescence quantification and labelling kinetics. Environ. Microbiol. *5*, 462–472.

Neu, T.R., Swerhone, G.D.W., Bockelmann, U., and Lawrence, J.R. (2005). Effect of CNP on composition and structure of lotic biofilms as detected with lectin-specific glycoconjugates. Aquat. Microb. Ecol. *38*, 283–294.

Nicol, G.W., Sven, L., Christa, S., and James, I.P. (2008). The influence of soil pH on the diversity, abundance and transcriptional activity of ammonia oxidizing archaea and bacteria. Environ. Microbiol. *10*, 2966–2978.

Nielsen, L.P., Christensen, P.B., Revsbech, N.P., and Sorensen, J. (1990). Denitrification and oxygen respiration in biofilms studied with a microsensor for nitrous oxide and oxygen. Microb. Ecol. *19*, 63–72.

Nielsen, J.L., Kragelund, C., and Nielsen, P.H. (2010). Ecophysiological Analysis of microorganisms in complex microbial systems by combination of fluorescence in situ hybridization with extracellular staining techniques. In Bioremediation: Methods and Protocols, Cummings, S.P., ed. (Humana Press, New York, NY, USA), pp. 117–128.

O'Brien, J.M., and Dodds, W.K. (2008). Ammonium uptake and mineralization in prairie streams: chamber incubation and short-term nutrient addition experiments. Freshwat. Biol. *53*, 102–112.

Odum, W.E. (1988). Comparative ecology of tidal freshwater and salt marshes. Ann. Rev. Ecol. Syst. *19*, 147–176.

Okabe, S., Satoh, H., and Watanabe, Y. (1999). In situ analysis of nitrifying biofilms as determined by in situ hybridization and the use of microelectrodes. Appl. Environ. Microbiol. *65*, 3182–3191.

Oguri, K., Kitazato, H., and Glud, R.N. (2006). Platinum octaetylporphyrin based planar optodes combined with an UV-LED excitation light source: an ideal tool for highresolution O_2 imaging in O_2 depleted environments. Mar. Chem. *100*, 95–107.

Okabe, S., Kindaichi, T., and Ito, T. (2005). Fate of ^{14}C-labeled microbial products derived from nitrifying bacteria in autotrophic nitrifying biofilms. Appl. Environ. Microbiology *71*, 3987–3994.

Olapade, O.A., and Leff, L.G. (2005). Seasonal response of stream biofilm communities to dissolved organic matter and nutrient enrichments. Appl. Environ. Microbiol. *71*, 2278–2287.

Olapade, O.A., and Leff, L.G. (2006). Influence of dissolved organic matter and inorganic nutrients on the biofilm bacterial community on artificial substrates in a northeastern Ohio, USA, stream. Can. J. Microbiol. *52*, 540–549.

Oviedo-Vargas, D., Royer, T.V., and Johnson, L.T. (2013). Dissolved organic carbon manipulation reveals coupled cycling of carbon, nitrogen, and phosphorus in a nitrogen-rich stream. Limnol. Oceanogr. *58*, 1196–1206.

Palmer, J., Flint, S., and Brooks, J. (2007). Bacterial cell attachment, the beginning of a biofilm. J. Ind. Microbiol. Biotechnol. *34*, 577–588.

Pande, S., Merker, H., Bohl, K., Reichelt, M., Schuster, S., de Figueiredo, L.F., Kaleta, C., and Kos, C. (2013). Fitness and stability of obligate cross-feeding interactions that emerge upon gene loss in bacteria. ISME J. *8*, 953–962.

Pastor, A., Peipoch, M., Cañas, L., Chappuis, E., Ribot, M., Gacia, E.,Riera, J.L., Martí, E., and Sabater, F. (2013). Nitrogen stable isotopes in primary uptake compartments across streams differing in nutrient availability. Environ. Sci. Technol. *47*, 10155–10162.

Pastor, A., Compson, Z.G., Dijkstra, P., Riera, J.L., Martí, E., Sabater, F., Hungate, B.A., and Marks, J.C. (2014). Stream carbon and nitrogen supplements during litter decomposition: contrasting patterns for two foundation species. Oecologia *176*, 1111–1121.

Peipoch, M., Gacia, E., Blesa, A., Ribot, M., Riera, J.L., and Martí, E. (2014). Contrasts among macrophyte riparian species in their use of stream water nitrate and ammonium: insights from ^{15}N natural abundance. Aquat. Sci. *76*, 203–215.

Pfeiffer, T., and Bonhoeffer, S. (2004). Evolution of cross-feeding in microbial populations. Am. Nat. *163*, E126-E135.

Polerecky, L., Franke, U., Werner, U., Grunwald, B., and de Beer, D. (2005). High spatial resolution measurement of oxygen consumption rates in permeable sediments. Limnol. Oceanogr. Meth. *3*, 75–85.

Poltak, S.R., and Cooper, V.S. (2010). Ecological succession in long-term experimentally evolved biofilms produces synergistic communities. ISME J. *5*, 369–378.

Price, K.J., and Carrick, H.J. (2014). Quantitative evaluation of spatiotemporal phosphorus fluxes in stream biofilms. Freshwat. Sci. *33*, 99–111.

Renslow, R.S., Majors, P.D., McLean, J.S., Fredrickson, J.K., Ahmed, B., and Beyenal, H. (2010). In situ effective diffusion coefficient profiles in live biofilms using pulsed-field gradient nuclear magnetic resonance. Biotechnol. Bioeng. *106*, 928–937.

Revsbech, N.P., Nielsen, L.P., Christensen, P.B., and Sorensen, J. (1988). Combined oxygen and nitrous oxide microsensor for denitrification studies. Appl. Environ. Microbiol. *54*, 2245–2249.

Ribot, M., von Schiller, D., Peipoch, M., Sabater, F., Grimm, N.B., and Martí, E. (2013). Influence of nitrate and ammonium availability on uptake kinetics of stream biofilms. Freshwat. Sci. *32*, 1155–1167.

Rickard, A.H., Gilbert, P., High, N.J., Kolenbrander, P.E., and Handley, P.S. (2003). Bacterial coaggregation: an integral process in the development of multi-species biofilms. Trends Microbiol. *11*, 94–100.

Riedel, K., Hentzer, M., Geisenberger, O., Huber, B., Steidle, A., Wu, H., Høiby, N., Givskov, M., Molin, S., and Eberl, L. (2001). N-acylhomoserine-lactone-mediated communication between *Pseudomonas aeruginosa* and *Burkholderia cepacia* in mixed biofilms. Microbiol. *147*, 3249–3262.

Rier, S.T., Kuehn, K.A., and Francoeur, S.N. (2007). Algal regulation of extracellular enzyme activity in stream microbial communities associated with inert substrata and detritus. J.N. Am. Benthol. Soc. *26*, 439–449.

Rier, S.T., Shirvinski, J.M., and Kinek, K.C. (2014). In situ light and phosphorus manipulations reveal potential role of biofilm algae in enhancing enzyme-mediated decomposition of organic matter in streams. Freshwat. Biol. *59*, 1039–1051.

Romani, A.M., and Sabater, S. (2000). Influence of algal biomass on extracellular enzyme activity in river biofilms. Microb. Ecol. *40*, 16–24.

Romaní, A.M., Giorgi, A., Acuña, V., and Sabater, S. (2004a) The influence of substratum type and nutrient supply on biofilm organic matter utilization in streams. Limnol. Oceanogr. *49*, 1713–1721.

Romaní, A.M., Guasch, H., Muñoz, I., Ruana, J., Vilalta, E., Schwartz, T., Emtiazi, F., and Sabater, S. (2004b). Biofilm structure and function and possible implications for riverine DOC dynamics. Microb. Ecol. *47*, 316–328.

Romaní, A.M., Fischer, H., Mille-Lindblom, C., andTranvik, L.J. (2006). Interactions of bacteria and fungi on decomposing litter: Differential extracellular enzyme activities. Ecology *87*, 2559–2569.

Romaní, A.M., Fund, K., Artigas, J., Schwartz, T., Sabater, S., and Obst, U. (2008). Relevance of polymeric matrix enzymes during biofilm formation. Microb. Ecol. *56*, 427–436.

Romaní, A.M., Artigas, J., and Ylla, I. (2012). Extracellular enzymes in aquatic biofilms: microbial interactions vs water quality effects in the use of organic matter. In Microbial Biofilms: Current Research and Applications, Lear, G., and Lewis, G., eds. (Caister Academic Press, Norwich, UK), pp. 153–174.

Romaní, A.M., Amalfitano, S., Artigas, J., Fazi, S., Sabater, S., Timoner, X., Ylla, I., and Zoppini, A. (2013). Microbial biofilm structure and organic matter use in mediterranean streams. Hydrobiol. *719*, 43–58.

Rotthauwe, J.-H., Witzel, K.-P., and Liesack, W. (1997). The ammonia monooxygenase structural gene *amoA* as a functional marker: molecular fine-scale analysis of natural ammonia oxidizing populations. Appl. Environ. Microbiol. *63*, 4704–4712.

Rubin, M.A., and Leff, L.G. (2007). Nutrients and other abiotic factors affecting bacterial communities in an Ohio River (USA). Microb. Ecol. *54*, 374–383.

Sala, M.M., Karner, M., Arin, L., and Marrasé, C. (2001). Measurement of ectoenzyme activities as an indication of inorganic nutrient imbalance in microbial communities. Aquat. Microb. Ecol. *23*, 310–311.

Sabater, S., Artigas, J., Romaní, A.M., Gaudes, A., Muñoz, I., and Urrea, G. (2011). Long-term moderate nutrient inputs enhance autotrophy in a forested Mediterranean stream. Freshwat. Biol. *56*, 1266–1280.

Schade, M., and Lemmer, H. (2006). In situ enzyme activities of filamentous scum bacteria in municipal activated sludge wastewater treatment plants. Act. Hydrochim. Hydrobiol. *34*, 480–490.

Scholz, O., and Marxsen, J. (1996). Sediment phosphatases of the Breitenbach, a first-order Central European stream. Arch. Hydrobiol. *135*, 433–450.

Scott, J.T., Back, J.A., Taylor, J.M., and King, R.S. (2008). Does nutrient enrichment decouple algal-bacterial production in periphyton? J.N. Am. Benthol. Soc. *27*, 332–344.

Scott, J.T., Lang, D.A., King, R.S., and Doyle, R.D. (2009). Nitrogen fixation and phosphatase activity in periphyton growing on nutrient diffusing substrata: evidence for differential nutrient limitation in stream periphyton. J.N. Am. Benthol. Soc. *28*, 57–68.

Sekar, R., Nair, K.V.K., Rao, V.N.R., and Venugopalan, V.P. (2002). Nutrient dynamics and successional changes in a lentic freshwater biofilm. Freshwat. Biol. *47*, 1893–1907.

Simon, K., Townsend, C., Biggs, B.J.F., and Bowden, W. (2005). Temporal variation of N and P uptake in 2 New Zealand streams. J. N. Am. Benthol. Soc. *24*, 1–18.

Singer, G., Besemer, K., Hödl, I., Chlup, A., Hochedlinger, G., Stadler, P., and Battin, T.J. (2006). Microcosm design and evaluation to study stream microbial biofilms. Limnol. Oceanogr. Meth. *4*, 436–447.

Singer, G., Besemer, K., Schmitt-Kopplin, P., Hödl, I., and Battin, T.J. (2010). Physical heterogeneity increases biofilm resource use and its molecular diversity in stream mesocosms. Plos One *5*, e9988.

Sinsabaugh, R.L., Golladay, S.W., and Linkins, A.E. (1991). Comparison of epilithic and epixylic biofilm development in a boreal river. Freshwat. Biol. *25*, 179–187.

Sinsabaugh, R.L., Van Horn, D.J., Shah, J.J.F., and Findlay, S. (2010). Ecoenzymatic stoichiometry in relation to productivity for freshwater biofilm and plankton communities. Microb. Ecology *60*, 885–893.

Sirová, D., Vrba, J., and Rejmánková, E. (2006). Extracellular enzyme activities in benthic cyanobacterial mats: comparison between nutrient-enriched and control sites in marshes of northern Belize. Aquat. Microb. Ecol. *44*, 11–20.

Sobczak, W.V. (1996). Epilithic bacterial responses to variations in algal biomass and labile dissolved organic carbon during biofilm colonization. J. N. Am. Benthol. Soc. *15*, 143–154.

Sobota, D.J., Johnson, S.L., Gregory, S.V., and Ashkenas, L.R. (2012). A stable isotope tracer study of the influences of adjacent land use and riparian condition on fates of nitrate in streams. Ecosystems *15*, 1–17.

Staal, M., Borisov, S.M., Rickelt, L.F., Klimant, I., and Kühl, M. (2011a). Ultrabright planar optodes for luminescence life-time based microscopic imaging of O_2 dynamics in biofilms. J. Microbiol. Meth. *85*, 67–74.

Staal, M., Prest, E.I., Vrouwenvelder, J.S., Rickelt, L.F., and Kühl, M. (2011b). A simple optode based method for imaging O_2 distribution and dynamics in tap water biofilms. Wat. Res. *45*, 5027–5037.

Stelzer, R.S., Heffernan, J., and Likens, G.E. (2003). The influence of dissolved nutrients and particulate organic matter quality on microbial respiration and biomass in a forest stream. Freshwat. Biol. *48*, 1925–1937.

Stewart, P.S., and Franklin, M.J. (2008). Physiological heterogeneity in biofilms. Nature Rev. Microbiol. *6*, 199–210.

Stewart, P.S. (2003). Diffusion in biofilms. Journal of bacteriology, *185*, 1485–1491.

Sutherland, I.W. (2001). The biofilm matrix–an immobilized but dynamic microbial environment. Trends Microbiol. *9*, 222–227.

Stout, L.M., Joshi, S.R., Kana, T.M., and Jaisi, D.P. (2014). Microbial activities and phosphorus cycling: an application of oxygen isotope ratios in phosphate. Geochim. Cosmochim. Acta *138*, 101–116.

Taherzadeh, D., Picioreanu, C., and Horn, H. (2012). Mass transfer enhancement in moving biofilm structures. Biophys. J. *102*, 1483–1492.

Tank, J.L., Meyer, J.L., Sanzone, D.M., Mulholland, P.J., Webster, J.R., Peterson, B.J., Wollheim, W.M., and Leonard, N.E. (2000). Analysis of nitrogen cycling in a forest stream during autumn using a ^{15}N-tracer addition. Limnol. Oceanogr. *45*, 1013–1029.

Tank, J.L., and Webster, J.R. (1998). Interaction of substrate and nutrient availability on wood biofilm processes in streams. Ecology *79*, 2168–2179.

Tank, J.L., and Dodds, W.K. (2003). Nutrient limitation of epilithic and epixylic biofilms in ten North American streams. Freshwat. Biol. *48*, 1031–1049.

Teissier, S., Torre, M., Delmas, F., and Garabétian, F. (2007). Detailing biogeochemical N budgets in riverine epilithic biofilms. J. N. Am. Benthol. Soc. *26*, 178–190.

Thamdrup, B. (2000). Bacterial manganese and iron reduction in aquatic sediments. In Advances in Microb. Ecol., Nelson, K.E., ed. (Springer, Berlin, Germany), pp. 41–84.

Thompson, A.J., and Sinsabaugh, R.L. (2000). Matric and particulate phosphatase and aminopeptidase activity in limnetic biofilms. Aquat. Microb. Ecol. *21*, 151–159.

Valle, A., Bailey, M.J., Whiteley, A.S., and Manefield, M. (2004). N-acyl-L-homoserine lactones (AHLs) affect microbial community composition and function in activated sludge. Environ. Microbiol. *6*, 424–433.

Van Ommen Kloeke, F., and Geesey, G.G. (1999). Localization and identification of populations of phosphatase-active bacterial cells associated with activated sludge flocs. Microb. Ecol. *38*, 201–214.

Von Schiller, D., Martí, E., Riera, J.L., and Sabater, F. (2007). Effects of nutrients and light on periphyton biomass and nitrogen uptake in Mediterranean streams with contrasting land uses. Freshwat. Biol. *52*, 891–906.

Von Schiller, D., Marti, E., and Riera, J.L. (2009). Nitrate retention and removal in Mediterranean streams bordered by contrasting land uses: a ^{15}N tracer study. Biogeosciences *6*, 181–196.

Wagner, M., Horn, M., and Daims, H. (2003). Fluorescence *in situ* hybridisation for the identification and characterisation of prokaryotes. Cur. Opin. Microbiol. *6*, 302–309.

Wang, S., Jiao, L.X., Yang, S., Jin, X., and Yi, W. (2012). Effects of organic matter and submerged macrophytes on variations of alkaline phosphatase activity and phosphorus fractions in lake sediment. J. Environ. Manag. *113*, 355–360.

Webster, J., and Valett, H. (2006). Solute dynamics. In Methods in Stream Ecology, Hauer, F.R., and Lamberti, G.A.E., eds. (Academic Press Inc., San Diego, CA, USA), pp. 169–185.

Wetzel, R.G. (2001). Limnology: Lake and River Ecosystems. (Academic Press Inc., San Diego, CA, USA).

Whitehead, K.A., and Verran, J. (2009). The effect of substratum properties on the survival of attached microorganisms on inert surfaces. In Marine and Industrial Biofouling, Flemming, H.-C., Murthy, P.S., Venkatesan, R., and Cooksey, K.E., eds (Springer, Berlin, Germany), pp. 13–33.

Wimpenny, J., Manz, W., and Szewzyk, U. (2000). Heterogeneity in biofilms. FEMS Microbiol. Rev. *24*, 661–671.

Wintermute, E.H., and Silver, P.A. (2010). Dynamics in the mixed microbial concourse. Gen. Develop. *24*, 2603–2614.

Woodruff, S.L., House, W.A., Callow, M.E., and Leadbeater, B.S.C. (1999). The effects of biofilms on chemical processes in surficial sediments. Freshwat. Biol. *41*, 73–89.

Xia, Y., Kong, Y., and Nielsen, P.H. (2007). In situ detection of protein-hydrolysing microorganisms in activated sludge. FEMS Microbiol. Ecol. *66*, 462–471.

Xia, Y., Kong, Y., and Nielsen, P.H. (2008). In situ detection of starch-hydrolyzing microorganisms in activated sludge. FEMS Microbiol. Ecol. *60*, 156–165.

Ylla, I., Romaní, A.M., and Sabater, S. (2007). Differential effects of nutrients and light on the primary production of stream algae and mosses. Fund. Appl. Limnol./Arch. Hydrobiol. *170*, 1–10.

Ylla, I., Borrego, C., Romaní, A.M., and Sabater, S. (2009). Availability of glucose and light modulates the structure and function of a microbial biofilm. FEMS Microbiol. Ecol. *69*, 27–42.

Ylla, I., Romaní, A.M., and Sabater, S. (2012). Labile and recalcitrant organic matter utilization by river biofilm under increasing water temperature. Microb. Ecol. *64*, 593–604.

Ylla, I., Canhoto, C., and and Romaní, A.M. (2014). Effects of warming on stream biofilm organic matter use capabilities. Microb. Ecology *68*, 132–145.

Zhou, X.H., Qiu, Y.Q., Shi, H.C., Yu, T., He, M., and Cai, Q. (2009). A New Approach to Quantify Spatial Distribution of Biofilm Kinetic Parameters by In Situ Determination of Oxygen Uptake Rate (OUR). Environ. Sci. Technol. *43*, 757–763.

Zhou, X.H., Yu, T., Shi, H.C., and Shi, H.M. (2001). Temporal and spatial inhibitory effects of zinc and copper on wastewater biofilms from oxygen concentration profiles determined by microelectrodes, Water Res. *45*, 953–959.

Zippel, B., and Neu, T.R. (2005). Growth and structure off phototrophic biofilms under controlled light conditions. Water Sci. Technol. *52*, 203–209.

Zak, D.R., Blackwood, C.B., and Waldrop, M.P. (2006). A molecular dawn for biogeochemistry. Trends Ecol. Evol. *21*, 288–295.

Part II

Biofilms and Pollution

Benthic Diatom Monitoring and Assessment of Freshwater Environments: Standard Methods and Future Challenges

6

Soizic Morin, Nora Gómez, Elisabet Tornés, Magdalena Licursi and Juliette Rosebery

Abstract

Since biofilms integrate the environmental effects of water chemistry, along with the physical and geomorphological characteristics of rivers and lakes, they have been widely applied in biomonitoring. In particular, diatoms are extensively used as reliable environmental indicators. Diatoms are microscopic, unicellular brown algae, which often dominate the algal biomass of biofilms. The shape and morphology of the siliceous skeleton, the frustules, unique to each taxon are used for taxonomical identification. The floras are diverse, in relation to their geographical location (climate, geology, relief) and to the quality of the aquatic environments they inhabit. Indeed, species are sensitive to the water physicochemical parameters and their presence/abundance is therefore correlated to water quality.

Diatom sensitivity or tolerance towards different environmental parameters has long been studied and used to implement bioassessment methods. Such methods evolved from indices of saprobity designed first for European streams, to developments of various diatom indicators worldwide, able to highlight different types of pollution (pH, salinity, nutrients, toxicants).

The objective of this chapter is to provide scientists and water managers with a broad overview of diatom tools helpful to monitor the ecological status of freshwater environments. We describe the applicability range and the limitations of the main existing methods, metrics (indices, traits) and types of surveys used, as well as the challenges faced by scientists to improve routine biomonitoring.

Introduction

Various biological metrics have been developed to assess the health of freshwater ecosystems. They reflect integrative effects of present and past conditions, whereas traditional chemical and physical measurements only apply to the current situation. Several studies have shown that community-based indicators are needed to evaluate water quality. Diatom biomonitoring has benefited, over other existing periphytic bioindicators, from a long research history. Indeed, typical water quality assessment in flowing watercourses is mainly based on benthic diatoms, i.e. diatoms growing on diverse natural (pebbles, macrophytes) or artificial (glass slides, ceramic tiles) substrates immerged in the water. However, biofilms are composed of diverse components other than diatoms. Although the planktonic microflora is used widely as biological quality element (especially in lacustrine environments), benthic green algae and cyanobacteria are poorly known and no consensual bioindicator exists.

Diatoms are a siliceous class of unicellular algae known to be very diverse (Guiry, 2012) and sensitive to chemical conditions. They are excellent bioindicators due to their short generation time and their varying ecological preferences. Moreover, the structural elements in their siliceous cell walls allow reliable taxonomic determination at specific and subspecific levels. They are distributed throughout the world in nearly all types of aquatic systems and usually account for the highest number of species among the primary producers in aquatic systems. They respond to environmental disturbances not only at the community level through changes in diversity, but also by shifts in dominant taxa. Over the last decades, a great number of diatom-based methods (reviewed in Prygiel *et al.*, 1999) have been proposed to assess water quality. No one single metric is applied worldwide, but most of the diatom indices used are minor adjustments of a common approach based on the knowledge of the ecological spectrum of species, combining sensitivity to pollution and indicator value.

This chapter aims to review the main characteristics of diatoms which have been used to develop bioassessment methods in freshwater environments. Various indicators are described, with respect to the types of pollution that can be addressed, and to the kinds of approach implemented. The pros and cons of existing systems, as well as future challenges in biomonitoring are presented.

Potential of diatoms for water quality assessment

Diatoms (Bacillariophyceae) are particularly interesting as indicators of water quality (Table 6.1). These unicellular brown microalgae (size ranging from < 10 to > 500 µm) occur in all aquatic environments and are found at almost all levels of pollution. Diatoms can account for up to 80% of the taxa present in streams, rivers, lakes and wetlands (McIntire *et al.*, 1996). They are used worldwide as biological markers to assess water quality: in Europe to support the implementation of the Water Framework Directive (Kelly *et al.*, 1998), but also in routine monitoring surveys in Canada, USA, Japan, Australia, South America, etc. (e.g. Lobo *et al.*, 2004; Gómez and Licursi, 2001; UNESCO-WHO-UNEP, 1996).

Diatoms occupy a diverse range of habitats: they can be planktonic (freely living in the water column) or fixed on diverse substrata (periphytic on hard surfaces, epipsammic on sand, epiphytic on macrophytes, see Chapter 1). Periphytic diatoms are relatively sedentary, and live their whole life cycle in the water implying that their community structure is tightly associated with local environmental conditions at the sample site. Thus, diatoms are generally collected from hard surfaces immerged (natural or introduced, like glass slides or ceramic tiles; e.g. Sekar *et al.*, 2004). Moreover, their sampling requires minimal effort (scraping or pipetting the surface, using corers), causes minimal disturbance to the sampling sites and it is possible at sites where other bioindicators, e.g. benthic invertebrates and fish are absent (Lear *et al.*, 2012).

Besides being easy to sample, diatoms have the great advantage that their frustules provide detailed features enabling reliable identification at species and subspecies levels. These siliceous walls are easily preserved and cleaned of organic material (with hydrogen peroxide, hydrochloric acid; AFNOR, 2003), to be permanently mounted on microscope slides in refringent resin for identification under light microscopy at high magnification (x1000), with oil immersion.

On the other hand, microscopic determination of diatom taxa requires precise identification than can only be reached by trained operators. Among the main diatom features to be taken into account when doing microscopic identification are: the general shape of the valve (the kind of symmetry), cell dimensions, length-to-breadth ratio, and ornamentations (presence or absence of one or two raphes, i.e. longitudinal slits on the valve; orientation of the striae; presence of specific features; stria density).

Diatom community structure reflects a gradient of general pollution in water quality (although causes are not always identifiable) integrated over time on a relatively short timescale, in relation to high sensitivity of individual species towards different levels of pollutions (Lowe and Pan, 1996; McCormick and Cairns, 1994; Stevenson and Pan, 1999; Whitton and Rott, 1996). In order to determine species sensitivity, the methodology most frequently used is a direct comparison between the physicochemical water quality and the species composition of the corresponding records in relative abundances of the taxa using appropriate multivariate analyses. Results are generally presented as species optima towards

Table 6.1 Advantages and drawbacks of diatoms as bioindicators of water quality

Advantages	Drawbacks
Sampling	
Biofilm presence generally visible to the naked eye on the substrata	
Quick and easy collection (scraping, pipetting, using corers for soft sediments and sand)	
Possible use of artificial substrates	Risks of vandalism or loss (e.g. floods) of artificial substrates
Taxon identification	
Numerous identification resources (books, articles, web)	Difficult and ever-changing systematics
	Time-consuming counting/identification of samples
	High quality microscope necessary
Bioassessment power	
Widespread distribution, even in hostile environments	Geographically specific distributions (endemism)
High species diversity (about 10 000 species known)	Heterogeneous biomass distribution (e.g. light dependent)
Sensitive to numerous kinds of pollution	Poorly sensitive to habitat disturbances
Conservation of the frustules (paleolimnological applications) and integrative power variable depending on the species	Integrative power lower than for higher organisms
Index calculation	
Dedicated software (e.g. Omnidia)	Applicability of indices generated for other geographic regions

different environmental parameters (Charles *et al.*, 2006; Ponader *et al.*, 2008), or species ecological profiles expressed as species probability of presence along ecological gradients (Potapova *et al.*, 2004).

Once the polluo-sensitivity or resistance of species to defined classes of quality are hierarchized (e.g. Van Dam *et al.*, 1994), species can be grouped and used as indicators for saprobic conditions (Sládeček, 1986), salinity (Ziemann, 1991), acidification (Birks *et al.*, 1990) and eutrophication in lakes and rivers (Steinberg and Schiefele, 1988; Hofmann, 1994). The relative abundances of the different species identified are reported and used, e.g. for calculation of water quality indices. Depending on the methods, a minimum of 400 individuals identified is required for reliable assessment (AFNOR 2000). Diatomists have created practical tools such as the software Ominida (Lecointe *et al.*, 1993). This software allows to efficiently compute diatom inventories from research and monitoring programmes, manage data and calculate indices.

Towards a harmonized way of using diatoms for biomonitoring

The need to monitor water quality led to the development of standardized sampling protocols and assessment methods, through single, simplified indices. Such indices were created by adapting the formula of Zelinka and Marvan (1961), basically combining the abundances of species and their individual ecological preferences, into a single score of water quality. Among them, DAipo (Diatom Assemblage Index to organic pollution; Watanabe *et al.*, 1986), IBD (Indice Biologique Diatomées, Coste *et al.*, 2009), IDEC (Indice Diatomées de l'Est du Canada; Lavoie *et al.*, 2006), IPS (Indice de Polluosensibilité Spécifique; Coste *in* Cemagref,

1982), TDI (Trophic Diatom Index; Kelly and Whitton, 1995), and IDP (Pampean Diatom Index; Gómez and Licursi, 2001) are successfully used in many monitoring programmes.

Recently, limitations in these approaches have been identified, linked to the importance of reliable autoecological taxon information and to the potential general application of indices developed in a particular geographical area (e.g. application of European indices to USA; Potapova and Charles, 2007). To overcome critical limitations concerning pertinent biomonitoring irrespective of regional differences in taxa, climate and other local constraints, approaches such as the use of a common metric for the different European countries (intercalibration exercises, Kelly *et al.*, 2009; Almeida *et al.*, 2014) have been developed.

Diatom assemblages respond rapidly (weeks) and sensitively to environmental change and provide highly informative assessment of the biotic integrity of aquatic ecosystems. To limit the biases linked to regional distribution of species, different characteristics can also be used in the monitoring such as taxa richness, diversity, biomass, autoecology of individual species, biotic indices, percentages of aberrant diatoms, percentages of motile diatoms, mortality, among others descriptors (Stevenson and Bahls, 1999). Other approaches are complementarily employed, based on non-taxonomical indicators, assuming that a given pressure selects for certain characteristics, whatever the location/site studied. The diagnosis therefore relies on the proportion of ecological characteristics derived from species distributions and on their classification in terms of ecological preferences (as described previously) or 'traits' (e.g. postures, growth forms, motility, or the ecological guilds defined by Passy, 2007).

Indices have been developed mainly to assess organic pollution, eutrophication or pH. But they fail when there is low nutrient enrichment or several superimposed anthropogenic influences. Therefore, attempts are made to develop diatom descriptors more specific to toxic pollution (see Chapter 8). They generally require a combination of structural and morphological descriptors (e.g. cell sizes, deformities).

Implementation of diatom-based assessment in rivers and streams: main approaches, advantages and limitations

Assessment of lotic systems has received more attention than lakes (King *et al.*, 2005), although the dynamics of phytoplankton as well as paleoecological reconstructions have been extensively investigated in these systems. However, lacustrine biomonitoring of contemporary water quality is increasingly applied, and is based on diatoms sampled in the littoral zone and with methods similar to those used in rivers and streams.

Diatom monitoring is mainly used for regulatory purposes, with very large scale programmes (from basin to country) and allows mapping water quality using simple biological indices. This approach gives information about the current state of rivers, and can also be used to monitor recovery processes following rehabilitation programmes. More specific approaches to determine the role of certain pressures in modifying the community *in situ* can also be performed, such as before/after impact studies (that can be studied in space or time), or the use of translocation experiments. Applied examples of these three complementary approaches are provided below.

Large scale monitoring programmes and data integration

In the last decades, numerous countries from the northern hemisphere have included diatoms in their biomonitoring programmes, while in the southern hemisphere their use is less frequent (Rimet, 2012).

Communicating the conditions of biological systems, and the impact of human activities on aquatic ecosystems, is the ultimate purpose of biological monitoring. As an example, the biomonitoring programme of the Matanza-Riachuelo River, a highly polluted Argentinean hydrographic system (Fig. 6.1) is presented. This basin has a surface area of 2240 km^2, and is inhabited by more than five million people. The deterioration of the water quality of the main watercourse and of most of its tributaries highlights a strong polluting load from household and industrial sewage waters. The urban pollution widely exceeds the diluting and self-depuration capacity of the river, as well

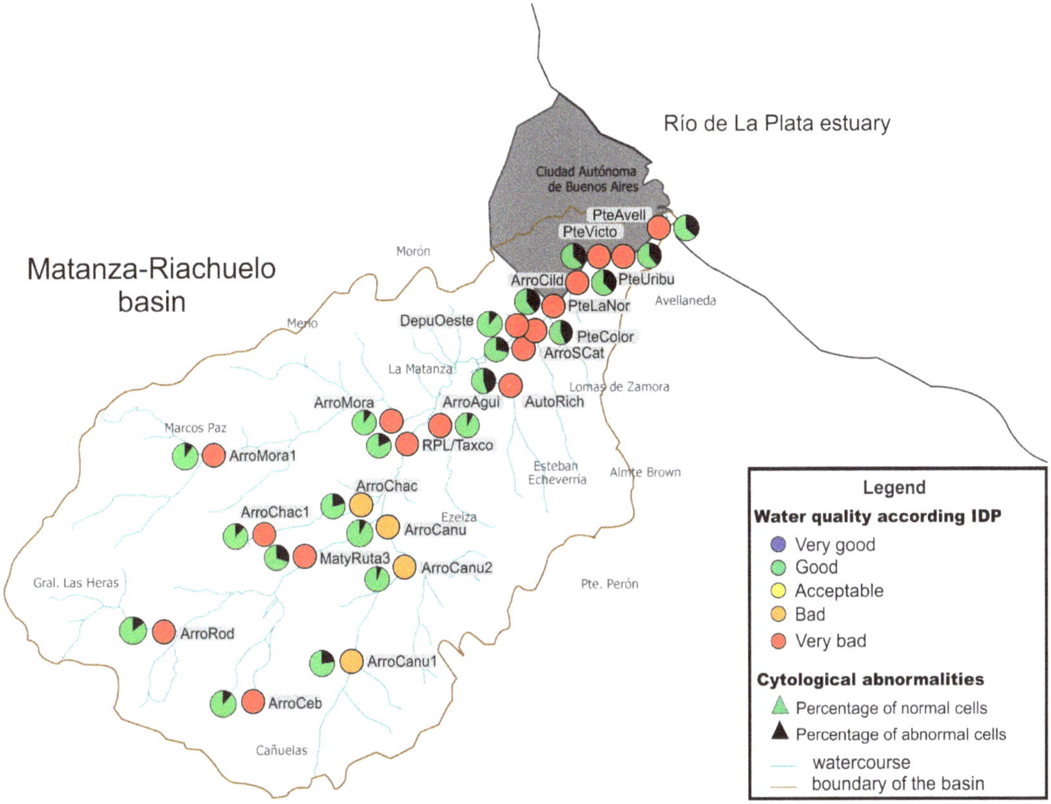

Figure 6.1 Water quality map corresponding to the 2010 monitoring results of Matanza-Riachuelo basin employing the Pampean Diatom Index (IDP) and percentage of cytologic abnormalities.

as toxic contaminants, chromium, copper, lead and cadmium concentrations above the guideline levels of water quality for protection of aquatic life. In this case, a regional index, the Pampean Diatom Index (IDP) (Gómez and Licursi, 2001), is employed to describe the different environmental conditions related to the eutrophication and the enrichment with organic matter through five water qualities codified with different colours (Fig. 6.1). Biomonitoring results expressed on maps constitute a useful tool for stakeholders as they provide quick visualization, through the use of different graphic codes and colours, of the evolution of the water quality in hydrographic basins through time and space. Likewise, the evaluation of cytological abnormalities (percentage of aberrant frustules and cytoplasmic content impaired) also contributes to the detection of changes of the water quality mainly associated to toxic pollution (Figs. 6.1 and 6.2). So, in the lower basin, strong symptoms of eutrophication and high amounts of organic matter and heavy metals can be diagnosed through high values of IDP, but also high percentages of cytological abnormalities.

Before/after impact studies (in space or time)

Diatoms have been used in experimental designs for comparing time series or differences in a treated area (or impacted) and control area, before or after the intervention or experimental treatment. The before/after impact assessment is a very suitable methodology for assessing whether or not a stress has changed the environment, to determine which components are adversely affected, and to estimate the magnitude of the effects. The simplest approach, referred to as the before–after design (BA), considers the time scale and involves collection of data prior to the beginning of activity and compares it with data recorded after the start of the activity (Smith, 2002). On the other hand the before-after-control-impact-paired

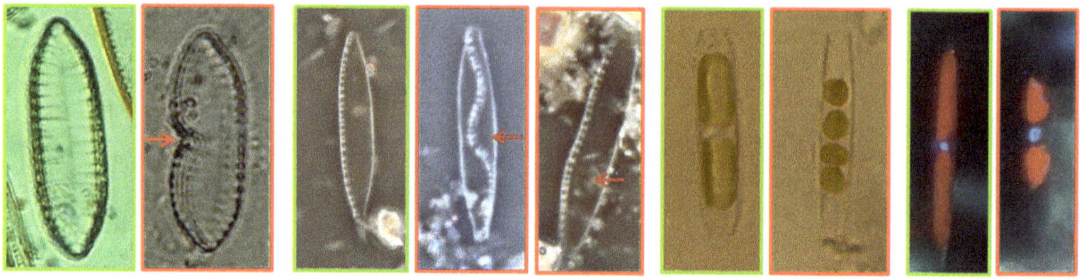

Figure 6.2 Normal cells (framed in green) and specimens with a deformed frustule or modified ornamentation (*Surirella angusta* and *Nitzschia palea*) and impaired cytoplasmic content (*Nitzschia* sp.) (framed in red, the arrows indicate the alterations) recorded in the Matanza-Riachuelo basin.

(BACIP) approach is a proper method, involving as well space scale, to evaluate the effects of the stressor being assessed; in this approach the 'impact area' is paired to another area referred to as 'control area' (Stewart-Oaten, 1996). Paired samples are collected a number of times, both before and after the perturbation, simultaneously (or nearly so) at both a 'Control' and 'Impact' location. The standard analytical approach, using the resulting BACIPS data, is to calculate the difference between 'Control' and 'Impact' values on each date (termed delta), and test whether the mean of these deltas changes from before to after the perturbation (Bence *et al.*, 1996). This type of approach allows the elucidation of potential effects that may occur in aquatic environments as a result of a stressor.

A study conducted in a Pampean watercourse (La Choza stream) is an example of application of this methodology in assessing the impact of a stressor (Artigas *et al.*, 2013). The diatom assemblages inhabiting the epipelic biofilm of this stream were exposed to a continuous surplus of inorganic nutrients; increasing concentrations of nitrogen and phosphorus in water 3-fold the basal concentration. Nutrient enrichment was achieved by the use of fertilizer bags distributed along the reach; the period of exposure was of 14 months. The changes in nutrient concentration were associated with a significant increase (BACIPS, $P < 0.001$) in diatom density and a decrease in species richness and diversity (Fig. 6.3). Changes in the relative proportions of the diatom taxa were also observed; while some taxa showed moderate to high variations (*Nitzschia palea*, *N. frustulum*, *Melosira varians*, *N. supralitorea* and *Caloneis bacillum*) others showed minor changes. Furthermore nutrient enrichment favoured shifting the proportion of stalked or filamentous to more motile growth forms (Fig. 6.4). The fertilization in La Choza caused a mild to moderate effect, not immediately felt, on the diatom assemblages. Taking into account the increase in urbanization and agricultural activity in the Pampean plain, it is likely that biodiversity can be seriously impaired if the entrance of nutrients to these ecosystems is not mitigated. A similar study carried out in an oligotrophic Mediterranean forest stream (Veraart *et al.*, 2008) reported that long-term nutrient addition has significant effects on the algal biomass and community composition, and this was detectable despite the low light availability resulting from the dense tree canopy. Results obtained from field experiments are extremely valuable, illustrating the expected effects of different pollution scenarios.

Active biomonitoring: use of translocation experiments

Over the last decades, artificial substrates (e.g. glass slides or plastic sheets for periphyton, plastic trays for epipelon) have been increasingly used for collecting diatom communities; consequently promoting the development of translocation experiments (Ivorra *et al.*, 1999; Morin *et al.*, 2010; Tolcach and Gómez, 2002; Sierra and Gómez, 2010). Basically, artificial substrates are immersed in a river site, and then transferred elsewhere to build scenarios concerning the responses of the community to changes in environmental conditions. Diatom communities from a reference (unpolluted) site can be transferred to an

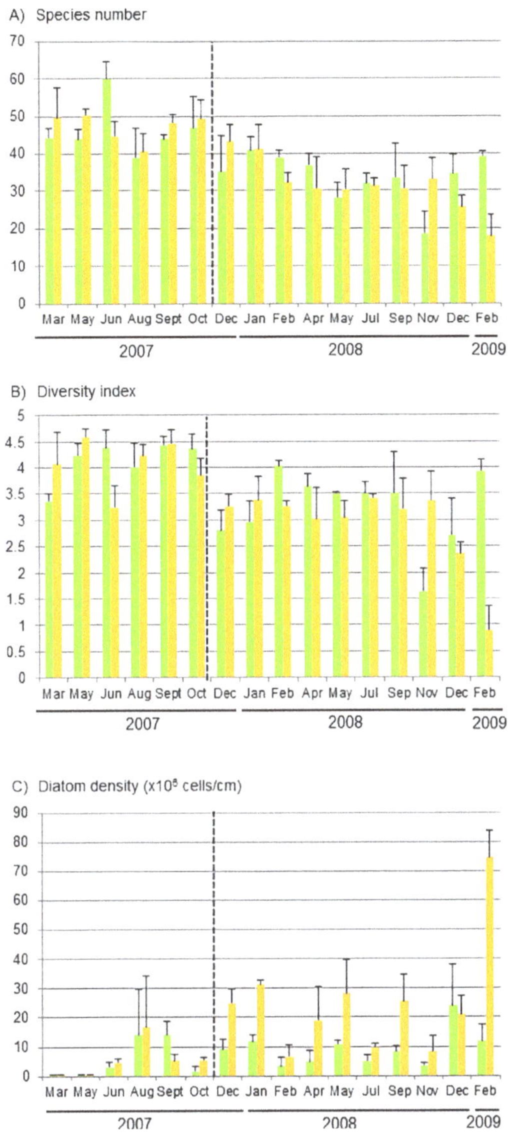

Figure 6.3 (A) Richness, (B) diversity and (C) densities in the control (green bars) and enriched (orange bars) reaches in La Choza stream. The dotted line indicates the onset of the fertilization period.

impacted location, to explore the effects of water quality degradation. Opposite translocations (polluted to unpolluted site) can be performed to study simulated improvement of water quality. Fig. 6.5 illustrates the recovery of community structure and water quality indices (using the IPS; Coste *in* Cemagref, 1982) in diatom communities sampled along a gradient of orthophosphates. Diatom composition differed significantly between upstream, intermediate and downstream sites. According to the gradient of eutrophication, IPS values decreased going downstream. Communities translocated from the two contaminated sites recovered a taxonomic composition closer to the reference (upstream) within 1 month, as well as increasing IPS values.

The use of translocation to assess the recovery potential of diatom communities has been recently questioned due to the fact that it is impossible to discriminate between the sole effect of water quality improvement and recolonization by immigrants from the upstream pool (Morin *et al.*, 2012a). However, translocation experiments remain powerful monitoring tools to assess potential gains in ecological health after remediation, especially in the case of connected hydrosystems.

Diatom-based assessment shifts towards fundamental ecology

The effects of anthropogenic modification of abiotic factors on natural variations in water quality variables and biotic relationships should be addressed. In contrast to studies on biofilm bacterial communities that have focused on large-scale biogeography (Martiny *et al.*, 2006; Battin *et al.*, 2007), information on the relative importance of natural and anthropogenically driven variations in diatom communities is limited to recent ecoregional approaches (see below). Recent works also aim at integrating the role of biotic factors, i.e. competition or facilitation within the community, in structuring the assemblages. Together, these works provide information that may be implemented in multimetric approaches. Finally, molecular approaches have also been developed to study the phylogeny of diatoms and more precisely define the concept of species in diatoms, and shifting towards molecular biomonitoring.

Spatial distributions

Many diatom species are known to have broad distributions; others seem limited to specific climatic zones or geographical regions, or are endemic to a particular habitat. Despite the importance of benthic diatoms as biomonitors, large and

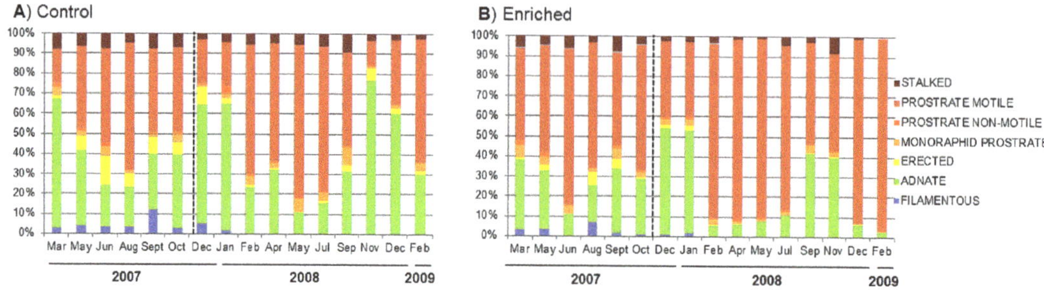

Figure 6.4 Relative abundance of diatom growth forms in the Control (A) and Enriched (B) reaches in La Choza stream. The dotted line indicates the onset of the fertilization period.

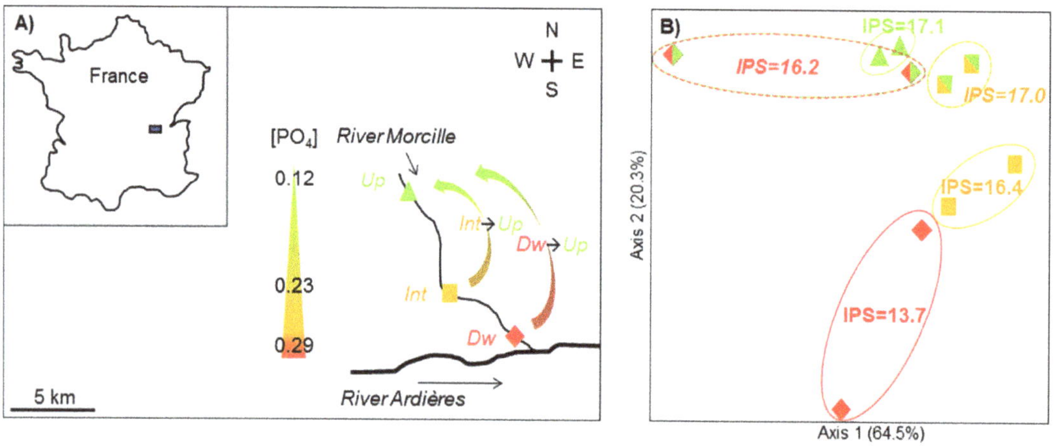

Figure 6.5 Translocation along a gradient of orthophosphates. (A) Location of the sample sites along the River Morcille. Diatom communities are sampled after 2 months at the three sites (Up: upstream, Int: intermediate, Dw: downstream), or after 1 month in their original site followed by 1 month upstream (Int → Up and Dw → Up). Average orthophosphate concentrations are shown in mg/l. (B) Principal component analysis based on relative abundances of the 40 dominant diatom taxa showing the discrimination between sampled communities, and corresponding average IPS values.

regional scale knowledge of their structure and function is still scarce. Their distributional patterns respond to a multitude of different factors, from the biogeochemical characteristics of water and its nutrient content, to geomorphological and physiographical features, but also to biotic interactions, that operate over a wide range of spatial and temporal scales (Menge and Olson, 1990). Water chemistry, light availability, variations in temperature, water velocity, substrata type and grazing are among the factors that potentially affect benthic diatoms (Stevenson *et al.*, 1996). The respective relevance of water quality variation and physiographical processes in a particular geographical area are expressed in a complex gradient, in which the interaction between local and broader-scale factors determines the composition of diatom communities (Leira and Sabater, 2005). In addition, the relative effects of anthropogenic variation over natural variation in determining the distribution of a given community should also be considered. Nutrient enrichment and human disturbances act to change local and large-scale factors reducing the regional differences (Tornés *et al.*, 2007). An obvious consequence of the overriding effect of human activities is that differences in relative abundance and composition of diatom communities are clearer among relatively undisturbed sites than among sites severely affected by nutrient enrichment. It has been proved then that this knowledge is crucial in the assessment of biological quality, which is based on the degree of deviation between expected and observed conditions (Tornés *et al.*, 2012).

Recently, an increasing body of research supports the idea that microorganisms exhibit biogeographical patterns with no strict evidence of ubiquitous and global distribution (Hillebrand *et al.*, 2001; Heino and Soininen, 2005). At the landscape scale, local species belong to broader metacommunities, shaped by dispersion, connectivity, biotic interactions and habitat area (Altermatt *et al.*, 2013; Chapter 4). In particular, historical processes (i.e. colonization, extinction, dispersion, migration) may determine global diversity patterns of diatoms (Vyverman *et al.*, 2007). Potapova and Charles (2002) demonstrated that large-scale spatial patterns of species dispersal, independent of local environmental characteristics, cannot be neglected in broad-scale studies. In their study almost one-third of the explainable variation in diatom species composition at the USA national scale was attributed to spatial factors. Similarly, another recent spatial concept applied to diatoms is nestedness (Soininen, 2008; Tornés and Ruhí, 2013). Nestedness is a metacommunity-based concept which quantifies the overlap in species composition between high and low diversity sites (Atmar and Patterson, 1993). Nested structures occur when assemblages of species-poor sites are subsets of the assemblages of species-rich sites. The extent to which environmental variables determine nestedness is still poorly understood. For example, Tornés and Ruhí (2013) identified hydrological stability as the main driver of nestedness of diatom communities in Mediterranean rivers.

Integrating the role of biotic interactions

Assembly rules, through general principles, try to explain the different processes leading to the presence of a particular species at a given site (Weiher and Keddy, 1999). The species sensitivity towards the environmental conditions and their ability to participate in spatial dispersion represent two important processes (see above). A third major process concerns biotic interactions, which have been very poorly studied to date. This consists in studying species co-occurrence in biofilms through the concepts of competition, passive coexistence and facilitation. But how to disentangle the relative importance of this biotic process in comparison to the others? In reality two types of approaches exist to unravel the causes: mathematically by comparing real communities to virtual ones with no biotic interactions, and practically by conducting laboratory experiments.

As diatoms are part of a complex three-dimensional matrix, we can assume that competition plays a significant role in community structure. However, to our knowledge, there is no strict evidence of competitive exclusion between diatom species from natural stream ecosystems. To date, only Heino and Soininen (2005) have tried to highlight such patterns for stream diatom communities, by a mathematical approach. From Finish data they calculated an index of species pairs that do not co-occur, i.e. chequerboard pairs (C-score, Stone and Roberts, 1990), which is a measure of the exclusion rate between species. They compared the results to those obtained from random communities, with no biotic interactions (Gotelli, 2000). Despite the higher number of chequerboard pairs in the real dataset, the particular structure of the environmental dataset used did not allow interpretation of the results as evidence of competitive exclusion.

Concerning positive biotic relationships, such as niche complementarity, Burkholder *et al.* (1990) by laboratory autoradiographic techniques, reported a direct comparison of phosphate uptake by adnate and by loosely attached diatoms in an intact biofilm matrix. They highlighted the fact that loosely attached cells took up significantly more radiolabel than did the underlying adnate cells, which were more isolated from the water column nutrient source. The results gave evidence of a physiological assimilation gradient among diatoms, where loosely attached cells can form a significant barrier to nutrient entry. It can then be concluded from those seminal works that such species could facilitate the persistence of underlying sensitive species. Moreover, in an experimental investigation of periphytic succession in recirculating laboratory streams, Passy and Larson (2011) examined the density and the relative abundance of diatoms across gradients of low to high nutrient supply and low to intermediate current velocity. They concluded that the mechanism of species succession, especially at a functional level, was a neutral coexistence where sensitive species were neither facilitated nor outcompeted by tolerant species but controlled by the environment.

Multimetric developments to better diagnose the type of stressors

Analysis of diatom assemblages can be based on taxonomic (e.g. composition and abundance, richness, diversity, species sensitivity, growth form, motility) or non-taxonomic measures (e.g. enzyme activity, chlorophyll content). The most common taxonomic-based techniques used for monitoring rivers worldwide belong to three basic approaches: indicator species or traits towards different environmental parameters, indices of community structure (e.g. diversity, richness, evenness), or biotic indices. Multimetric indices can combine these three types of approaches into a unitless measure, to be able to respond both to specific sources of stress and to general perturbations (Karr and Chu, 1997). Ideally, the series of indicators should represent key information about structure, function and composition (Dale and Beyeler, 2001). Then, such indices integrate metrics that must show clear relationships with different types of stressors, and should reduce uncertainty and increase robustness of assessment in comparison to single metrics by combining different types of metrics indicative of different environmental conditions and community features. Several studies in Europe (Hering *et al.*, 2006) and in USA (Barbour *et al.*, 1999; Karr and Chu, 1999) have proved the multimetric index to be a valuable approach for assessing the ecological status of water bodies. The first attempt to settle a diatom-based multimetric index was that of Hill *et al.* (2000) who proposed a Periphytic Index of Biotic Integrity (PIBI) for the Mid-Appalachian region (USA). PIBI includes diatom and non-diatom criteria, linked to the relative abundance of some genera, the percentage of acidophilic or eutraphentic diatoms, the percentage of motile diatoms. These metrics were well correlated with both water quality and stream depth and width. Then Griffith *et al.* (2002) proposed for the Southern Rockies ecoregion of Colorado (USA) diatom metrics rather linked to riparian disturbances (percentage of the species *A. minutissimum* for example) or metal pollution (number of diatom cells). More recently Delgado *et al.* (2010) developed a diatom multimetric index (MDIAT) as a combination of metric values (existing European diatom-based indices and percentage of local reference taxa) responding to organic and nutrient stressors.

Progress in taxonomic identification using molecular tools

Molecular biology is increasingly used for diatom phylogenetic analyses (Bruder and Medlin, 2007; Medlin *et al.*, 2008; Medlin, 2010; Medlin and Kaczmarska, 2004), but classifications based on molecular studies and current systematics based on morphological features rarely cross (Cox, 2009). Resulting identifications are difficult to compare, with in some cases generic boundaries incorrectly drawn (Medlin *et al.*, 2008). Challenge in correct identification concerns species with very subtle morphological differences (Behnke *et al.*, 2004) or original species descriptions covering more than one genotype (Evans *et al.*, 2008; Mann *et al.*, 2008; Sarno *et al.*, 2005) and vice-versa with species of large phenotypic plasticity, e.g. leading to identification of the extremes as different taxa (Mann *et al.*, 2010). Primers specific to diatoms have been proposed by Valiente Moro *et al.* (2009) but they are not specific enough to characterize the real diversity of samples (Morin *et al.*, 2012b). Next-generation sequencing was recently attempted to inventory taxonomic diversity in diatom communities (Kermarrec *et al.*, 2013a). Although taxonomic assignation was not always stringent, barcoding approaches offer promising perspectives for high throughput screening of diatom diversity, and may represent a powerful tool for biomonitoring in the future.

Today, molecular tools are rather used to support taxonomy, particularly in the case of cryptic species where they are needed to refine identifications (e.g. Evans *et al.*, 2008; Mann *et al.*, 2008). The combination of molecular techniques with microscopic observations allows progress in classification, especially when dealing with complicated species complexes (Kermarrec *et al.*, 2013b).

Conclusions

Diatom-based biomonitoring is widely used for regulatory purposes, at large scales or for many local applications. Here we reviewed methods differing in their objectives (evaluation of the general water quality, diagnostic of local impacts) and highlighted the recent inflexions in diatom research, towards increasing consideration of fundamental ecology and multi-proxy approaches. Molecular tools are also considered to potentially contribute to a better assessment of water quality, as they may improve our knowledge of diatom taxonomy and help to refine taxa ecological requirements.

Besides diatoms, diverse other periphytic components are likely to be employed, although no consensual bioindicator exists. Classifications have been proposed for monitoring eutrophication based on non-diatom benthic algae in Europe (Rott *et al.*, 1999; Whitton, 1999). The Periphyton Index of Trophic status (PIT: Schneider and Lindstrøm, 2011) developed in Norway combines non-diatom indicator species with non-autotrophic taxa and provides accurate eutrophic assessment in waters where diatoms are hardly found. Recent developments in molecular approaches also offer promising perspectives, e.g. for the use of benthic cyanobacteria as powerful bioindicators (Loza *et al.*, 2013).

References

AFNOR (2000). Qualité de l'Eau. Détermination de l'indice biologique diatomées (IBD) – Norme NF T90-354 -63p.

AFNOR (2003). Qualité de l'Eau. Guide pour l'échantillonnage en routine et le pré-traitement des diatomées benthiques de rivières – Norme NF EN 13946 -16p.

Almeida, S.F.P., Elias, C., Ferreira, J., Tornés, E., Puccinelli, C., Delmas, F., Dörflinger, G., Urbanič, G., Marcheggiani, S., Rosebery, J., *et al.* (2014). Water quality assessment of rivers using diatom metrics across Mediterranean Europe: a methods intercalibration exercise. Sci. Total Environ. *476–477*, 768–776.

Altermatt; F., Seymour, M., and Martinez, N. (2013). River network properties shape α-diversity and community similarity patterns of aquatic insect communities across major drainage basins. J. Biogeogr. *40*, 2249–2260.

Artigas, J., García-Berthou, E., Bauer, D.E., Castro, M.I., Cochero, J., Colautti, D., Cortelezzi, A., Donato, J.C., Elosegi, A., Feijoó, C., *et al.* (2013). Global pressures, specific responses: effects of nutrient enrichment in streams from different biomes. Environ. Res. Lett. *8*, 014002.

Atmar, W., and Patterson, B. (1993). The measure of order and disorder in the distribution of species in fragmented habitat. Oecologia. *96*, 373–382.

Battin, T.J., Sloan, W.T., Kjelleberg, S., Daims, H., Head, I.M., Curtis, T.P., and Eberl, L. (2007). Microbial landscapes: new paths to biofilm research. Nat. Rev. Microbiol. *5*, 76–81.

Barbour, M.T., Gerritsen, J., Snyder, B.D., and Stribling, J.B. (1999). Rapid bioassessment protocols for use in streams and wadeable rivers: periphyton, benthic macroinvertebrates and fish, second edition. EPA-841-B-99-002. (US Environmental Protection Agency, Office of Water, Washington, DC, USA).

Behnke, A., Friedl, T., Chepurnov, V.A., and Mann, D.G. (2004). Reproductive compatibility and rDNA sequence analyses in the *Sellaphora pupula* species complex (Bacillariophyta). J. Phycol. *40*, 193–208.

Bence, J.R., Stewart-Oaten, A., and Schoeter, S.C. (1996). Estimating the size of an effect from a before-after-control-impact paired series design: the predictive approach applied to a power plant study. In Detecting Ecological Impacts: Concepts and Application in Coastal Habitats, Schmitt, R.J., and Osenberg, C.W., eds. (Academic Press Inc., San Diego, CA, USA), pp. 133–150.

Birks, H.J.B., Une, J.M., Juggins, S., Stevenson, A.C., and Braack, C.J.F.T. (1990). Diatom pH reconstruction. Phil. Trans. R. Soc. Lond. B, 263–278.

Bruder, K., and Medlin, L.K. (2007). Molecular assessment of phylogenetic relationships in selected species/genera in the naviculoid diatoms (Bacillariophyta). I. The genus *Placoneis*. Nova Hedw. *85*, 331–352.

Burkholder, J.M., Wetzel, R.G., and Klomparens, K.L. (1990). Direct comparison of phosphate uptake by adnate and loosely attached microalgae within an intact biofilm matrix. Appl. Environ. Microbiol. *56*, 2882–2890.

Cemagref (1982). Etude des méthodes biologiques d'appréciation quantitative de la qualité des eaux (Rapport Q.E. Lyon A.F. – Bassin Rhône-Méditerranée-Corse).

Charles, D.F., Acker, F.W., Hart, D.D., Reimer, C.W., and Cotter, P.B. (2006). Large-scale regional variation in diatom-water chemistry relationships: Rivers of the eastern United States. Hydrobiologia *561*, 27–57.

Coste, M., Boutry, S., Tison-Rosebery, J., and Delmas, F. (2009). Improvements of the Biological Diatom Index (BDI): Description and efficiency of the new version (BDI-2006). Ecol. Indic. *9*, 621–650.

Cox, E.J. (2009). What's in a name? – Diatom classification should reflect systematic relationships. Acta Bot. Croat. *68*, 443–454.

Dale, V.H., and Beyeler, S.C. (2001). Challenges in the development and use of ecological indicators. Ecol. Indic. *1*, 3–10.

Delgado, C., Pardo, I., and García, L. (2010). A multimetric diatom index to assess the ecological status of coastal Galician rivers (NW Spain). Hydrobiologia. *644*, 371–384.

Evans, K.M., Wortley, A.H., Simpson, G.E., Chepurnov, V.A., and Mann, D.G. (2008). A molecular systematic approach to explore diversity within the *Sellaphora pupula* species complex (Bacillariophyta) J. Phycol. *44*, 215–231.

Gómez, N., and Licursi, M. (2001). The Pampean Diatom Index (IDP) for assessment of rivers and streams in Argentina. Aquat. Ecol. *35*, 173–181.

Gotelli, N.J. (2000). Null model analysis of species co-occurrence patterns. Ecology. *81*, 2606–2621.

Griffith, M.B., Hill, B.H., Herlihy, A.T., and Kaufmann, P.R. (2002). Multivariate analysis of periphyton assemblages in relation to environmental gradients in Colorado rocky mountain streams. J. Phycol. *38*, 83–95.

Guiry, M.D. (2012). How many species of algae are there? J. Phycol. *48*, 1057–1063.

Heino, J., and Soininen, J. (2005). Assembly rules and community models for unicellular organisms: patterns in diatoms of boreal streams. Freshwat. Biol. *50*, 567–577.

Hering, D., Feld, C.K., Moog, O., and Ofenböck, T. (2006). Cook book for the development of a multimetric Index for biological condition of aquatic ecosystems: experiences from the European AQEM and STAR projects and related initiatives. Hydrobiologia. *566*, 311–324.

Hill, B.H., Herlihy, A.T., Kaufmann, P.R., Stevenson, R.J., McCormick, F.H., and Johnson, C.B. (2000). Use of periphyton assemblage data as an index of biotic integrity. J. N. Am. Benthol. Soc. *19*, 50–67.

Hillebrand, H., Watermann, F., Karez, R., and Berninger, U.G. (2001). Differences in species richness patterns between unicellular and multicellular organisms. Oecologia. *126*, 114–124.

Hofmann, G. (1994). Aufwuchs diatoms in Seen und ihre Eignung als Indikatoren der Trophie (Bibliotheca Diatomologica 30).

Ivorra, N., Hettelaar, J., Tubbing, G.M.J., Kraak, M.H.S., Sabater, S., and Admiraal, W. (1999). Translocation of microbenthic algal assemblages used for *in situ* analysis of metal pollution in rivers. Arch. Environ. Contam. Toxicol. *37*, 19–28.

Karr, J.R., and Chu, E.W. (1997). Biological monitoring and assessment: using multimetric indexes effectively (University of Washington, EPA 235-R97–001).

Karr, J.R., and Chu, E.W. (1999). Restoring life in Running Waters: Better Biological Monitoring (Island Press, Washington, DC, USA).

Kelly, M., Bennett, C., Coste, M., Delgado, C., Delmas, F., Denys, L., Ector, L., Fauville, C., Ferréol, M., Golub, M., *et al.* (2009). A comparison of national approaches to setting ecological status boundaries in phytobenthos assessment for the European Water Framework Directive: results of an intercalibration exercise. Hydrobiologia *621*, 169–182.

Kelly, M., Cazaubon, A., Coring, E., Dell'Uomo, A., Ector, L., Goldsmith, B., Guasch, H., Hürlimann, J., Jarlman, A., Kawecka, B., *et al.* (1998). Recommendations for the routine sampling of diatoms for water quality assessments in Europe. J. Appl. Phycol. *10*, 215–224.

Kelly, M.G., and Whitton, B.A. (1995). The Trophic Diatom Index: a new index for monitoring eutrophication in rivers. J. Appl. Phycol. *7*, 433–444.

Kermarrec, L., Franc, A., Rimet, F., Chaumeil, P., Humbert, J.F., and Bouchez, A. (2013a). Next-generation sequencing to inventory taxonomic diversity in eukaryotic communities: a test for freshwater diatoms. Mol. Ecol. Resour. *13*, 607–619.

Kermarrec, L., Bouchez, A., Rimet, F., and Humbert, J.-F. (2013b). First Evidence of the Existence of semi-cryptic species and of a phylogeographic structure in the Gomphonema parvulum (Kützing) Kützing complex (Bacillariophyta). Protist. *164*, 686–705.

King, L., Clarke, G., Bennion, H., Kelly, M., and Yallop, M. (2005). Sampling littoral diatoms in lakes for ecological status assessments: a literature review (Science Report SC030103/SR1, Environment Agency, Bristol, UK).

Lavoie, I., Campeau, S., Grenier, M., and Dillon, P.J. (2006). A diatom-based index for the biological assessment of eastern Canadian rivers: an application of correspondence analysis (CA). Can. J. Fish. Aquat. Sci. *63*, 1793–1811.

Lear, G., Dopheide, A., Ancion, P., Roberts, K., Washington, V., Smith, J., and Lewis, G.D. (2012). Biofilms in freshwater: their importance for the maintenance and monitoring of freshwater health. In Microbial Biofilms: Current Research and Applications, Lear, G., and Lewis, G.D., eds. (Caister Academic Press, Norfolk, UK), pp. 129–152.

Lecointe, C., Coste, M., and Prygiel, J. (1993). Omnidia – Software for taxonomy, calculation of diatom indices and inventories management. Hydrobiologia. *269/270*, 509–513.

Leira, M., and Sabater, S. (2005). Diatom assemblages distribution in catalan rivers, NE Spain, in relation to chemical and physiographical factors. Wat. Res. *39*, 73–82.

Lobo, E.A., Callegaro, V.L.M., Hermany, G., Bes, D., Wetzel, C.A., and Oliveira, M.A. (2004). Use of epilithic diatoms as bioindicators from lotic systems in southern Brazil, with special emphasis on eutrophication. Acta Limnol. Bras. *16*, 25–40.

Lowe, R.L., and Pan, Y. (1996). Benthic algal communities as biological monitors. In Algal Ecology: Freshwater Benthic Ecosystems, R.J. Stevenson, M.L. Bothwell, and R.L. Lowe, eds. (Aquatic Ecology Series, Academic Press Inc., San Diego, CA, USA), pp. 705–739.

Loza, V., Perona, E., and Mateo, P. (2013). Molecular fingerprinting of cyanobacteria from river biofilms as a water quality monitoring tool. Appl. Environ. Microbiol. *79*, 1459–1472.

Mann, D.G., Thomas, S.J., and Evans, K.M. (2008). Revision of the diatom genus *Sellaphora*: a first account of larger species in the British Isles. Fottea. *8*, 15–78.

Mann, D.G., Williams, D.W., Kociolek, J.P., and Cox, E.J. (2010). Discovering diatom species: is a long history of disagreements about species-level taxonomy now at an end? Plant Ecol. Evol. *143*, 251–264.

Martiny, J.B., Bohannan, B.J., Brown, J.H., Colwell, R.K., Fuhrman, J.A., Green, J.L., Horner-Devine M.C., Kane, M., Krumins, J.A., Kuske, C.R., *et al.* (2006).

Microbial biogeography: putting microorganisms on the map. Nat. Rev. *4*, 102–112.

McCormick, P.V., and Cairns, J. (1994). Algae as indicators of environmental change. J. Appl. Phycol. *6*, 509–526.

McIntire, C.D., Gregory, S.V., Steinman, A.D., and Lamberti, G.A. (1996). Modelling algal communities: an example from stream ecology. In Algal Ecology: Freshwater Benthic Ecosystems, Stevenson, R.J., Bothwell, M.L., and Lowe, R.L., eds. (Aquatic Ecology Series, Academic Press Inc., San Diego, CA, USA), pp. 669–704.

Medlin, L., Jung, I., Bahulikar, R., Mendgen, K., P., K., and Kooistra, W.H.C.F. (2008). Evolution of the Diatoms. VI. Assessment of the new genera in the araphids using molecular data. Nova Hedw. *133*, 81–100.

Medlin, L.K. (2010). Pursuit of a natural classification of diatoms: An incorrect comparison of published data. Eur. J. Phycol. *45*, 155–166.

Medlin, L.K., and Kaczmarska, I. (2004). Evolution of the diatoms: V. Morphological and cytological support for the major clades and a taxonomic revision. Phycologia *43*, 245–270.

Menge, B.A., and Olson, A.M. (1990). Role of scale and environmental factors in regulation of community structure. Trends Ecol. Evol. *5*, 52–57.

Morin, S., Lambert, A.-S., Artigas, J., Coquery, M., and Pesce, S. (2012a). Diatom immigration drives biofilm recovery after chronic copper exposure. Freshwat. Biol. *57*, 1658–1666.

Morin, S., Roubeix, V., Batisson, I., Winterton, P., and Pesce, S. (2012b). Characterization of freshwater diatom communities: Comparing taxonomic and genetic-fingerprinting approaches. J. Phycol. *48*, 1458–1464.

Morin, S., Pesce, S., Tlili, A., Coste, M., and Montuelle, B. (2010). Recovery potential of periphytic communities in a river impacted by a vineyard watershed. Ecol. Indic. *10*, 419–426.

Passy, S.I. (2007). Diatom ecological guilds display distinct and predictable behavior along nutrient and disturbance gradients in running waters. Aquat. Bot. *86*, 171–178.

Passy, S., and Larson, C. (2011). Succession in stream biofilms is an environmentally driven gradient of stress tolerance. Microb. Ecol. *62*, 414–424.

Ponader, K.C., Charles, D.F., Belton, T.J., and Winter, D.M. (2008). Total phosphorus inference models and indices for coastal plain streams based on benthic diatom assemblages from artificial substrates. Hydrobiologia *610*, 139–152.

Potapova, M.G., and Charles, D.F. (2002). Benthic diatoms in USA rivers: distributions along spatial and environmental gradients. J. Biogeogr. *29*, 167–187.

Potapova, M., and Charles, D.F. (2004). Potential use of rare diatoms as environmental indicators in USA rivers, in: Proceedings of the 17th International Diatom Symposium (Biopress, Bristol, UK), pp. 281–295.

Potapova, M., and Charles, D.F. (2007). Diatom metrics for monitoring eutrophication in rivers of the United States. Ecol. Indic. 7, 48–70.

Prygiel, J., Coste, M., and Bukowska, J. (1999). Review of the major diatom-based techniques for the quality assessment of rivers – State of the art in Europe. In Use of Algae for Monitoring Rivers III, J. Prygiel, B.A. Whitton, and J. Bukowska, eds. (Agence de l'Eau Artois Picardie, Douai), p224–238.

Rimet, F. (2012). Recent views on river pollution and diatoms. Hydrobiologia. *683*, 1–24.

Rott, E., Pipp, E., Pfister, P., van Dam, H., Orther, K., Binder, N., and Pall, K. (1999). Indikationslisten für Aufwuchsalgen in Österreichischen Fliessgewassern. Teil 2: Trophieindikation. vol 2. Bundesministerium für Land- und Forstwirtschaft. Wien.

Sarno, D., Kooistra, W.H.C.F., Medlin, L.K., Percopo, I., and Zingone, A. (2005). Diversity in the genus *Skeletonema* (Bacillariophyceae), 2. An assessment of the taxonomy of *S. costatum*-like species, with the description of four new species. J. Phycol. *41*, 151–176.

Schneider, S., and Lindstrøm, E.-A. (2011). The periphyton index of trophic status PIT: a new eutrophication metric based on non-diatomaceous benthic algae in Nordic rivers Hydrobiologia. *665*, 143–155.

Sekar, R., Venugopalan, V.P., Satpathy, K.K., Nair, K.V.K., and Rao, V.N.R. (2004). Laboratory studies on adhesion of microalgae to hard substrates Hydrobiologia. *512*, 109–116.

Sládeček, V. (1986). Diatoms as indicators of organic pollution. Acta Hydroch. Hydrob. *14*, 555–566.

Smith, E.P. (2002). BACI design. In Encyclopedia of Environmetrics, A.H. El-Shaarawi and W.W. Piegorsch, eds. (John Wiley & Sons Ltd, Chichester, UK), pp. 141–148.

Soininen, J. (2008). The ecological characteristics of idiosyncratic and nested diatoms. Protist. *159*, 65–72.

Steinberg, C., and Schiefele, S. (1988). Biological indication of trophy and pollution of running waters. Z. Wasser Abwass. For. *21*, 227–234.

Stevenson, R.J., and Bahls, L. (1999). Periphyton protocols. In Rapid Bioassessment Protocols for use in Streams and Wadeable Rivers: Periphyton, Benthic Macroinvertebrates and Fish, 2nd edn, M.T. Barbour, J. Gerritsen, B.D. Snyder, and J.B. Stribling, eds. (US Environmental Protection Agency, Office of Water, Washington, DC, USA), p1–22.

Stevenson, R.J., Bothwell, M.L., and Lowe, R.L. (1996). Algal ecology: Freshwater Benthic Ecosystems (Academic Press Inc., San Diego, CA, USA).

Stevenson, R.J., and Pan, Y. (1999). Assessing environmental conditions in rivers and streams with diatoms. In The Diatoms – Applications for the Environmental and Earth Sciences, E.F. Stoermer and J.P. Smol, eds. (Cambridge University Press, Cambridge, UK), pp. 11–40.

Stewart-Oaten, A. (1996). Problems in the analysis of environmental monitoring data. In Detecting Ecological Impacts: Concepts and Application in Coastal Habitats, Schmitt, R.J., and Osenberg, C.W., eds. (Academic Press Inc., San Diego, CA, USA), pp. 109–132.

Stone, L., and Roberts, A. (1990). The checkerboard score and species distribution. Oecologia *85*, 74–79.

Tolcach, E.R., and Gómez, N. (2002). The effect of translocation of microbenthic communities in a polluted lowland stream. Ver. Int. Verein. Limnol. *28*, 254–258.

Tornés, E., Cambra, J., Gomà, J., Leira, M., Ortiz, R., and Sabater, S. (2007). Indicator taxa of benthic diatom communities: a case study in Mediterranean streams. Ann. Limnol. *43*, 1–11.

Tornés, E., Leira, M., and Sabater, S. (2012). Is the biological classification ofbenthic diatom communities concordant with ecotypes? Hydrobiologia *695*, 43–55.

Tornés, E., and Ruhí, A. (2013). Flow intermittency decreases nestedness and specialisation of diatom communities in Mediterranean rivers. Freshwat. Biol. *58*, 2555–2566.

UNESCO-WHO-UNEP (1996). Water Quality Assessments (Chapman & Hall, Cambridge, UK).

Valiente Moro, C., Crouzet, O., Rasconi, S., Thouvenot, A., Coffe, G., Batisson, I., and Bohatier, J. (2009). New design strategy for development of specific primer sets for PCR-based detection of Chlorophyceae and Bacillariophyceae in environmental samples. Appl. Environ. Microbiol. *75*, 5729–5733.

van Dam, H., Mertens, A., and Sinkeldam, J. (1994). A coded checklist and ecological indicator values of freshwater diatoms from the Netherlands. Neth. J. Aquat. Ecol. *28*, 117–133.

Veraart, A.J., Romaní, A.M., Tornés, E., and Sabater, S. (2008). Algal response to nutrient enrichment in forested oligotrophic stream. J. Phycol. *44*, 564–572.

Sierra, M.V., and Gómez, N. (2010). Assessing the disturbance caused by an industrial discharge using field transfer of epipelic biofilm. Sci. Total Environ. *408*, 2696–2705.

Vyverman, W., Verleyen, E., Sabbe, K., Vanhoutte, K., Sterken, M., Hodgson, D.A., Mann, D.G., Juggins, S., Vijver, B.V.D., and Jones, V. (2007). Historical processes constrain patterns in global diatom diversity. Ecology. *88*, 1924–1931.

Watanabe, T., Asai, K., Houki, A., Tanaka, S., and Hizuka, T. (1986). Saprophilous and eurysaprobic diatom taxa to organic water pollution and Diatom Assemblage Index (DAIpo). Diatom. *2*, 23–73.

Weiher, E., and Keddy, P. (1999). Ecological assembly rules: perspectives, advances, retreats (Cambridge University Press, Cambridge, UK).

Whitton, B.A., and Rott, E. (1996). Use of Algae for monitoring rivers II (Studia Student. G.m.b.H).

Whitton, B.A. (1999). Perspective on the use of phototrophs to monitor nutrients in running waters Aquat. Conserv.: Mar. Freshwat. Ecosyst. *9*, 545–549.

Zelinka, M., and Marvan, P. (1961). Zurprazisierung der biologischen Klassifikation des Reinheitfliessendergewasser. Arch. Hydrobiol. *57*, 389–407.

Ziemann, H. (1991). Veränderungen der Diatomeenflora der Werra unter dem Einfluß des Salzgehaltes. Acta Hydroch. Hydrob. *19*, 159–174.

The Use of Biofilms to Assess the Effects of Chemicals on Freshwater Ecosystems

7

Helena Guasch, Joan Artigas, Berta Bonet, Chloe Bonnineau, Oriol Canals, Natàlia Corcoll, Arnaud Foulquier, Julio López-Doval, Sandra Kim Tiam, Soizic Morin, Enrique Navarro, Stephane Pesce, Lorenzo Proia, Humbert Salvadó and Alexandra Serra

Abstract

Nowadays, biofilms are one of the principal targets of community ecotoxicology in aquatic ecosystems with a high potential for future use in ecotoxicology. A large set of methods derived from biofilm ecology has successfully been applied in ecotoxicology providing a diverse and comprehensive toolbox. Our ability to quantify the effects of pollution on different biofilm components, allows the direct effects of pollutants on the most sensitive community and their indirect effects on the rest of biofilm components to be evaluated. Biofilms are also a site for biotransfomation and/or transfer of chemicals to other aquatic organisms, supporting a more generalized use of biofilms in environmental chemistry. Investigations aiming to describe processes at biofilm scale, like nutrient dynamics and those including simple food chains, have recently been applied, providing the opportunity of upscaling the effects of pollutants on biofilms to food webs and ecosystems. Finally, biofilm ecotoxicology should now focus on providing the theoretical background for understanding the complex set of responses of natural communities to pollution. This knowledge should also be the basis for guiding the selection of the most appropriate tools and the development of new approaches for a better detection of the impact of pollution on aquatic life.

Introduction

The increasing worldwide contamination of freshwater systems with thousands of industrial and natural chemical compounds is one of the key environmental problems facing humanity. Developing and refining tools to assess the impact of these pollutants on aquatic life is still a challenging issue (Hering *et al.*, 2010). In spite of the inherent complexity of natural systems, the basis for using natural biofilms to assess acute and chronic effects of pollution is rather simple. It is expected that the effects of toxicity will first trigger a biochemical response, e.g. by the activation of detoxification mechanisms, causing thereafter physiological alterations, such as a reduction in photosynthetic activity and respiration, and leading finally to a reduction in the growth of the most sensitive species and the selection of the most tolerant species causing a shift in the structure (i.e. species composition) of the biofilm community. Together with the analysis of water chemistry, and the prevailing environmental conditions, a set of biofilm parameters (i.e. endpoints) may be used to assess the effects of pollutions under real-exposure scenarios (Fig. 7.1).

It has been shown that the use of biofilms in ecotoxicology is rather common, either in field or laboratory investigations. The majority of studies deal with metals and pesticides, but several investigations have recently been focused on emerging compounds (Guasch *et al.*, 2012). Here we aim to update previous reviews and to provide a critical overview of the most common endpoints used to assess the biological and ecological effects of pollution, the results obtained in different exposure scenarios, and future trends in the use of biofilms in ecotoxicology.

Biofilm ecotoxicology – a multi-component approach

The biological composition of biofilms is very broad, including several types of communities: algal, bacterial, fungal, protozoan and microinvertebrate communities, each of them including a large list of species involved in many ecological processes.

Owing to the prominent role that algae play in biofilms growing in illuminated surfaces, the majority of ecotoxicological investigations have focused on the algal component of biofilms (reviewed in Corcoll *et al.*, 2012a), while fewer studies have focused on the bacterial component (reviewed in Proia *et al.*, 2012a). Studies dealing with other biofilm components, such as fungi or protozoa, in spite of their important role, have received little attention.

Effects of pollutants on the autotrophic component of biofilms

If light is available, algae and other phototrophic organisms become the main component of biofilms. Effects of pollutants have been investigated on both the function and structure of the autotrophic component of biofilms (Table 7.1).

Among the functional descriptors used, photosynthesis-related parameters are some of the most relevant endpoints for assessing toxicity towards algae. Pulse amplitude modulated (PAM) fluorescence techniques were developed to measure among other parameters, photosynthetic capacity and efficiency, and non-photochemical photosynthetic processes. These functional endpoints are largely applied to evaluate the effects of chemicals on biofilms for their sensitivity to a large panel of chemicals, especially those targeting photosystem II, like herbicides or certain metals (e.g. copper) (Serra *et al.*, 2009; Ricart *et al.*, 2009; Laviale *et al.*, 2011). Fluorescence techniques are easy to apply (for more details, see review by Corcoll *et al.*, 2012a) and are even useful for assessing the impact of physical stressors, such as ultraviolet radiation (Navarro *et al.*, 2008). The analysis of accessory pigments (e.g. β-carotene, diatoxanthin, diadinoxanthin, pheophytin, etc.) has also been shown to suitably detect early toxicity of compounds targeting directly and/or indirectly the photosynthetic apparatus (Laviale *et al.*, 2010; Corcoll *et al.*, 2012b,c; Bonnineau *et al.*, 2013).

Chronic exposure to contaminants exerts a selection pressure on the community that may be reflected by physiological changes at species level (modifying its tolerance against contaminants) or by changes in the abundance and composition of algal communities (i.e. biomass, species composition). PAM fluorescence, or high-performance liquid chromatography (HPLC), is used to quantify the relative distribution of algal groups (green, blue and brown algae) within a biofilm based on photosynthetic pigments (e.g. Corcoll *et al.*, 2012a,b). In biofilms, taxonomic community identification is generally performed for diatoms (Chapter 6), a highly diverse, cosmopolitan class of brown algae. Shifts in structure have led to classifications based on species sensitivity/tolerance to contaminants (Morin *et al.*, 2009, 2012, 2014; Ricart *et al.*, 2009). Specific morphological endpoints, e.g. teratologies (Falasco *et al.*, 2009) or cell sizes (Luís *et al.*, 2011), have also proved to detect metal pollution successfully. Algal taxonomy has been largely used to study toxicant-induced selection in biofilm communities, due to its tradition but also its high sensitivity. More recently, molecular tools using DNA sequences have been described as promising tools to assess the prevalence of specific gene sequences in tolerant communities and their taxonomic affinities in natural biofilms (Eriksson *et al.*, 2009).

Quantitative real-time polymerase chain reaction (qPCR) techniques have been used with success in the field of ecotoxicology in order to assess the effects of various contaminants on different diatom species (planktonic and benthic). For instance, after the exposure of *Thalassiosira pseudonana* to PAHs, Bopp and Lettleri (2007) observed strong up-regulation of *lacsA*, which is involved in the fatty acid metabolism and repression of *sil3*, contributing to the formation of the silica shell; highlighting then a possible impact of such compounds on these functions. In a different study, Guo *et al.* (2013) reported up-regulation of

Table 7.1 Summary of biofilm endpoints (in bold) and methods (in italics) in ecotoxicology

Functional responses at molecular, cell, community or ecosystem level	Changes in biomass	Effects on the structure and architecture of the community
Autotrophic organisms		
Photosynthesis: *PAM, ^{14}C-HCO_3 uptake*	**Chlorophyll concentration**: *spectrophotometry*	**Algal groups**: *microscope, HPLC*
Tolerance induction: *toxicity assays, DNA sequences*	**Algal density**: *microscope, flow cytometry*	**Species composition**: *microscope*
		Diatom cell size, teratoforms: *microscope, flow cytometry*
		Genetic diversity: *fingerprinting techniques*
Bacteria		
C uptake: *^{3}H-thymidne incorporation*	**Bacterial density**: *microscope, flow cytometry*	**Genetic diversity**: *fingerprint, FISH, CARD-FISH, NGS*
Respiration: *substrate-induced respiration*		
Physiological profile: *MicroRespTM*		
Denitrification		
Antibiotic resistance genes		
Fungi		
Respiration: *substrate- induced respiration*	**Fungal density**: *microscope (mycelium growth)*	**Species composition**: *microscope*
Reproduction: *sporulation*	**Biomass**: *ergosterol concentration*	**Genetic diversity**: *fingerprint, NGS*
Extracellular degradation of organic matter: *EEA by enzymatic assays or qPCR*		
Protozoa		
Duplication rate: *dynamics of cell density*	**Cell density**: *microscope*	**Cell damage**: *lysosomal membrane stability, cytoplasmatic vacuolization, etc.*
Grazing activity and endocytotic rate: *clearance assays, intake of particles*		**Species composition**: *microscope*
		Genetic diversity: *fingerprint*
Whole biofilm		
CR, GPP and NPP: *O_2 change, MicroResp*	**AFDM**: *weight of organic material after burning biomass*	**3D structure**: *confocal microscopy*
PO4 and NH4 uptake: *nutrient addition*	**DW**: *weight of the whole biofilm after drying biomass*	**Accumulation and bio accumulation**: *intracellular/total metal concentration, total concentration of chemicals*
Antioxidant response: *antioxidant enzyme activities (AEA)*		**Contaminant transfer**: *food web experiments*
Extracellular degradation of organic matter: *EEA by enzymatic assays*		
Leaf litter breakdown: *biomass changes*		

PAM, pulse amplitude modulated fluorescence; HPLC, high-performance liquid chromatography; fingerprint, DGGE (denaturing gradient gel electorphoresis); T-RFLP, terminal restriction fragment length polymorphism; FISH, fluorescence *in situ* hybridization; NGS, next-generation sequencing; MicroResp, basal/substrate induced respiration; EEA, extracellular enzyme activity; AEA, antioxidant enzyme activity; qPCR, quantitative polymerization chain reaction; AFDM, ash-free dry mass; CR, community respiration; GPP, gross primary production; NPP, net primary production; DW, dry weight.
The endpoints and methods used to assess the effects of chemicals can be specific to the different biofilm communities: phototrophic organisms; bacteria; fungi and protozoa, or affect the whole biofilm. These methods provide information about functional attributes (from molecular and physiological responses to biofilm-mediated ecosystem functions), changes in biomass, effects on the community structure (e.g. community composition) and architecture (3D structure) of biofilms or accumulation and trophic transfer of chemicals.

heat shock protein 70/90 (HSP70 and HSP90) on the diatom *Ditylum brightwellii* after copper and nickel exposure but not after exposure to endocrine-disrupting chemicals (BPA, PCB, and endosulfan), revealing that these genes are differentially involved in the defence response against various environmental stressors. Moreover, gene expression is an early and sensitive biomarker of toxicant exposure. Actually, qPCR tools are able to reveal toxic effects, whereas other endpoints like growth inhibition are not (Bopp and Lettleri, 2007; Kim Tiam *et al.*, 2012). They were also shown to respond at environmental concentrations. Indeed Kim Tiam *et al.* (2012) observed early differential expression of genes involved in regulation of mitochondrial metabolism (*cox1*, *nad5*, 12S) and photosynthesis (*psaA*, *d1*) on the diatom *Eolimna minima* after exposure to cadmium concentrations of 10µg/l.

Effects of pollutants on bacteria

Given the generally close link between bacterial and algal production in stream biofilms (Scott *et al.*, 2008), effects of toxicants on biofilm bacterial communities can be either direct, or indirect by following changes in the autotrophic component (Ricart *et al.*, 2009; Proia *et al.*, 2011). The functional response of biofilm bacteria to environmental stressors can be evaluated using a large set of global descriptors, including bacterial growth (Lawrence *et al.*, 2007), bacterial production, by measuring incorporation of radiolabelled thymidine (Paulson *et al.*, 2000; Blanck *et al.*, 2003), and bacterial survival rates (Ricart *et al.*, 2010) (Table 7.1). Toxicants can also affect biogeochemical processes associated with bacterial metabolism, such as organic matter decomposition and nutrient cycling. Such effects on biofilm bacterial communities can be assessed through the measurement of extracellular enzyme activities (EEA) involved in carbon, nitrogen or phosphorus acquisition (Ricart *et al.*, 2009; Tlili *et al.*, 2010; Fechner *et al.*, 2012), or through the measurement of gas production to evaluate basal or substrate-induced respiration (Tlili *et al.*, 2011a,b), denitrification (Chénier *et al.*, 2006; Wang *et al.*, 2014) or community-level physiological profile (Lawrence *et al.*, 2004, 2007; Boivin *et al.*, 2006; Tlili *et al.*, 2011b). The potential of biofilm bacterial communities to degrade or mineralize organic compounds (e.g. pesticides, pharmaceuticals or endocrine disruptors) can also be viewed as a promising ecotoxicological tool (Paje *et al.*, 2002; Pesce *et al.*, 2009; Writer *et al.*, 2011a, 2011b). In addition to their functional impact, toxicants may affect the structure and diversity of biofilm bacteria. Those effects can be assessed quantitatively, by determining bacterial cell densities using microscopy (Proia *et al.*, 2011, 2012b) or flow cytometry (Villeneuve *et al.*, 2011), and semi-quantitatively, by using fluorescence *in situ* hybridization (FISH) and catalysed reported deposition-fluorescence *in situ* hybridization (CARD-FISH) to detect the impact of toxicants on community composition at a broad phylogenetic level (Brummer *et al.*, 2000; Lawrence *et al.*, 2007; Proia *et al.*, 2013a). Toxicant effects on the bacterial community composition can also be evaluated by using molecular fingerprint techniques (Dorigo *et al.*, 2010; Tlili *et al.*, 2010). New perspectives are now given by next-generation sequencing (NGS) that provide a more detailed characterization of community composition and allow taxonomic identification of bacterial community members, as shown by recent studies aimed at assessing bacterial diversity on river biofilms using NGS-based approaches (Besemer *et al.*, 2012; Hall *et al.*, 2012; Bricheux *et al.*, 2013) (Table 7.1).

In the last decade, ecotoxicology has also been focused on investigating the fate and effects of antibiotics in nature. As an example, the prevalence of antibiotic resistance genes in bacteria of stream biofilms has recently been demonstrated (e.g. Dutour *et al.*, 2002; Fox *et al.*, 2008; Marti *et al.*, 2013), as well as the effects of real mixtures of antibiotics detected in the bacterial compartment of highly impacted river biofilm (Proia *et al.*, 2013a).

Effects of pollutants on fungi

Evaluation of chemical stress in aquatic fungal communities has been mostly performed in leaf biofilms, because of the great fungal biomass accrual (ca. 98% of total microbial biomass) and strong toxicant adsorption potential in this substratum. Responses of leaf fungal communities to toxicants are mostly evaluated through the litter breakdown (Moreirinha *et al.*, 2011; Artigas *et al.*,

2012; Flores *et al.*, 2014), a key ecosystem process used as an indicator of functional stream integrity (Gessner and Chauvet, 2002). Metals (e.g. copper and zinc) and organic pesticides (e.g. azole fungicides) can depress litter decomposition (Duarte *et al.*, 2008; Artigas *et al.*, 2012) above a certain threshold concentration. Toxicant effects may be based on the respiration (substrate-induced respiration) and reproduction (sporulation) activities of the fungal community (Tlili *et al.*, 2010; Moreirinha *et al.*, 2011). Functional descriptors, such as cellulolytic (cellobiohydrolase), hemicellulolytic (β-xylosidase) and ligninolytic (phenol oxidase) extracellular enzyme activities, have been used to determine toxicant impairment on fungal capacities to degrade organic matter and alter carbon cycling in rivers (Artigas *et al.*, 2012). Methodological approaches based on gene regulation encoding for extracellular enzymes (e.g. quantitative real-time PCR, Solé *et al.*, 2012) have become promising tools to advance in the understanding of molecular mechanisms controlling microbial activities involved in carbon cycling and mitigation of environmental pollution (e.g. pesticide degradation). From a structural point of view, the density and taxonomic composition of aquatic hyphomycete communities (dominant in submerged leaves) are shown to be sensitive to heavy metals (Duarte *et al.*, 2008) and organic pesticides (Bundschuh *et al.*, 2011). Genetic approaches (including fingerprint, and NGS-techniques) are considered as useful tools to identify toxicant effects in aquatic hyphomycete communities (Moreirinha *et al.*, 2011; Artigas *et al.*, 2012; Tolkkinen *et al.*, 2013; Flores *et al.*, 2014), but *in situ* approaches are lacking regarding the literature. In parallel, the use of stable isotope probing techniques (optimized for soil microbial communities, Park *et al.*, 2006) are promising tools to identify populations capable of degrading pollutants and, therefore, of comprehending the adaptation potential of fungal communities in contaminated ecosystems including their use in bioremediation (Table 7.1).

Effects of pollutants on protozoa

As unicellular organisms associated to biofilms, protozoa are closely in contact with the surrounding environment and show high sensitivity to aquatic pollution. Compared to other aquatic consumers, protozoa communities have a faster physiological response and succession process (i.e. the replacement of species over time) due to their higher growth rate (Salvadó *et al.*, 1995; Nicolau *et al.*, 2001; Zhou *et al.*, 2008; Madoni, 2011). Indeed, protozoa are also affected by pollutants. Heavy metals (Niederlehner and Cairns, 1992; Madoni, 2000; Holtze *et al.*, 2003; Díaz *et al.*, 2006; Martín-González *et al.*, 2005; Rico *et al.*, 2009; Ancion *et al.*, 2013), ammonia (Niederlehner and Cairns, 1990), pesticides (Shi *et al.*, 2013), polycyclic aromatic hydrocarbons, PAHs (Lara *et al.*, 2007) and nanoparticles (Mortimer *et al.*, 2010) among other pollutants (Bringmann and Kühn, 1980; Nalecz-Jawecki *et al.*, 1993; Selivanovskaya *et al.*, 1997) have been demonstrated to affect protozoa. The effects of each pollutant vary depending on its concentration and its exposure time (Cairns and Pratt, 1993) and by the specific capability of each species to acclimatize, to recover its population and to bioaccumulate the pollutant (Martín-González *et al.*, 2006). In that sense, the study of structural and functional attributes of the protozoa community provides several useful endpoints for assessing pollution in aquatic ecosystems. Effects of pollutants have been observed on protozoa richness (Gracia *et al.*, 1994; Fernandez-Leborans and Novillo, 1995; Nicolau *et al.*, 2005) or species composition (Fernandez-Leborans and Novillo, 1995; Canals *et al.*, 2013), e.g. the stalked ciliate *Opercularia* spp is normally associated to stressed or polluted ecosystems. In addition to classical endpoints, such as mortality (Bergquist and Bovee, 1976; Salvadó *et al.*, 1997) or duplication rate (Salvadó *et al.*, 1997; Gomiero *et al.*, 2012), effects of pollutants on cell viability (e.g. Nalecz-Jawecki *et al.*, 1993; Salvadó *et al.*, 1997; Mortimer *et al.*, 2010), grazing activity or endocytotic rate (K_e) (Nicolau *et al.*, 2001; Gomiero *et al.*, 2012) have also been measured. Finally, effects of toxicity at cellular level, such as lysosomal membrane stability (Gomiero *et al.*, 2012), cytoplasmatic vacuolization and mitocondrial degeneration have also been observed (Martín-Gonzalez *et al.*, 2006). In addition to classical methods based on microscopic analyses, fingerprinting techniques, such as denaturing gradient gel electrophoresis (DGGE) and terminal restriction fragment length polymorphism (T-RFLP), are gaining greater prominence, as these approaches are increasing our knowledge of the complexity of biofilm protozoa communities (Dopheide *et al.*, 2008, 2009). Nevertheless, combining microscopic and molecular analyses is recommended to obtain further information.

Ecotoxicological responses of the whole biofilm

Biofilms are not only an assemblage of aquatic organisms but ubiquitous complex structures with a large proportion of non-living organic and inorganic matter with a high adsorption capacity. While many ecotoxicological investigations focus on the effects of chemicals on specific compartments of the biofilm, endpoints providing information about the effects of chemical exposure on the whole biofilm, such as bioaccumulation, oxidative stress or nanoparticle toxicity, are also interesting (Table 7.1). Investigations describing processes at biofilm scale, like primary production and nutrient dynamics, provide the opportunity of upscaling the effects of pollutants on biofilms to ecosystem functioning.

Accumulation of pollutants in natural biofilms

Total concentrations of chemicals in water fluctuate in time, and do not always reflect the integrated exposure to water chemicals of organisms living in that environment, thus complicating the establishment of direct relationships to toxicity. Monitoring chemical bioaccumulation may overcome this problem because it can represent real bioavailability and exposure. Thus, the accumulation of pollutants in biofilms can be considered the first step in the exposure of microbial organisms living in the biofilm matrix and of those placed at higher trophic levels. In addition, it can also be considered as a detoxification pathway (see Chapter 10).

Bioaccumulation of chemicals in biofilms is influenced by several interacting physical and chemical parameters of the environment like current velocity, temperature, pH, nutrients and organic matter concentration in water or the hydrophobicity of each compound (Headley *et al.*, 1998; Sabater *et al.*, 2002; Meylan *et al.*, 2004; Lundqvist *et al.*, 2012), but also by biological proprieties of the biofilm, such as its age, thickness or EPS composition (Headley *et al.*, 1998; Lawrence *et al.*, 2001). Bioaccumulation kinetics of chemicals are rather complex and depend on the substance's chemical properties, as well as on uptake mechanisms that may be passive and/or active. Metal bioaccumulation in biofilms has been studied extensively, and is described as a two-step process. Metals are first adsorbed extracellularly (in the EPS or onto cell surfaces), before being absorbed into cells by uptake mechanisms (Holding *et al.*, 2003). Intracellular and total metal content in biofilms can be measured easily, to improve the description of exposure (Meylan *et al.*, 2003; Morin *et al.*, 2008a; Serra *et al.*, 2009). In spite of the expected variability in bioaccumulation capacity of biofilms, a large number of studies reported a strong relation between metal bioaccumulation and changes in the structure, composition and function of algal and bacterial communities living in biofilms (Duong *et al.*, 2008; Morin *et al.*, 2008b; Ancion *et al.*, 2010; Bonet *et al.*, 2012; Corcoll *et al.*, 2012c). Studies reporting herbicide bioaccumulation in biofilms are rather numerous (Headley *et al.*, 1998; Lawrence *et al.*, 2001). However, the investigations of the link to toxicity are scarce, probably because of the highly complex and diverse toxicokinetics of these compounds and the impossibility to separate between intracellular and extracellular accumulation. More recently, several authors have reported the bioaccumulation of pharmaceuticals and endocrine disruptors in biofilms (Writer *et al.*, 2011a, 2013; Wunder *et al.*, 2011). However, their link to toxicity on biofilm is still not confirmed. As many compounds susceptible to provoking deleterious impacts on the biota are likely to be accumulated in biofilms (e.g. Lawrence *et al.*, 2001; Sabater, 2003), measuring toxicant concentrations in this 'natural passive sampler' – the biofilm – may be a valuable alternative to traditional chemical monitoring. This measure would provide ecologically relevant information about the potential risk of contaminants for the aquatic ecosystem and may be especially useful and reliable for those compounds not undergoing metabolization into the biofilm (e.g. metals).

Detecting biofilm under oxidative stress

Chemical contamination in biofilm is likely to induce direct or indirect oxidative stress by enhancing reactive oxygen species (ROS) production or impairing cellular antioxidant responses. The resulting excess in ROS can provoke lipid peroxidation, membrane disruption, alteration in cell structures and mutagenesis (Scandalios, 1993; Mittler, 2002; Edreva, 2005; Wolfe-Simon *et al.*, 2005; Lesser, 2006). Though oxidative stress can be specifically induced by some toxicant (e.g. copper), it can also result from general metabolism alteration and thus indicates a low 'health' status of biofilm. Therefore, the detection of oxidative stress damage and response within the whole biofilm community is expected to provide information on biofilm stress status and its ability to cope with further oxidative stress (Bonnineau *et al.*, 2013)

Lipid peroxide quantification is a common measure of cellular oxidative damage that can be estimated at community level. For instance, Vera *et al.* (2012) used the thiobarbituric acid-reactive substances (TBARS) assay to show how exposure to an environmentally relevant concentration of a glyphosate formulation provoked oxidative damage in the biofilm community.

Nevertheless, most of the recent work has been focused on biofilm antioxidant capacity, rather than on oxidative damage. In fact, to keep the oxidative balance under control, organisms have non-enzymatic mechanisms (e.g. glutathione, carotenoids and phenolics; Okamoto *et al.*, 2001) as well as enzymatic mechanisms (e.g. glutathione-S-transferase: GST, catalase: CAT, ascorbate peroxidase: APX, glutathione reductase: GR and superoxide dismutase: SOD activities). In particular, several authors have proposed using antioxidant enzyme activities (AEAs) as biomarkers of pollution due to their capacity to respond to both organic and inorganic pollutants (Valavanidis *et al.*, 2006; Guasch *et al.*, 2010a,b; Maharana *et al.*, 2010; Bonnineau *et al.*, 2011; Bonet *et al.*, 2012, 2013, 2014). In biofilms, AEAs are defined as a global indicator of the 'health' status of the whole biofilm, then considered as a black box. AEA measurement at community level is expected to reflect the tendency (activation or inhibition) observed in the majority of individuals and species within the community (Bonnineau *et al.*, 2012).

Biofilm AEAs have been used at different scales in both laboratory and field studies, mainly to determine the antioxidant response of the community to a specific chemical. For instance, in several studies, AEAs have been found to be more sensitive to contaminant than traditional biomarkers such as photosynthetic parameters (Dewez *et al.*, 2005; Guasch *et al.*, 2010b; Bonet *et al.*, 2013, 2014). Measuring AEA response throughout a gradient of oxidative stress can also provide information on the antioxidant capacity of the community. Indeed, AEAs are expected to increase with increasing oxidative stress until ROS overcomes the cell defence system and AEAs eventually decrease due to cellular damage. From this unimodal (bell shape) pattern of response, a range of oxidative stress levels by which AEAs increased can be defined; within this range the community is expected to be able to alleviate oxidative stress. This range defines the antioxidant capacity of a community and is influenced by various parameters (e.g. biofilm age, pre-exposure to contamination). For instance, chronic exposure of biofilm to the herbicide oxyfluorfen led to an increase in biofilm CAT capacity. Biofilms chronically exposed to oxyfluorfen were able to respond to higher concentrations of oxyfluorfen by an increase in CAT activity while in non-adapted biofilms (those not previously exposed), CAT activity decreased in response to acute exposure to high levels of oxyfluorfen, probably because of oxidative damage (Bonnineau *et al.*, 2013).

Since oxidative stress can be greatly influenced by environmental parameters, such as light or temperature (Butow *et al.*, 1997; Aguilera *et al.*, 2002; Li *et al.*, 2010), both laboratory and field studies are needed to better understand AEAs responses and interpret their variations (Bonnineau *et al.*, 2013). For instance, Bonet *et al.* (2012) showed that, under controlled conditions (microcosm study), APX clearly decreased due to Zn exposure while, in the field, the inhibition of GST was shown to be a biomarker of Zn exposure (Bonet *et al.*, 2013, 2014). These differences were attributed to variations in environmental parameters and a specific effort has been made to better understand field variability. In an annual monitoring, Bonet *et al.* (2013) observed that AEA followed the seasonality of the system, changing as a response to light and water temperature fluctuations. However, seasonality was not observed in the polluted site, where Zn masked this pattern of variation.

Detecting oxidative stress in biofilms provides information on oxidative damage (e.g. Lipid peroxidation), antioxidant responses and antioxidant capacity of the community. These markers of oxidative stress can be used to detect alteration within biofilm community due to pollutant exposure but also to environmental variations (e.g. climate change) (Bonet *et al.*, 2013).

Differential biofilm gene expression

In aquatic ecosystems, biofilm ecotoxicology has been used to investigate contaminant effects at different levels of biological organization, from species composition to biogeochemical processes. New approaches suggest going even deeper within biofilm and investigating structure and function at molecular level. Differential gene expression has been studied until now on single species (i.e. diatom cultures). Nevertheless, the tools developed for diatoms are extremely promising and in the future could be expanded to the whole biofilm. The qPCR tools have been tested with less success at the community level (i.e. a biofilm composed of different diatom species and other organisms), using these specific gene sequences principally because of the lack of available nucleotide sequences of such organisms in genomic databases (Tiam, unpublished data).

Biofilms in nanoparticle ecotoxicology

Experiments with biofilms are an optimal target for assessing the environmental risks related to new emerging toxicants such as nanoparticles. As biofilms grow on submerged surfaces, they are especially exposed to engineered nanoparticles (ENPs). Nanotechnology development is leading to a proliferation of products that are likely to become a source of many different engineered nanoparticles (ENP) in the environment, where their fate, behaviour and effects are mostly unknown. Their nano-size allows these materials to interact at molecular scale with organisms present in the environment. Among other effects in rivers, ENPs may impact on photosynthetic organisms. The ENPs have direct and indirect toxicological effects on different organisms present in biofilms (Navarro *et al.*, 2008). The physical characteristics of ENPs facilitate their transport in suspension. In addition, their large density, as that of metallic nanomaterials and their surface properties, which may enhance agglomeration processes, may provoke sedimentation under reduced hydrodynamics. This process will deposit ENPs in biofilms where biouptake may take place, thus leading to toxic effects. In addition, since biofilm ecotoxicological testing can be done under the controlled conditions of micro or mesocosms (artificial channels), certain methodological problems associated to ENP experimentation, mostly related to the lack of control and characterization of ENPs and the environmental conditions prevailing during the exposure of the organisms, will be avoided (Handy *et al.*, 2012).

Biofilm ecotoxicology – link between pollution and ecosystem health

Biofilms are considered biological entities which play a key role in ecosystem functioning, and are in turn very sensitive to chemical exposure. Investigations aiming to describe processes at biofilm scale like nutrient dynamics and those including simple food chains, are common in ecological research but less used in ecotoxicology. These approaches have recently been applied, providing the opportunity of upscaling the effects of pollutants on biofilms to food webs and ecosystems.

Upscaling biofilm responses to ecosystem processes

Biofilm communities are composed of many microbial species with a key role in ecosystem functioning offering important insights regarding mechanisms occurring from the single cell level to biogeochemical processes at a larger scale by mediating processes, such as oxygen production, nutrient uptake and organic matter transformation (Battin *et al.*, 2003; see Chapter 5). In fluvial systems, for example, combining community-scale (such as mesocosm experiments) and

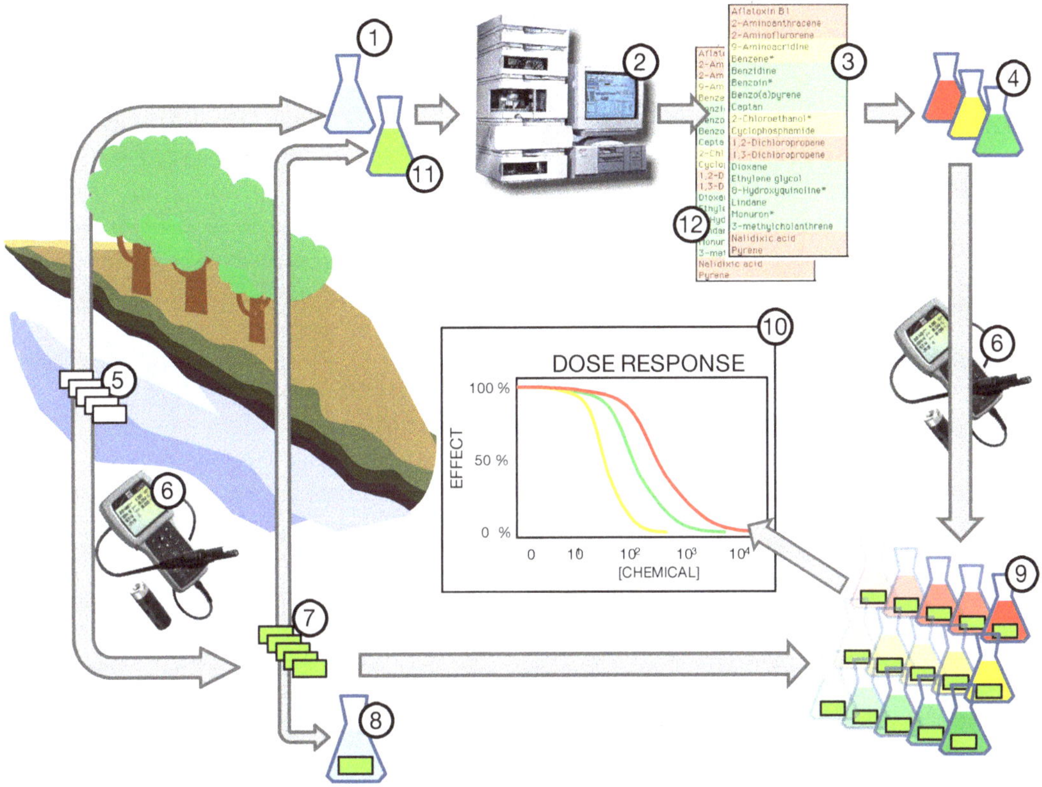

Figure 7.1 Conceptual framework for the use of freshwater biofilms in aquatic ecotoxicology. Water samples (1) from a river affected by chemical pollution are analysed (2). The chemicals present in the water (3) are classified according to their mode of action (different shaded colours), and a reduced number of chemicals representing the different modes of action (represented by different colours) are selected (4). In parallel, artificial substrata (5) are placed at the sampling points, and during a few weeks or months (depending on the type of community and local conditions), the environmental variables are monitored (6). This information can be used later during the experimental design, to study their role as modulating factors of the measured ecotoxicity. After the colonization time, the substrata containing the freshwater biofilms are sampled (7), some are preserved for analytical purposes (8), and others are used for dose–response experiments (9). Results are plotted and modelled (10) and the relevant concentrations measured (as the Effective Concentration reducing by 50% the biological parameter measured: EC_{50}). In addition, biofilms can be processed to extract (11) and analyse (2) the chemicals adsorbed and/or bioaccumulated. The list of the detected chemicals (12) can complement the information obtained from water samples (3).

whole-reach scale (field or ecosystem) measurements together with the characterization of the structure and function of biofilms contribute to a better understanding of the effects of human activities on ecosystem processes.

Rosi-Marshall *et al.* (2013) in a field study, using Nutrient Diffusing Substrata (NDS) found that a mix of pharmaceuticals produced changes in biofilm metabolism parameters, consisting of a decrease in community respiration (CR) and gross primary production (GPP). Another study used field translocation between impacted and non-impacted sites to assess the effects of industrial discharge on biofilm functioning. These authors found a rapid shift (one week) from autotrophic- to heterotrophic-dominated metabolism when the communities from the non-impacted site were translocated to impacted sites. In contrast, when communities were transferred back to the reference site, the recovery to autotrophy took up to four weeks (Sierra and Gómez, 2010).

Hill *et al.* (1997) used chambers to measure benthic metabolism in a rocky mountain stream which had elevated metal pollution and found that GPP and net primary productivity (NPP) decreased with increasing metal concentrations by one order of magnitude from the reference site to the most impacted site. The effects of different pharmaceuticals on biofilm metabolism and nutrient uptake were assessed *in vitro* and *in situ* (using NDS in the field) in a central Indiana river (USA). The *in vitro* experiments showed that ammonium uptake was reduced after exposure to nicotine and caffeine, and nitrate uptake was increased by nicotine exposure, while no effects were observed on microbial metabolism. On the other hand, an *in situ* experiment showed that nicotine increased microbial respiration (Bunch and Bernot, 2011). Nutrient uptake was also used to assess the effects of metals on fluvial biofilms. Serra *et al.* (2009) found a slight decrease in phosphate uptake after chronic copper exposure of the biofilm to 26 μg/l in artificial channels. In another mesocosm study, Proia *et al.* (2011) showed that triclosan (60 μg/l) inhibited biofilm phosphate uptake up to 71% and uptake rates did not recover until two weeks after the end of exposure. The negative effect of triclosan on biofilm capacity to uptake phosphate was confirmed in other investigations using microcosms and revealed the persistence of this effect over time (Proia *et al.*, 2013b; Guasch *et al.*, *in press*).

These studies exemplify how classical biofilm processes, such as nutrient uptake and community metabolism commonly investigated in ecosystem ecology, are sensitive tools for assessing the ecotoxicological effects of pollutants on freshwater communities and ecosystems. In addition, addressing these endpoints allows the ecological relevance of the observed effects at different levels, from community to whole ecosystem scales, to be increased.

Biofilms in ecotoxicological food web studies

Studies based on food-web relationships between biofilms and their grazers provide a high degree of environmental realism. Due to the increased complexity, these studies allow us to assess the responses of communities and within communities and the evaluation of the direct and indirect effects of pollutants at different trophic levels (Culp *et al.*, 2000; Geiszinger *et al.*, 2009).

These investigations are mainly performed in experimental conditions in order to ensure control of environmental variables (Fig. 7.2), but some field studies also exist. Literature reviews provide interesting experimental models (Culp *et al.*, 1996; Ledger *et al.*, 2009). In most cases,

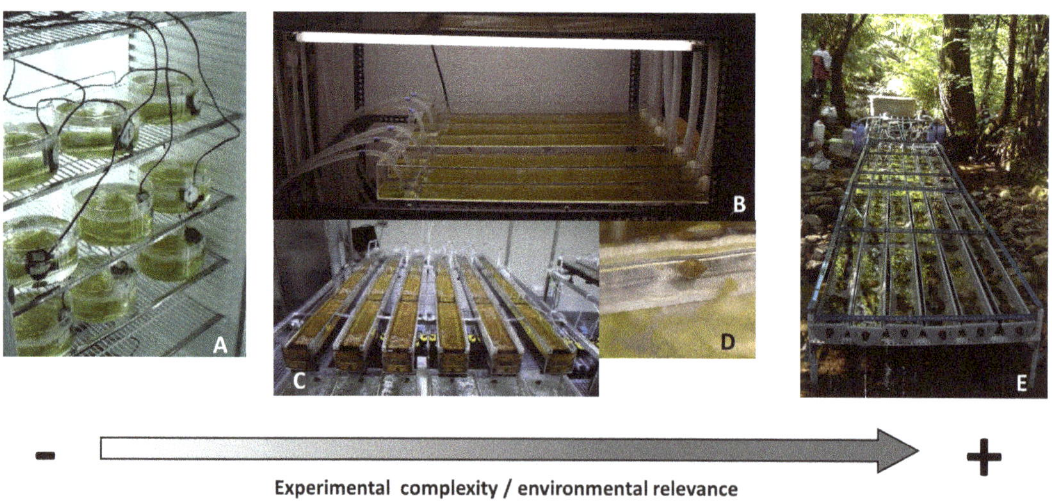

Figure 7.2 Experimental settings of biofilm ecotoxicology. Biofilm communities growing on artificial substrata (e.g. sandblasted glass substrata) are exposed to chemicals under controlled conditions. Exposure can be done in (A) crystallizing dishes (1.5 l volume) with recirculating water; (B) recirculating indoor channels (1 m long); (C) one flow through indoor channels (2 m long). (D) detail of a snail (grazer) placed on top of biofilms in a grazing experiment. (E) one flow through outdoor channels (5 m long).

food-web experiments involve biofilms and a grazer in order to study biomagnification and transfer of the test substance from primary producer to consumer. Other experiments address the possible additional effects of grazing pressure on a chemically stressed biofilm or the possible indirect effects of pollutants on grazers or biofilms due to toxicant-induced alterations of ecological relevance. Generally speaking, insects and molluscs have been used as grazers. Food-web experiments with biofilms have been applied in order to study the ecotoxicological effect of pesticides (Muñoz *et al.*, 2001; Real *et al.*, 2003; López-Doval *et al.*, 2010; Lundqvist *et al.*, 2012), metals (Irwing *et al.*, 2003; Conley *et al.*, 2011; Xie and Buchwalter, 2011; Kim *et al.*, 2012; Li *et al.*, 2012), nanoparticles (Kulacki *et al.*, 2012) and emerging pollutants (Evans-White and Lamberti, 2009), among other compounds.

Several authors demonstrated the importance of biofilms in the introduction of toxicants in the food web by means of food-web experiments. In the case of zinc, bioaccumulation in biofilm, metal transfer and bioaccumulation in the grazer *Centroptilum triangulifer* were shown (Kim *et al.*, 2012). Irwing *et al.* (2003) demonstrated that mayflies grazing on biofilms contaminated with cadmium showed significant inhibition in growth and feeding in comparison to those exposed to contaminated water. Xie and Buchwalter (2011), using biochemical responses in the mayfly *C. triangulifer*, confirmed that cadmium is more toxic by ingestion of contaminated biofilm than by direct exposure to contaminated water. Experiments with food webs demonstrated that high nutritional quality and quantity of available biofilm diminish the toxicological response of mayflies to selenium (Conley *et al.*, 2011). Bioavailability of pollutants is modulated by the influence of environmental factors on biofilm, as demonstrated with food-web experiments. Increasing levels of phosphate enhanced bioaccumulation of copper in biofilms and dietary toxicity to the amphipod *Hyalella azteca* (Li *et al.*, 2012). In an experiment with freshwater snails and biofilms, Lundqvist *et al.* (2012) reported that dissolved organic matter in water interferes in the sorption of pesticides (carbofuran, lindane and chlorpyrifos) to biofilms and is, therefore, a factor that can modulate bioavailability and bioaccumulation of insecticides.

The presence or absence of grazers can interfere in the effects of toxicants on functional or structural characteristics of biofilms. Muñoz *et al.* (2001) studied the effects of atrazine in a single food web and described reduction of carbon incorporation and algal diversity in biofilm due to the interaction of grazers (*Physa acuta*) with the herbicide. Evans-White and Lamberti (2009) observed that toxicants in combination with grazers increased chlorophyll concentration and algal diversity. On the contrary, similar experiments with food webs did not find interactive effects of grazing and toxicants on biofilm (Real *et al.*, 2003; López-Doval *et al.*, 2010). Indirect effects on the structure and function of biofilms have been observed as a consequence of the changes in the physiology and behaviour of *P. acuta* induced by the toxicant (Evans-White and Lamberti, 2009).

Overall, it is reasonable to expect that grazing may influence the response of biofilms to toxic exposure. Communities suffering both grazing pressure and the effects of toxic substances will have less ability to overcome grazing effects than non-exposed communities, because toxicity will limit algae regrowth and facilitate the extinction of the less abundant species after grazing. This interaction may have remarkable ecological implications since grazing pressure will magnify the negative effects that toxicants exert on ecosystem processes, such as primary production and nutrient cycling (Fig. 7.3).

Environmental factors modulating biofilm response to pollutants

In the field, environmental conditions are highly variable and organisms are rarely under optimal conditions. There is growing awareness that these abiotic parameters can strongly constrain ecosystem responses to anthropogenic contamination (Fisher *et al.*, 2013). Nevertheless, their influence is rarely taken into account in single-species ecotoxicological tests. Indeed, single species have a limited range of acclimation to environmental parameters and studies performed at community level appear to be better suited for investigating the influence of environmental factors on contamination effects (Clements *et al.*, 2009). In

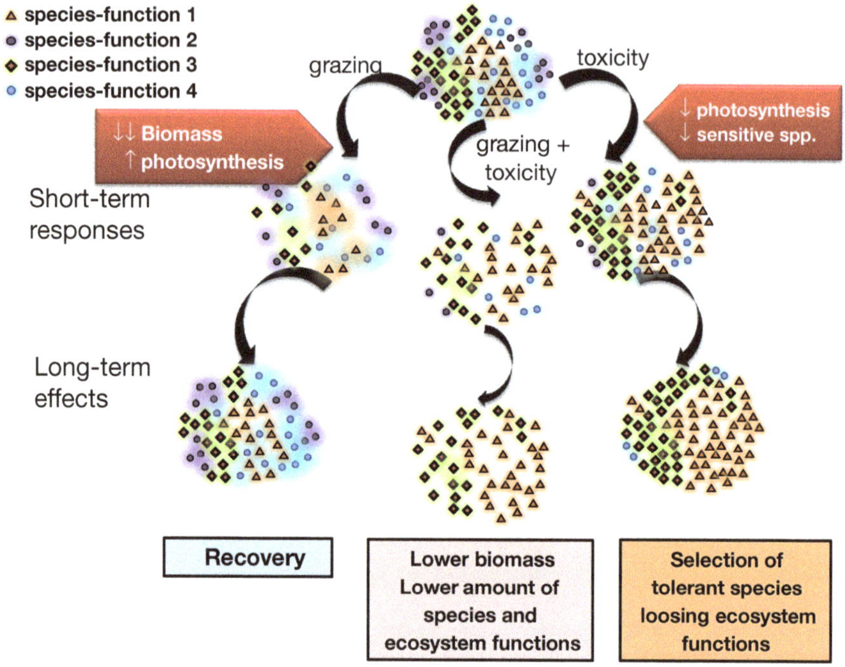

Figure 7.3 General model of the individual and combined effects of grazing and toxicity on biofilms. Based on a simple four-species biofilm model, it is expected that toxicity will constrain the ability of the community to recover from grazing pressure. In addition to the selection pressure exerted by toxicity, causing a reduction in activity (i.e. photosynthesis) and an increase in the relative abundance of the most tolerant species (sp1 and sp3), the reduction in population size caused by the non-selective effect of grazing on biofilms, will increase the risk of extinction for the less abundant species. Overall, grazing and toxicity will have cumulative negative effects on biofilms causing a reduction in the number of species, the biomass and ecosystem functions.

aquatic ecosystems, environmental parameters, such as light intensity, flow regime or temperature, strongly influence biofilm structure and function (Chapter 1) and these factors can have a critical effect on biofilm community response to contamination (Fig. 7.4).

Light intensity and regime is highly variable in the field due to seasonal variations and/or changes in riparian vegetation. Nevertheless, light is the first energy source for the autotrophic component of biofilm and therefore modulates not only biofilm structure and function but also biofilm response to herbicides and metals, as shown by several authors at both laboratory and field scale (Guasch *et al.*, 2003; Laviale *et al.*, 2010; Bonnineau *et al.*, 2012; Bonet *et al.*, 2013). Not only was biofilm grown under high light intensity more sensitive to the herbicide atrazine (field study, Guasch *et al.*, 2003), but it was also more tolerant to glyphosate (laboratory study, Bonnineau *et al.*, 2012).

While flow regime can affect chemical bioavailability (Osorio *et al.*, 2014), this highly variable abiotic factor can also modulate biofilm structure and function (Graba *et al.*, 2013). Therefore, the flow regime under which biofilm is grown is also susceptible to alter the capacity of biofilm to cope with chemical toxicity. For instance, a simulated drought event in artificial streams reduced biofilm capacity to recover from a subsequent 48 hour exposure to a bactericide (87 μg/l of triclosan) at both structural (high bacterial mortality) and functional level (reduced phosphate uptake) (Proia *et al.*, 2013b). Villeneuve *et al.* (2011) also showed that biofilms grown under a turbulent flow regime have a higher sensitivity to pesticides than biofilms grown under a laminar flow regime.

The influence of other factors like sediment deposition (Magbanua *et al.*, 2013), temperature (Larras *et al.*, 2013), nutrient concentration (Tlili *et al.*, 2010) or salinization (Rotter *et al.*, 2013)

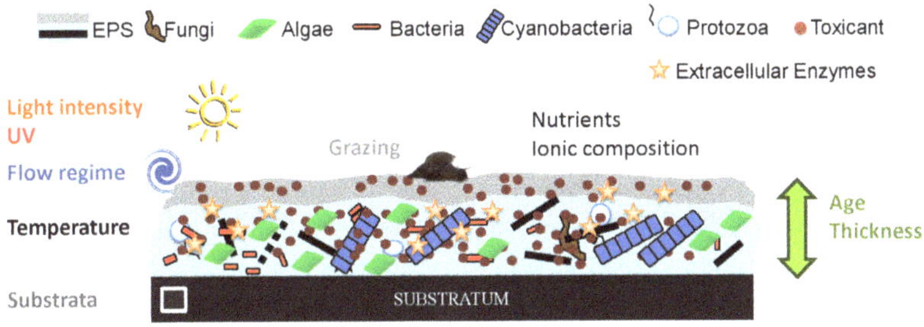

Modulating environmental factors	Biofilm	Toxicant effect
⇕	Mortality	⇧
⇕	Biomass	⇩
■ ⇕	Species composition	Change
	Photosynthesis	⇩
	Oxidative stress	⇧
⇕	Food quality for grazers	⇩

Figure 7.4 Interactions between environmental factors and contamination in river biofilms. The main abiotic factors modulating biofilm structure and function are indicated in a schematic view of a biofilm (adapted from Romani, 2010). Environmental parameters and toxicants are likely to affect similar biofilm parameters, as indicated in the table; the expected negative impact of a toxicant is indicated by an arrow.

on biofilm response to pollutants has also been investigated (Fig 7.4).

These previous studies have shown how environmental parameters can constrain community capacity to respond to a pollutant but also to recover from contamination exposure. To better understand ecosystem responses to contamination, it is essential to take into account these parameters in toxicity assessment. Since the influence of abiotic factors on biofilm structure and function has been intensively investigated in ecology, the use of biofilm in ecotoxicology appears then as a realistic approach, in which environmental parameters can be integrated into toxicity assessment.

Conclusions and future recommendations

Biofilms are nowadays one of the principal targets of community ecotoxicology with a high potential for future uses in ecotoxicology. A large set of methods derived from biofilm ecology have successfully been applied in ecotoxicology providing a diverse and comprehensive toolbox. On the one hand, our ability to quantify the effects of pollution on different biofilm components, allows us to evaluate the direct effects of pollutants on the most sensitive community (e.g. algae in the case of herbicides or bacteria for antibiotics) and also their indirect effects on the rest of biofilm components and on higher trophic levels because all of them are closely related through biological interactions. For example, the model presented for biofilms exposed to toxicants under grazing pressure exemplifies the advantage of using complex biological models like biofilms and their grazers to improve our ability to predict the effects of pollution in multiple-stress scenarios (Fig. 7.3). On the other hand, enormous progress has been made regarding sensitivity. The application of early warning systems, for example the study of AEAs in whole biofilms, may allow us to detect early responses of the community by the activation of mechanisms of defence towards toxicity. In terms of analytical chemistry, different methods have been refined to quantify low concentrations of a large panel of chemicals in biota, including biofilm samples. In addition to metals, recent investigations have shown that many organic pollutants have a tendency to adsorb and/or be uptaken in biofilms, acting as 'natural passive samplers'. Biofilms are also a site for biotransfomation and/or transfer of chemicals to other aquatic organisms, supporting a more generalized use of biofilm samples in environmental chemistry. This methodological progress is also visible in terms of new applications like the use of biofilms to investigate nanoparticle toxicity.

The set of biofilm endpoints described (Table 7.1) provides a powerful toolbox covering the expected responses of biofilms to pollution at different temporal scales: from early responses to acute exposure (e.g. by the activation of mechanisms of detoxification, the inhibition of photosynthesis or respiration), to long-term effects after chronic exposure (e.g. extinction of the most sensitive species and changes in the whole community structure). It is important to highlight the potential that different molecular approaches may have on our ability to detect the effects of pollution on the diversity of species of the different biofilm components (Table 7.1). In contrast to the study of some of the biofilm components, such as algae, with a long tradition in taxonomy (i.e. the use of diatom species composition as biological indicators; see Chapter 6), assessing the effects of toxicity on the species composition of other biofilm components is less common (i.e. bacteria). In this regard, the application of molecular tools may contribute to overcome this limitation, understood, however, as a complement of rather than a substitute for microscope observation classical taxonomy.

Based on the principles of ecotoxicology and their progress as a scientific discipline, there has been an increasing interest in linking chemical pollution with ecosystem health. In addition to biofilm endpoints, which are biomarkers of exposure, biofilm ecology provides an opportunity to link exposure with ecosystem functioning. Classical biofilm processes, such as nutrient uptake and community metabolism commonly investigated in ecosystem ecology, allows the ecological relevance of the observed effects from community to whole ecosystem integrity to be increased.

It is also important to point out that biofilm ecotoxicology has also benefited from the fast progress of genetics. As an example, metagenomics is envisaged as a promising approach for targeting the effect of contaminants on specific biofilm functions, and the microorganism responsible behind them.

While biofilm ecotoxicology studies have inherited the methods and basics of biofilm ecology and community ecotoxicology, a general framework to formulate a hypothesis about the response of this model of an aquatic community to human perturbations is still lacking. As shown in this review, a large set of methods has been refined and validated. Bearing this in mind, biofilm ecotoxicology should now focus on providing the theoretical background for understanding the complex set of responses of natural communities to pollution. This knowledge should also be the basis to guide the selection of the most appropriate tools and the development of new approaches for a better detection of the impact of pollution on aquatic life.

Acknowledgements

This study has been carried out with financial support from the French National Research Agency (ANR) in the framework of the Investments for the Future Programme, within the Cluster of Excellence COTE (ANR-10-LABX-45) and ANR CESA PoToMAC (ANR-11-CESA-0022). Financial support was also provided by the Spanish Science and Education Ministry (project CTM2009-14111-CO2–01) and the Spanish Economy and Competitiveness Ministry (project CGL2013–43822-R).

References

Aguilera, J., Bischof, K., Karsten, U., Hanelt, D., and Wiencke, C. (2002). Seasonal variation in ecophysiological patterns in macroalgae from an Arctic fjord. II. Pigment accumulation and biochemical defence systems against high light stress. Mar. Biol. *140*, 1087–1095.

Ancion, P.Y., Lear, G., and Lewis, G.D. (2010). Three common metal contaminants of urban runoff (Zn, Cu and Pb) accumulate in freshwater biofilm and modify embedded bacterial communities. Environ. Pollut. *158*, 2738–2745.

Ancion, P.Y., Lear, G., Dopheide, A., and Lewis, G.D. (2013). Metal concentrations in stream biofilm and sediments and their potential to explain biofilm microbial community structure. Environ. Pollut. *173*, 117–124.

Artigas, J., Majerholc, J., Foulquier, A., Margoum, C., Volat, B., Neyra, M., and Pesce, S. (2012). Effects of the fungicide tebuconazole on microbial capacities for litter breakdown in streams. Aquat. Toxicol. *122–123*, 197–205.

Battin, T.J., Kaplan, L.A., Newbold, J.D., and Hansen, C.M.E. (2003). Contributions of microbial biofilms to ecosystem processes in stream mesocosms. Nature *426*, 439–442.

Bergquist, B.L., and Bovee, E.C. (1976). Cadmium: quantitative methodology and study of its effect upon the locomotor rate of Tetrahymena pyriformis. Acta Protozool. *15*, 47.

Besemer, K., Peter, H., Logue, J.B., Langenheder, S., Lindström, E.S., Tranvik, L.J., and Battin, T.J. (2012). Unraveling assembly of stream biofilm communities. ISME J. *6*, 1459–1468.

Blanck, H., Admiraal, W., Cleven, R.F.M.J., Guasch, H., Van den Hoop, M.A.G.T., Ivorra, N., Nyström, B., Paulsson, M., Petterson, R.P., Sabater, S., and Tubbing, G.M.J. (2003). Variability in zinc tolerance, measured as incorporation of radio-labeled carbon dioxide and thymidine, in periphyton communities sampled from 15 European river stretches. Arch. Environ. Contam. Toxicol. *44*, 17–29.

Boivin, M.E.Y., Massieux, B., Breure, A.M., Greve, G.D., Rutgers, M., and Admiraal, W. (2006). Functional recovery of biofilm bacterial communities after copper exposure. Environ. Pollut. *140*, 239–246.

Bonet, B., Corcoll, N., and Guasch, H. (2012). Antioxidant enzyme activities as biomarkers of Zn pollution in fluvial biofilms. Ecotox. Environ. Safe. *80*, 172–178.

Bonet, B., Corcoll, N., Acuña, V., Sigg, L., Behra, R., and Guasch, H. (2013). Seasonal changes in antioxidant enzyme activities of freshwater biofilms in a metal polluted Mediterranean stream. STOTEN. *444*, 60–72.

Bonet, B., Corcoll, N., Tlili, A., Morin, S., and Guasch, H. (2014). Antioxidant enzyme activities in biofilms as biomarker of Zn pollution in a natural system: an active bio-monitoring study. Ecotox. Environ. Safe. *103*, 82–90.

Bonnineau, C., Bonet, B., Corcoll, N., and Guasch, H. (2011). Catalase in fluvial biofilms: a comparison between different extraction methods and example of application in a metal-polluted river. Ecotoxicology *20*, 293–303.

Bonnineau, C., Moeller, A., Barata, C., Bonet, B., Proia, L., Sans-Piché, F., Schmitt-Jansen, M., Guasch, H., and Segner, H. (2012). Advances in the Multibiomarker Approach for Risk Assessment in Aquatic Ecosystems. In Emerging and Priority Pollutants in Rivers Bringing Science into River Management Plans, Guasch, H., Ginebreda, A., and Geiszinger, A. eds. The Handbook of Environmental Chemistry. (Springer, Berlin/Heidelberg, Germany), pp. 147–179.

Bonnineau, C., Tlili, A., Faggiano, L., Montuelle, B., and Guasch, H. (2013). The use of antioxidant enzymes in freshwater biofilms: Temporal variability vs. toxicological responses. Aquat. Toxicol. *136–137*, 60–71.

Bopp, S.K., and Lettleri, T. (2007). Gene regulation in the marine diatom Thalassiosira pseudonana upon exposure to polycyclic aromatic hydrocarbons (PAHs). Gene. *396*, 293–302.

Bricheux, G., Morin, L., Le Moal, G., Coffe, G., Balestrino, D., Charbonnel, N., Bohatier, J., and Forestier, C. (2013). Pyrosequencing assessment of prokaryotic and eukaryotic diversity in biofilm communities from a French river. Microbiologyopen. *2*, 402–414.

Bringmann, G., and Kühn, R. (1980). Comparison of the toxicity thresholds of water pollutants to bacteria, algae and protozoa in the cell multiplication inhibition test. Water Res. *14*, 231–241.

Brummer, I.H., Fehr, W., and Wagner-Dobler, I. (2000). Biofilm community structure in polluted rivers: abundance of dominant phylogenetic groups over a complete annual cycle. Appl. Environ. Microbiol. *66*, 3078–3082.

Bunch, A.R., and Bernot, M.J. (2011). Distribution of nonprescription pharmaceuticals in central Indiana streams and effects on sediment microbial activity. Ecotoxicology *20*, 97 – 109.

Bundschuh, M., Zubrod, J.P., Kosol, S., Maltby, L., Stang, C., Duester, L., and Schulz, R. (2011). Fungal composition on leaves explains pollutant-mediated indirect effects on amphipod feeding. Aquat. Toxicol. *104*, 32–37.

Butow, B.J., Wynne, D., and Tel-Or, E. (1997). Antioxidative protection of Peridinium gatunense in Lake Kinneret: seasonal and daily variation. J. Phycol. *33*, 780–786.

Cairns, Jr.J., and Pratt, J.R. (1993). Trends in ecotoxicology. STOTEN (Proceedings of the 2nd European Conference on Ecotoxicology) *134*, 7–22.

Canals, O., Salvadó, H., Auset, M., Hernández, C., and Malfeito, J.J. (2013). Microfauna communities as performance indicators for an A/O Shortcut Biological Nitrogen Removal moving-bed biofilm reactor. Water Res. *47*, 3141–3150.

Chénier, M.R., Beaumier, D., Fortin, N., Roy, R., Driscoll, B.T., Lawrence, J.R., and Greer, C.W. (2006). Influence of nutrient inputs, hexadecane and temporal variations on denitrification and community composition of river biofilms. Appl. Environ. Microbiol. 72, 575–584.

Clements, W.H., and Rohr, J.R. (2009). Community responses to contaminants: Using basic ecological principles to predict ecotoxicological effects. Environ. Toxicol. Chem. *28*, 1789–1800.

Conley, J.M., Funk, D.H., Cariello, N.J., and Buchwalter, D.B. (2011). Food rationing affects dietary selenium bioaccumulation and life cycle performance in the mayfly Centroptilum triangulifer. Ecotoxicology *20*, 1840–1851.

Corcoll, N., Ricart, M., Franz, S., Sans-Piché, F., Schmitt-Jansen, M., and Guasch, H. (2012a). The Use of Photosynthetic Fluorescence Parameters from Autotrophic Biofilms for Monitoring the Effect of Chemicals in River Ecosystems. In Emerging and Priority Pollutants in Rivers, Guasch, H., Ginebreda, A., and Geiszinger, A. eds. (Springer, Berlin/Heidelberg, Germany), pp. 85–115.

Corcoll, N., Bonet, B., Leira, M., Montuelle, B., Tlili, A., and Guasch, H. (2012b). Light History Influences the Response of Fluvial Biofilms to Zn Exposure. J. Phycol. *48*, 1411–1423.

Corcoll, N., Bonet, B., Morin, S., Tlili, A., Leira, M., and Guasch, H. (2012c). The effect of metals on photosynthesis processes and diatom metrics of biofilm from a metal-contaminated river: A translocation experiment. Ecol. Indic. *18*, 620–631.

Culp, J.M., Podemski, C.L., Cash, K.J., and Lowell, R.B. (1996). Utility of field-based artificial streams for assessing effluent effects on riverine ecosystems. J. Aquat. Ecosyst. Health. *5*, 117–124.

Culp, J.M., Lowell, R.B., and Cash, K.J. (2000). Integrating mesocosm experiments with field and laboratory studies to generate weight-of-evidence risk assessments for large rivers. Environ. Toxicol. Chem. *19*, 1167–1173.

Dewez, D., Geoffroy, L., Vernet, G., and Popovic, R. (2005). Determination of photosynthetic and enzymatic biomarkers sensitivity used to evaluate toxic effects of copper and fludioxonil in alga Scenedesmus obliquus. Aquat. Toxicol. *74*, 150–159.

Díaz, S., Martín-González, A., and Gutiérrez, J.C. (2006). Evaluation of heavy metal acute toxicity and bioaccumulation in soil ciliated protozoa. Environ. Int. *32*, 711–717.

Dopheide, A., Lear, G., Stott, R., and Lewis, G. (2008). Molecular characterization of ciliate diversity in stream biofilms. Appl. Environ. Microbiol. *74*, 1740–1747.

Dopheide, A., Lear, G., Stott, R., and Lewis, G. (2009). Relative diversity and community structure of ciliates in stream biofilms according to molecular and microscopy methods. Appl. Environ. Microbiol. *75*, 5261–5272.

Dorigo, U., Bérard, A., Rimet, F., Bouchez, A., and Montuelle, B. (2010). In situ assessment of periphyton recovery in a river contaminated by pesticides. Aquat. Toxicol. *98*, 396–406.

Duarte, S., Pascoal, C., Alves, A., Correia, A., and Cássio, F. (2008). Copper and zinc mixtures induce shifts in microbial communities and reduce leaf litter decomposition in streams. Freshwater Biol. *53*, 91–101.

Duong, T.T., Morin, S., Herlory, O., Feurtet-Mazel, A., Coste, M., and Boudou, A. (2008). Seasonal effects of cadmium accumulation in periphytic diatom communities of freshwater biofilms. Aquat. Toxicol. *90*, 19–28.

Dutour, C., Bonnet, R., Marchandin, H., Boyer, M., Chanal, C., Sirot, D., and Sirot, J. (2002). CTX-M-1, CTX-M-3 and CTX-M-14 beta-lactamases from Enterobacteriaceae isolated in France. Antimicrob. Agents. Chemother. *46*, 534–537.

Edreva, A. (2005). Generation and scavenging of reactive oxygen species in chloroplasts: a submolecular approach. Agr. Ecosyst. Environ. *106*, 119–133.

Evans-White, M.A., and Lamberti, G.A. (2009). Direct and indirect effects of a potential aquatic contaminant on grazer–algae interactions. Environ. Toxicol. Chem. *28*, 418–426.

Eriksson, K.M., Antonelli, A., Nilsson, R.H., Clarke, A.K., and Blanck, H. (2009). A phylogenetic approach to detect selection on the target site of the antifouling compound irgarol intolerant periphyton communities. Environ. Microbiol. *11*, 2065–2077.

Falasco, E., Bona, F., Badino, G., Hoffmann, L., and Ector, L. (2009). Diatom teratological forms and environmental alterations: a review. Hydrobiologia *623*, 1–35.

Fechner, L.C., Dufour, M., and Gourlay-Francé, C. (2012). Pollution-induced community tolerance of freshwater biofilms: measuring heterotrophic tolerance to Pb using an enzymatic toxicity test. Ecotoxicology *21*, 2123–2131.

Feng, H., Huang, X., Zhang, Q., Wei, G., Wang, X., and Kang, Z. (2012). Selection of suitable inner reference genes for relative quantification expression of microRNA in wheat. Plant Physiol. Biochem. *51*, 116–122.

Fernandez-Leborans, G., and Novillo, A. (1995). The effects of cadmium on the successional stages of a freshwater protozoa community. Ecotox. Environ. Safe. *31*, 29–36.

Fischer, B.B., Pomati, F., and Eggen, R.I.L. (2013). The toxicity of chemical pollutants in dynamic natural systems: The challenge of integrating environmental factors and biological complexity. STOTEN. *449*, 253–259.

Flores, L., Banjac, Z., Farré, M., Larrañaga, A., Mas-Martí, E., Muñoz, I., Barceló, D., and Elosegi, A. (2014). Effects of a fungicide (imazalil) and an insecticide (diazinon) on stream fungi and invertebrates associated with litter breakdown. STOTEN. *476–477*, 532–541.

Fox, R.E., Zhong, Z., Krone, S., and Top, E.M. (2008). Spatial structure and nutrients promote invasion of IncP-1 plasmids in bacterial populations. ISME J. *2*, 1024–1039.

Geiszinger, A., Bonnineau, C., Faggiano, L., Guasch, H., Lopez-Doval, J.C., Proia, L., Ricart, M., Ricciardi, F., Romani, A., Rotter, S., *et al.* (2009). The relevance of the community approach linking chemical and biological analyses in pollution assessment. Trends Anal. Chem. (TrAC). *28*, 619–626.

Gessner, M.O., and Chauvet, E. (2002). A case for using litter breakdown to assess functional stream integrity. Ecol. Appl. *12*, 498–510.

Gomiero, A., Sforzini, S., Dagnino, A., Nasci, C., and Viarengo, A. (2012). The use of multiple endpoints to assess cellular responses to environmental contaminants in the interstitial marine ciliate Euplotes crassus. Aquat. Toxicol. *114–115*, 206–216.

Gracia, M.P., Salvadó, H., Rius, M., and Amigó, J.M. (1994). Effects of copper on ciliate communities from activated sludge plants. Acta Protozool. 33, 219–226.

Guasch, H., Admiraal, W., and Sabater, S. (2003). Contrasting effects of organic and inorganic toxicants on freshwater periphyton. Aquat. Toxicol. *64*, 165–175.

Guasch, H., Serra, A., Corcoll, N., Bonet, B., and Leira, M. (2010a). Metal ecotoxicology in fluvial biofilms: potential influence of water scarcity. In Sabater, S., and

Barceló, D. (eds), Water Scarcity in the Mediterranean: Perspectives Under Global Change (Springer, Berlin, Germany), pp. 41–54.

Guasch, H., Atli, G., Bonet, B., Corcoll, N., Leira, M., and Serra, A. (2010b). Discharge and the response of biofilms to metal exposure in Mediterranean rivers. Hydrobiologia *657*, 143–157.

Guasch, H., Bonet, B., Bonnineau, C., Corcoll, N., López-Doval, J., Muñoz, I., Ricart, M., Serra, A., and Clements, W. (2012). How to link field observations with causality? Field and experimental approaches linking chemical pollution with ecological alterations. In Emerging and Priority Pollutants in Rivers. Guasch, H., Ginebreda, A., and Geiszinger, A., eds (Springer, Berlin/Heidelberg, Germany), pp. 181–218.

Guasch, H., Ricart, M., López-Doval, J., Bonnineau, C., Proia, L., Morin, S., Muñoz, I., Romaní, A.M., and Sabater, S. (2015). Interactive effects of natural biotic stressors and chemicals on periphyton communities: the influence of grazing on triclosan toxicity. Freshwater Biol. In press.

Guo, R., Lee, M.A., and Ki, J.S. (2013). Different transcriptional responses of heat shock protein 70/90 in the marine diatom Ditylum brightwellii exposed to metal compounds and endocrine-disrupting chemicals. Chemosphere. *92*, 535–543.

Hall, E.K., Besemer, K., Kohl, L., Preiler, C., Scheider, T., Riedel, K., Wanek, W., and Battin, T. (2012). Effects of resource chemistry on the composition and function of stream hyporheic biofilms. Front. Microbiol. *14*, 3–35.

Handy, R.D., van den Brink, N., Chappell, M., Muhling, M., Behra, R., Dusinska, M., Simpson, P., Ahtiainen, J., Jha, A.N., Seiter, J., Bednar, A., Kennedy, A., Fernandes, T.F., and Riediker, M. (2012). Practical considerations for conducting ecotoxicity test methods with manufactured nanomaterials: what have we learnt so far? Ecotoxicology *21*, 933–972.

Headley, J.V., Gandrass, J., Kuballa, J., Peru, K.M., and Gong, Y. (1998). Rates of sorption and artitioning of contaminants in river biofilm. Environ. Sci. Technol. *32*, 3968–3973.

Hering, D., Borja, A., Carstensen, J., Carvalho, L., Elliott, M., Feld, C.K., Heiskanen, A.S., Johnson, R.K., Moe, J., Pont, D., Solheim, A.L., and van de Bund, W. (2010). The European Water Framework Directive at the age of 10: A critical review of the achievements with recommendations for the future. STOTEN. *19*, 4007–4019.

Hill, B.H., Lazorchak, J.M., McCormick, F.H., and Willingham, W.T. (1997). The effects of elevated metals on benthic community metabolism in a rocky mountain stream. Environ. Pollut. *95*, 183–190.

Holding, K.L., Gill, R.A., and Carter, J. (2003). The relationship between epilithic periphyton (Biofilm) bound metals and metals bound to sediments in freshwater systems. Environ. Geochem. Hlth. *25*, 87–93.

Holtze, M.S., Ekelund, F., Rasmussen, L.D., Jacobsen, C.S., and Johnsen, K. (2003). Prey-predator dynamics in communities of culturable soil bacteria and protozoa: differential effects of mercury. Soil Biol. Biochem. *35*, 1175–1181.

Irwing, E.C., Baird, D.J., and Culp, J.M. (2003). Ecotoxicological responses of the Mayfly *Baetis tricaudatus* to dietary and waterborne cadmium: implications for toxicity testing. Environ. Toxicol. Chem. *22*, 1058–1064.

Kim, K.S., Funk, D.H., and Buchwalter, D.B. (2012). Dietary (periphyton) and aqueous Zn bioaccumulation dynamics in the mayfly Centroptilum triangulifer. Ecotoxicology *21*, 2288–2296.

Kim Tiam, S., Feurtet-Mazel, A., Delmas, F., Mazzella, N., Morin, S., Daffe, G., and Gonzalez, P. (2012). Development of q-PCR approaches to assess water quality: Effects of cadmium on gene expression of the diatom Eolimna minima. Water Res. *46*, 934–942.

Kulacki, K.J., Cardinale, B.J., Keller, A.A., Bier, R., and Dickson, H. (2012). How do stream organisms respond to, and influence, the concentration of Titanium dioxide nanoparticles? A mesocosm study with algae and herbivores. Environ. Toxicol. Chem. *31*, 2414–2422.

Lara, E., Berney, C., Ekelund, F., Harms, H., and Chatzinotas, A. (2007). Molecular comparison of cultivable protozoa from a pristine and a polycyclic aromatic hydrocarbon polluted site. Soil Biol. Biochem. *39*, 139–148.

Larras, F., Lambert, A.-S., Pesce, S., Rimet, F., Bouchez, A., and Montuelle, B. (2013). The effect of temperature and a herbicide mixture on freshwater periphytic algae. Ecotox. Environ. Safe. *98*, 162–170.

Laviale, M., Prygiel, J., and Créach, A. (2010). Light modulated toxicity of isoproturon toward natural stream periphyton photosynthesis: a comparison between constant and dynamic light conditions. Aquat. Toxicol. *97*, 334–342.

Laviale, M., Morin, S., and Créach, A. (2011). Short term recovery of periphyton photosynthesis after pulse exposition to the PSII inhibitors atrazine and isoproturon. Chemosphere. *84*, 731–734.

Lawrence, J.R., Kopf, G., Headley, J.V., and Neu, T.R. (2001). Sorption and metabolism of selected herbicides in river biofilm communities. Can. J. Microbiol. *47*, 634–641.

Lawrence, J.R., Chénier, M., Roy, R., Beaumier, D., Fortin, N., Swerhone, G.D.W., Neu, T.R., and Greer, C.W. (2004). Microscale and molecular assessment of the impacts of nickel, nutrients, and oxygen level on river biofilm communities. Appl. Environ. Microbiol. *70*, 4326–4339.

Lawrence, J.R., Swerhone, G.D.W., Topp, E., Korber, D.R., Neu, T.R., and Wassenaar, L.I. (2007). Structural and functional responses of river biofilm communities to the nonsteroidal anti-inflammatory diclofenac. Environ. Toxicol. Chem. *26*, 573–582.

Ledger, M.E., Harris, R.M.L., Armitage, P.D., and Milner, A.M. (2009). Realism of model ecosystems: an evaluation of physicochemistry and macroinvertebrate assemblages in artificial streams. Hydrobiologia *617*, 91–99.

Lesser, M.P. (2006). Oxidative stress in marine environments: biochemistry and physiological ecology. Annu. Rev. Physiol. *68*, 253–278.

Li, L., Zhao, J., and Tang, X. (2010). Ultraviolet irradiation induced oxidative stress and response of antioxidant system in an intertidal macroalgae *Corallina officinalis* L.J. Environ. Sci. *22*, 716–722.

Li, M., Costello, D.M., and Burton, G.M. (2012). Interactive effects of phosphorus and copper on Hyalella azteca via periphyton in aquatic ecosystems. Ecotox. Environ. Safe. *83*, 41–46.

López-Doval, J.C., Ricart, R., Guasch, H., Romaní, A.M., Sabater, S., and Muñoz, I. (2010). Does Grazing Pressure Modify Diuron Toxicity in a Biofilm Community? Arch. Environ. Con. Tox. *58*, 955–962.

Luís, A.T., Teixeira, P., Almeida, S.F.P., Matos, J.X., and da Silva, E.F. (2011). Environmental impact of mining activities in the Lousal area (Portugal): Chemical and diatom characterization of metal-contaminated stream sediments and surface water of Corona stream. STOTEN, *409*, 4312–4325.

Lundqvist, A., Bertilsson, S., and Goedkoop, W. (2012). Interactions with DOM and biofilms affect the fate and bioavailability of insecticides to invertebrate grazers. Ecotoxicology *21*, 2398–2408.

Madoni, P. (2000). The acute toxicity of nickel to freshwater ciliates. Environ. Pollut. *109*, 53–59.

Madoni, P. (2011). Protozoa in wastewater treatment processes: A minireview. Ital. J. Zool. *78*, 3–11.

Magbanua, F.S., Townsend, C.R., Hageman, K.J., Lange, K., Lear, G., and Lewis, G.D., and Matthaei, C.D. (2013). Understanding the combined influence of fine sediment and glyphosate herbicide on stream periphyton communities. Water Res. *47*, 5110–5120.

Maharana, D., Jena, K., Pise, N.M., and Jagtap, T.G. (2010). Assessment of oxidative stress indices in a marine macro brown alga Padina tetrastromatica (Hauck) from comparable polluted coastal regions of the Arabian Sea, West coast of India. J. Environ. Sci. *22*, 1413–1417.

Marti, E., Jofre, J., and Balcazar, J.L. (2013). Prevalence of antibiotic resistance genes and bacterial community composition in a river influenced by a wastewater treatment plant. PLoS One *8*, e78906.

Martín-González, A., Borniquel, S., Díaz, S., Ortega, R., and Gutiérrez, J.C. (2005). Ultrastructural alterations in ciliated protozoa under heavy metal exposure. Cell Biol. Int. *29*, 119–126.

Martín-González, A., Díaz, S., Borniquel, S., Gallego, A., and Gutiérrez, J.C. (2006). Cytotoxicity and bioaccumulation of heavy metals by ciliated protozoa isolated from urban wastewater treatment plants. Res. Microbiol. *157*, 108–118.

Meylan, S., Behra, R., and Sigg, L. (2003). Accumulation of copper and zinc in periphyton in response to dynamic variations of metal speciation in freshwater. Environ. Sci. Technol. *37*, 5204–5212.

Meylan, S., Behra, R., and Sigg, L. (2004). Influence of metal speciation in natural freshwater on bioaccumulation of copper and zinc in periphyton: a microcosm study. Environ. Sci. Technol. *38*, 3104–3111.

Mittler, R. (2002). Oxidative stress, antioxidant and stress tolerance. Plant. Sci. *7*, 405–410.

Moreirinha, C., Duarte, S., Pascoal, C., and Cássio, F. (2011). Effects of cadmium and phenanthrene mixtures on aquatic fungi and microbially mediated leaf litter decomposition. Arch. Environ. Contam. Toxicol. *61*, 211–219.

Morin, S., Duong, T.T., Dabrin, A., Coynel, A., Herlory, O., Baudrimont, M., Delmas, F., Durrieu, G., Schäfer, J., Winterton, P., Blanc, G., and Coste, M. (2008a). Long-term survey of heavy-metal pollution, biofilm contamination and diatom community structure in the Riou Mort watershed, South-West France. Environ. Pollut. *151*, 532–542.

Morin, S., Duong, T.T., Herlory, O., Feurtet-Mazel, A., and Coste, M. (2008b). Cadmium toxicity and bioaccumulation in freshwater biofilms. Arch. Environ. Contam. Toxicol. *54*, 173–186.

Morin, S., Bottin, M., Mazzella, N., Macary, F., Delmas, F., Winterton, P., and Coste, M. (2009). Linking diatom community structure to pesticide input as evaluated through a spatial contamination potential (Phytopixal): A case study in the Neste river system (South-West France). Aquat. Toxicol. *94*, 28–39.

Morin, S., Cordonier, A., Lavoie, I., Arini, A., Blanco, S., Duong, T.T., Tornés, E., Bonet, B., Corcoll, N., Faggiano, L., *et al.* (2012). Consistency in diatom response to metal-contaminated environments. In Emerging and Priority Pollutants in Rivers: Bringing Science into River Management Plans (Handbook of Environmental Chemistry), Guasch, H., Ginebreda. A., and Geiszinger, A., eds. (Springer, Heidelberg, Germany), pp. 117–146.

Morin, S., Corcoll, N., Bonet, B., Tlili, A., and Guasch, H. 2014. Diatom responses to zinc contamination along a Mediterranean river. Plant Ecol. Evol. *147*, 325–332.

Mortimer, M., Kasemets, K., and Kahru, A. (2010). Toxicity of ZnO and CuO nanoparticles to ciliated protozoa *Tetrahymena thermophila*. Toxicology *269*, 182–189.

Muñoz, I., Real, M., Guasch, H., Navarro, E., and Sabater, S. (2001). Effects of atrazine on periphyton under grazing pressure. Aquat. Toxicol. *55*, 239–249.

Nalecz-Jawecki, G., Demkowicz-Dobrzahski, K., and Sawicki, J. (1993). Protozoan *Spirostomum ambiguum* as a highly sensitive bioindicator for rapid and easy determination of water quality. STOTEN. *134*, 1227–1234.

Navarro, E., Robinson, C.T., and Behra, R. (2008). Increased tolerance to ultraviolet radiation (UVR) and cotolerance to cadmium in UVR-acclimatized freshwater periphyton. Limnol. Oceanogr. *53*, 1149–1158.

Nicolau, A., Dias, N., Mota, M., and Lima, N. (2001). Trends in the use of protozoa in the assessment of wastewater treatment. Res. Microbiol. *152*, 621–630.

Nicolau, A., Martins, M.J., Mota, M., and Lima, N.l. (2005). Effect of copper in the protistan community of activated sludge. Chemosphere. *58*, 605–614.

Niederlehner, B.R., and Cairns, J. Jr. (1990). Effects of ammonia on periphytic communities. Environ. Pollut. *66*, 207–221.

Niederlehner, B.R., and Cairns, J. Jr. (1992). Community response to cumulative toxic impact: effects of acclimation on Zinc tolerance of Aufwuchs. Can. J. Fish. Aquat. Sci. *49*, 2155–2163.

Okamoto, O.K., Pinto, E., Latorre, L.R., Bechara, E.J.H., and Colepicolo, P. (2001). Antioxidant Modulation in Response to Metal-Induced Oxidative Stress in Algal Chloroplasts. Arch. Environ. Contam. Toxicol. *40*, 18–24.

Osorio, V., Proia, L., Ricart, M., Pérez, S., Ginebreda, A., Cortina, J.L., Sabater, S., and Barceló, D. (2014). Hydrological variation modulates pharmaceutical levels and biofilm responses in a Mediterranean river. STOTEN. *472*, 1052–1061.

Paje, M.L.F., Kuhlicke, U., Winkler, M., and Neu, T.R. (2002). Inhibition of lotic biofilms by Diclofenac. Appl. Microbiol. Biotechnol. *59*, 488–492.

Park, J., Congeevaram, S., Ki, D.W., and Tiedje, J.M. (2006). Use of stable isotope probing in selectively isolating target microbial community genomes from environmental samples for enhancing resolution in ecotoxicological assessment. Mol. Cell. Toxicol. *2*, 11–14.

Paulsson, M., Nyström, B., and Blanck, H. (2000). Long-term toxicity of zinc to bacteria and algae in periphyton communities from the river Göta Älv, based on a microcosm study. Aquat. Toxicol. *47*, 243–257.

Pesce, S., Martin-Laurent, F., Rouard, N., and Montuelle, B. (2009). Potential for microbial diuron mineralisation in a small wine-growing watershed: from treated plots to lotic receiver hydrosystem. Pest. Manag. Sci. *65*, 651–657.

Proia, L., Morin, S., Peipoch, M., Romaní, A.M., and Sabater, S. (2011). Resistance and recovery of river biofilms receiving short pulses of Triclosan and Diuron. Sci. Total Environ. *409*, 3129–3137.

Proia, L., Cassiò, F., Pascoal, C., Tlili, A., and Romaní, A.M. (2012a). The use of attached microbial communities to assess ecological risks of pollutants in river ecosystems. The role of heterotrophs. In Emerging and Priority Pollutants in Rivers: Bringing Science into River Management Plans, Guasch, H., Ginebreda, A., and Geiszinger, A. eds (Springer Verlag, Berlin/Heidelberg, Germany) pp. 55–83.

Proia, L., Romaní, A.M., and Sabater, S. (2012b). Nutrients and light effects on stream biofilms: a combined assessment with CLSM, structural and functional parameters. Hydrobiologia 695, 281–291.

Proia, L., Lupini, G., Osorio, V., Pérez, S., Barceló, D., Schwartz, T., Amalfitano, S., Fazi, S., Romaní, A.M., and Sabater, S. (2013a). Response of biofilm bacterial communities to antibiotic pollutants in a Mediterranean river. Chemosphere 92, 1126–1135.

Proia, L., Vilches, C., Boninneau, C., Kantiani, L., Farré, M., Romaní, A.M., Sabater, S., and Guasch, H. (2013b). Drought episode modulates the response of river biofilm to triclosan. Aquat. Toxicol. *127*, 36–45.

Real, M., Muñoz, I., Guasch, G., Navarro, E., and Sabater, S. (2003). The effect of copper exposure on a simple aquatic food chain. Aquat. Toxicol. *63*, 283–291.

Ricart, M., Barceló, D., Geiszinger, A., Guasch, H., de Alda, M.L., Romaní, A.M., Vidal, G., Villagrasa, M., and Sabater, S. (2009). Effects of low concentrations of the phenylurea herbicide diuron on biofilm algae and bacteria. Chemosphere. *76*, 1392–1401.

Ricart, M., Guasch, H., Alberch, M., Barceló, D., Bonnineau, C., Geiszinger, A., Farré, M.L., Ferrer, J., Ricciardi, F., Romaní, A.M., Morin, S., Proia, L., Sala, L., Sureda, D., and Sabater, S. (2010). Triclosan persistence trough wastewater treatment plants and its potential toxic effects on river biofilms. Aquat. Toxicol. *100*, 346–353.

Rico, D., Martín-González, A., Díaz, S., de Lucas, P., and Gutiérrez, J.C. (2009). Heavy metals generate reactive oxygen species in terrestrial and aquatic ciliated protozoa. Comp. Biochem. Physiol. C. Toxicol. Pharmacol. *149*, 90–96.

Romani, A.M. (2010). Freshwater biofilms. In Biofouling, Durr, S., and Thomason, J.C., eds. (Blackwell Publishing Ltd., Oxford, UK), pp. 137–153.

Rosi-Marshall, E.J., Kincaid, D.W., Bechtold, H.A., Royer, T.V., Rojas, M., and Kelly, J.J. (2013). Pharmaceuticals suppress algal growth and microbial respiration and alter bacterial communities in stream biofilm. Ecol. Appl. *23*, 583–593.

Rotter, S., Heilmeier, H., Altenburger, R., and Schmitt-Jansen, M. (2013). Multiple stressors in periphyton – comparison of observed and predicted tolerance responses to high ionic loads and herbicide exposure. J. Appl. Ecol. *50*, 1459–1468.

Sabater, S., Navarro, E., and Guasch, H. (2002). Effects of copper on algal communities at different current velocities. J. Appl. Phycol. *14*, 391–398.

Sabater, S., Buchaca, T., Cambra, J., Catalan, J., Guasch, H., Ivorra, N., Munoz, I., Navarro, E., Real, M., and Romani, A.M. (2003). Structure and function of benthic algal communities in an extremely acid river. J. Phycol. *39*, 481–489.

Salvadó, H., Gracia, M.P., and Amigó, J.M. (1995). Capability of ciliated protozoa as indicators of effluent quality in activated sludge plants. Water Res. *29*, 1041–1050.

Salvadó, H., Gracia, M.P., Amigó, J.M., and Rius, M. (1997). Effects of Cadmium on growth and motility in *Euplotes aediculatus* isolated from Activated Sludge. Bull. Environ, Contam. Toxicol. *58*, 838–844.

Scandalios, J.G. (1993). Oxygen Stress and Superoxide Dismutases. Plant. Physiol. *101*, 7–12.

Scott, J.T., Back, J.A., Taylor, J.M., and King, R.S. (2008). Does nutrient enrichment decouple algal–bacterial production in periphyton? J. N. Am. Benthol. Soc. *27*, 332–344.

Selivanovskaya, S.Y., Petrov, A.M., Egorova, K.V., and Naumova, R.P. (1997). Protozoan and metazoan communities treating a simulated petrochemical industry wastewater in a rotating disc biological reactor. World J. Microb. Biot. *13*, 511–517.

Serra, A., Corcoll, N., and Guasch, H. (2009). Copper accumulation and toxicity in fluvial periphyton: the influence of exposure history. Chemosphere. *74*, 633–641.

Shi, Y., Lu, Y., Meng, F., Guo, F., and Zheng, X. (2013). Occurrence of organic chlorinated pesticides and their ecological effects on soil protozoa in the agricultural soils of North Western Beijing, China. Ecotoxicol. Environ. Safe. *92*, 123–128.

Sierra, V.M., and Gómez, N. (2010). Assessing the disturbance caused by an industrial discharge using field transfer of epipelic biofilm. STOTEN. *408*, 2696–2705.

Solé, M., Müller, I., Pecyna, M.J., Fetzer, I., Harms, H., and Schlosser, D. (2012). Differential Regulation by Organic Compounds and Heavy Metals of Multiple Laccase Genes in the Aquatic Hyphomycete Clavariopsis aquatica. Appl. Environ. Microb. *78*, 4732–4739.

Tlili, A., Bérard, A., Roulier, J.L., Volat, B., and Montuelle, B. (2010). PO43- dependence of the tolerance of autotrophic and heterotrophic biofilm communities to copper and diuron. Aquat. Toxicol. *98*, 165–177.

Tlili, A., Marechal, M., Montuelle, B., Volat, B., Dorigo, U., and Bérard, A. (2011a). Use of the MicroRespTM method to assess pollution induced community tolerance to metals for lotic biofilms. Environ. Pollut. *159*, 18–24.

Tlili, A., Corcoll, N., Bonet, B., Morin, M., Montuelle, B., Bérard, A., and Guasch, H. (2011b). In situ spatio-temporal changes in pollution-induced community tolerance to zinc in autotrophic and heterotrophic biofilm communities. Ecotoxicology *20*, 1823–1839.

Tolkkinen, M., Mykrä, H., Markkola, A.M., Aisala, H., Vuori, K.M., Lumme, J., Pirttilä, and A.M., and Muotka, T. (2013). Decomposer communities in human-impacted streams: species dominance rather than richness affects leaf decomposition. J. Appl. Ecol. *50*, 1142–1151.

Valavanidis, A., Vlahogianni, T., Dassenakis, M., and Scoullos, M. (2006). Molecular biomarkers of oxidative stress in aquatic organisms in relation to toxic environmental pollutants. Ecotoxicol. Environ. Saf. *64*, 178–189.

Vera, M.S., Fiori, E.D., Lagomarsino, L., Sinistro, R., Escaray, R., Iummato, M.M., Juárez, A., Molina, M. del C.R. de, Tell, G., and Pizarro, H. (2012). Direct and indirect effects of the glyphosate formulation Glifosato Atanor® on freshwater microbial communities. Ecotoxicology *21*, 1805–1816.

Villeneuve, A., Montuelle, B., and Bouchez, A. (2011). Effects of flow regime and pesticides on periphytic communities: evolution and role of biodiversity. Freshwater Biol. *56*, 2245–2259.

Wang, S.Y., Bernhardt, E.S., and Wright, J.P. (2014). Urban stream denitrifier communities are linked to lower functional resistance to multiple stressors associated with urbanization. Hydrobiologia *726*, 13–23.

Wolfe-Simon, F., Grzebyk, D., Schofield, O., and Falkowski, P.O. (2005). The role and evolution of Superoxide dismutases in algae. J. Phycol. *41*, 453–465.

Writer, J.H., Ryan, J.N., and Barber, L.B. (2011a). Role of biofilms in sorptive removal of steroidal hormones and 4-nonylphenol compounds from streams. Environ. Sci. Technol. *45*, 7275–7283.

Writer, J.H., Barber, L.B., Ryan, J.N., and Bradley, P.M. (2011b). Biodegradation and attenuation of steroidal hormones and alkylphenols by stream biofilms and sediments. Environ. Sci. Technol. *45*, 4370–4376.

Writer, J.H., Antweiler, R.C., Ferrer, I., Ryan, J.N., and Thurman, E.M. (2013). In-stream attenuation of neuro-active pharmaceuticals and their metabolites. Environ. Sci. Technol. *47*, 9781–9790.

Wunder, D.B., Bosscher, V.A., Cok, R.C., and Hozalski, R.M. (2011). Sorption of antibiotics to biofilm. Water Res. *45*, 2270–2280.

Xie, L., and Buchwalter, D.B. (2011). Cadmium exposure route affects antioxidant responses in the mayfly Centroptilum triangulifer. Aquat. Toxicol. *105*, 199–205.

Zhou, Q., Zhang, J., Fu, J., Shi, J., and Jiang, G. (2008). Biomonitoring: an appealing tool for assessment of metal pollution in the aquatic ecosystem. Anal. Chim. Acta *606*, 135–150.

Biofilm Development in Sewer Networks

8

Oriol Gutierrez, Guangming Jiang, Keshab Sharma and Zhiguo Yuan

Abstract

Wastewater collection systems, or sewers, are crucial sections of the urban water cycle where complex microbial, chemical and physicochemical processes take place. This chapter aims to give an overview of the diversity and importance of biofilms and bioreactions occurring in sewers, paying special attention to its detrimental effects. Sewer biofilms can be divided in two main classes:

1 Submerged biofilms: including activities of sulfate-reducing bacteria (SRB) responsible for the formation of sulfide (H_2S, an odorous, toxic and corrosion-inducer compound), Methanogenic Archaea (MA) responsible for the formation of methane (CH_4, an explosive and potent greenhouse gas) and the Fermentation processes that increase the two previous biofilms metabolism.
2 Unsubmerged biofilms: activities of biofilms growing on the gas phase of sewers that causes loss of concrete mass, cracking of the sewer pipes and ultimately, structural collapse. This process is known as microbially induced concrete corrosion (MICC).

The structure of sewer-biofilms and mechanisms for the control of its harmful effects are described.

Introduction

Wastewater collection systems, or sewers, consist of an underground network of physical structures-installations composed of pipelines, pump stations, manholes and channels that convey the wastewater from its source to the point where it is discharged. The discharge point is usually a Wastewater Treatment Plant (WWTP) but may also be natural environments (Fig. 8.1). Sewer systems are crucial in protecting public health as these prevent the spread of diseases by avoiding population exposure to the contaminated wastewater. By definition, wastewater is water that has been adversely affected in quality by anthropogenic influence, either from domestic households (such as showers, toilets and washing machines) or from industrial processes (Metcalf and Eddy, 2003). Thus composition of wastewater, although contains more than 99% of water, varies widely depending upon the source. Sewage can contain suspended solids (that can create sludge deposits and anaerobic conditions in sewers), biodegradable organics (proteins, carbohydrates and fats that can lead to the depletion of oxygen in water and develop septic conditions), pathogens, nutrients (nitrogen and phosphorus), priority pollutants (organic and inorganic compounds with acute toxicity, and heavy metals from industrial processes) and dissolved inorganics (such as calcium, sodium and sulfate).

Sewers are very important assets of the urban water systems. For instance, in Spain only, the sewer networks have an extension of around 89,900 kilometres (equivalent to twice the equator distance) that collect the wastewater generated from 8110 municipalities, covering 86% of the total population. Similarly, the length of sewer pipes are 1,200,000 and 117,000 km in the USA and Australia, respectively (AWA, 2011; US EPA, 1991). The total asset value of these networks is estimated to be about one trillion dollars in the

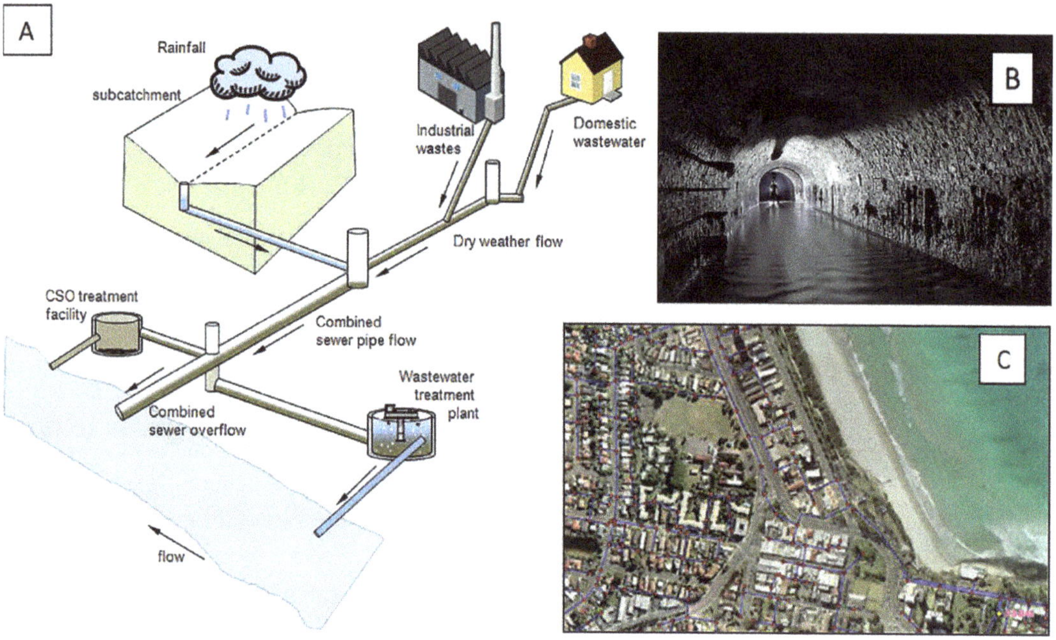

Figure 8.1 (A) Scheme of the urban sewer system; (B) picture of sewer trunk main in Paris (France), courtesy of sub-urban.com; and (C) distribution of a sewer network in the domestic suburb of Burleigh Heads, courtesy of Gold Coast City Council (Australia).

USA and \$100 billion in Australia (Brongers, 2001).

Depending on the topography, sewers are classified to include two types of pipes: gravity sections and pressure mains (Fig. 8.2). Gravity sewers are used to collect wastewater from multiple sources and convey the wastewater by gravity to a central location. Gravity pipes have sufficient slopes to keep the wastewater flowing naturally through the system without having excessive solid deposition. Whenever wastewater has to be transported to a higher location and flow under gravity is not possible, pump stations need to be employed. Pump stations are normally installed at low elevation points of the sewer network in order to pump the sewage up through a pressure main to another gravity pipe, to convey wastewater over a hill, and/or in case of nearly flat terrains up to a treatment facility. Depending on the topography of each catchment, gravity pipes or pressure mains may be predominant in a sewer network.

Sewers have been traditionally considered only as a system for hydraulic transport of sewage. However processes occurring in sewer systems are

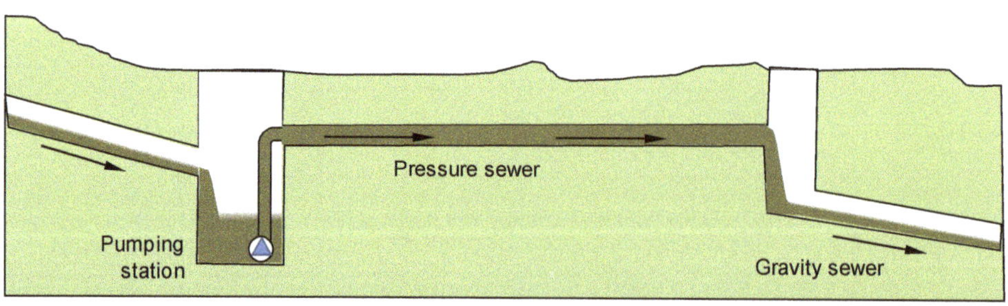

Figure 8.2 Sewer network types of pipe depending on topography.

much broader than solely the hydraulic processes. Sewers are 'reactors' where complex microbial, chemical and physicochemical processes take place. The complexity of those reactions depends very much on the inherent sewer features such as:

1 Wastewater matrix, which includes a diversity of microorganisms and pollutants (such as organic matter, nutrients and particles) varying with the location and time within the network.
2 Presence of different sewer phases such as suspended wastewater phase, biofilms, sediments and surface of the sewer in contact with the sewer atmosphere (Fig. 8.3).
3 Microbial processes that occur under changing environmental conditions such as aerobic (presence of dissolved oxygen in wastewater), anoxic (presence of nitrate or nitrite), anaerobic (absence of oxidant compounds in the wastewater) and at different level of redox potentials.

Fig. 8.3 presents the most important processes occurring in sewer systems. It can be seen that some processes are biologically mediated by biofilms. The diversity and importance of bioreactions in sewers is going to be described in sections below.

Biofilms and biological reactions have important impacts on sewers functioning. Microorganisms are widely present in the wastewater and are exposed to a range of substrates that contain fractions both of organic matter and inorganic compounds (Metcalf and Eddy, 2003). Sewer biofilms grow attached to sewer surfaces including walls, sediments and other physical supports (Hvitved-Jacobsen, 2002). They take the form of a concentrated layer of microorganisms with self-imbedded matrix of extracellular polymeric substances that hold together in a shape of slime. According to Hvitved-Jacobsen (2002), sewer biofilms typically have a water content of 70–90%, 50–90% of organic matter and a relatively high content of carbohydrates and proteins. Different types of microorganisms will prevail in particular sections of sewers depending on hydraulic features and wastewater composition. Hydraulic features are related to the reaeration zones of the pipes and include turbulence flow of sewage, ventilation of the systems and depth of the wastewater column. Wastewater composition characteristics are related to substrate availability, dissolved organic matter, temperature, pH and redox of sewage.

This chapter aims to give an overview of the most common biofilms found in sewers, paying special attention to its detrimental effects.

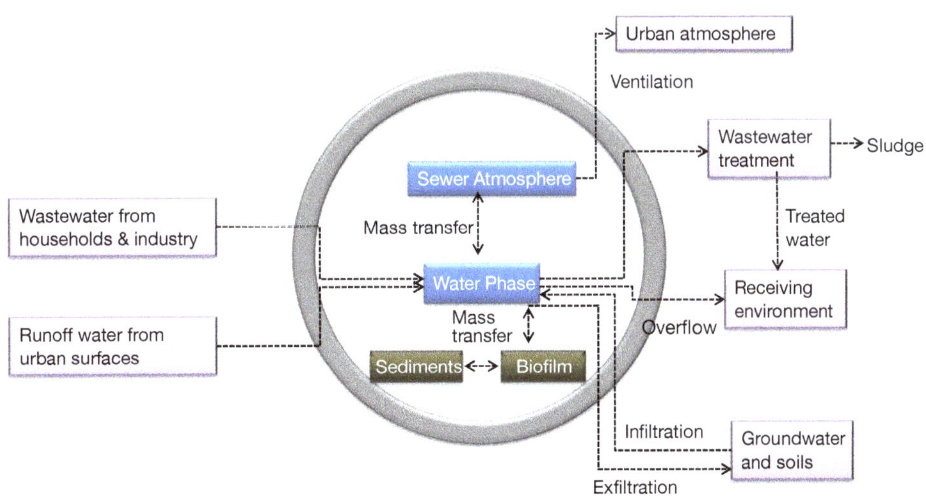

Figure 8.3 Wastewater flows and processes occurring in sewers.

Submerged sewer biofilms dynamics

As widely explained in this book, biofilms need the presence of water, therefore sewer biofilms develop mainly in submerged conditions. Anaerobic sewer bioprocesses, related to both the sulfur and the carbon cycles, are very important in sewers. The interaction of those processes and the aerobic transformation of wastewater are crucial for the performance of urban wastewater systems. To alleviate and control the sewer corrosion and odour problems, various liquid-phase technologies have been used to reduce the formation/emission of H_2S or CH_4 (WERF, 2007). The most commonly used chemicals for liquid-phase technologies include oxygen, nitrate, magnesium hydroxide, iron salts, and caustic shocking to deactivate sewer biofilms (Gutierrez *et al.*, 2008, 2009, 2014; Jiang and Yuan, 2013a; Jiang *et al.*, 2011a; Zhang *et al.*, 2009). The dynamics of submerged sewer biofilms under different environmental conditions are presented in this section.

Anaerobic sewer biofilms

Anaerobic conditions develop under the absence of dissolved oxygen leading to sewage septicity. Anaerobic conditions are very common in sewers and promote the development of biofilms which are the cause of the majority of process-related problems. The main anaerobic bioprocesses in sewers are:

- anaerobic sulfate respiration
- methane generation
- fermentation.

Anaerobic sulfate respiration

In situations where there is lack of oxygen and nitrate, sulfate (SO_4^{2-}) becomes the most thermodynamically efficient electron acceptor. The anaerobic sulfate respiration is carried out by a functional group of bacteria commonly called sulfate-reducing bacteria (SRB) under anaerobic conditions as a dissimilatory reduction reaction (equations 8.1–8.5). SRB obtain energy by oxidizing organic compounds and/or molecular hydrogen (H_2) while reducing sulfate (SO_4^{2-}) to hydrogen sulfide (H_2S-HS^-) during its metabolism. *Desulfovibrio* and *Desulfotomaculum* are the dominant genera of SRB (Muyzer and Stams, 2008).

SRB grow inside the biofilm in sewer walls and take the sulfate and organic matter present in sewage for their metabolism (Fig. 8.4). Sulfate concentrations between 15–30 mg S-SO_4^{2-}/l are typical in domestic sewage but those can go higher depending on the source of drinking water or if the water comes from mineral sources with strong presences of salts like sulfate (Pikaar *et al.*, 2014).

Within sewer networks, sulfide is formed mainly in completely filled rising main sections, more prone to septicity than partially filled gravity sections where reareation from the gas phase can occur (Hvitved-Jacobsen, 2002). Sulfide concentration in rising mains depends upon the sewer features such as the hydraulic retention time (HRT, time that sewage remains in a certain pipe) and the area/volume ratio (A/V ratio, the inner surface area to volume ratio, i.e. 2/r, where r is the radius of the pipe, relative surface of sewer inner-walls to the volume of sewage that contains). The production and accumulation of hydrogen sulfide is a major concern for wastewater utilities and health of residents nearby the sewers. The problems related with sulfide depend upon the extent by which H_2S escapes from the liquid phase of sewers (where is formed) to the sewer

Sulfate-reducing reactions	ΔGo' (kJ/reaction)*	Equation no.
$4H_2 + SO_4^{2-} + H^+ \rightarrow HS^- + 4H_2O$	−151.9	8.1
$Acetate^- + SO_4^{2-} \rightarrow 2HCO_3^- + HS^-$	−47.6	8.2
$Propionate^- + 0.75SO_4^{2-} \rightarrow Acetate^- + HCO_3^- + 0.75HS^- + 0.25H^+$	−37.7	8.3
$Butyrate^- + 0.5SO_4^{2-} \rightarrow 2\ Acetate^- + 0.5HS^- + 0.5H^+$	−27.8	8.4
$Lactate^- + 0.5SO_4^{2-} \rightarrow Acetate^- + HCO_3^- + 0.5HS^-$	−80.2	8.5

*Thauer *et al.* (1977).

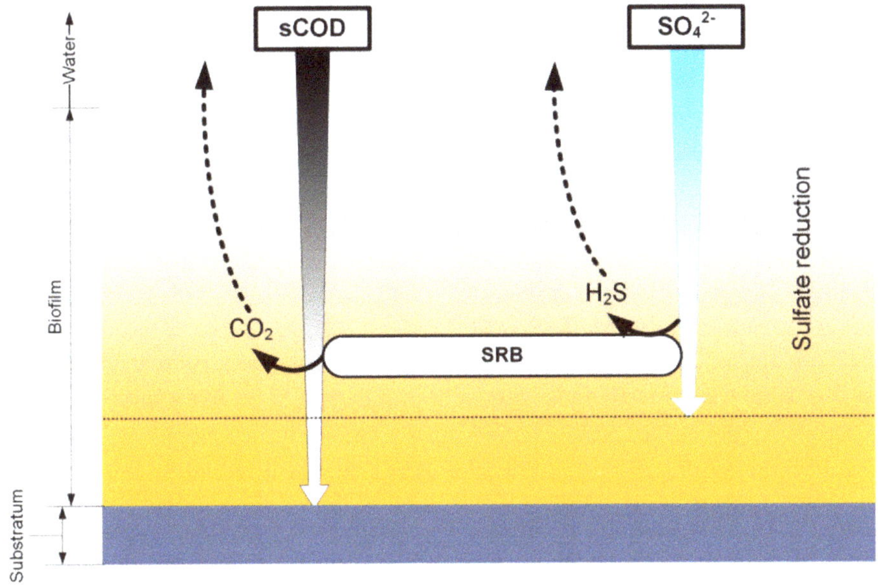

Figure 8.4 Conceptual representation of the anaerobic sulfate respiration of sewer biofilms. sCOD: Organic matter dissolved in wastewater expressed as soluble chemical oxygen demand.

atmosphere, the rate of H_2S stripping or transfer to the sewer atmosphere that takes place under turbulent conditions (drop structures, manholes, wetwells) and in gravity sewers. The extent of transfer depends mainly on the pH of sewage. H_2S is a weak diabasic acid which dissociates As Per the equation 8.6.

$$H_2S \leftrightarrow H^+ + HS^-$$
$$pK_a = -\log K_a = 7.1 \qquad (8.6)$$

As long as H_2S remains dissolved in wastewater, it does not present any harm and will be eventually oxidized chemical or biochemically. The ionic form (HS^-) is soluble in water and remains dissolved. On the other hand the molecular form (H_2S) is more volatile, and can be easily released from the sewer liquid phase. When H_2S is released to the sewer headspace and build-up there, it causes major detrimental effects including odour nuisance and toxicity. H_2S is a flammable and poisonous gas with a characteristic odour of rotten eggs. Its odour concentration threshold is very low, 0.0047 ppmv in gas, and is potentially dangerous because its smell is quickly lost as the concentration increases. Exposure to lower concentrations (10–100 ppmv) can result in eye irritation, a sore throat and cough, nausea, shortness of breath, and fluid in the lungs. These effects are believed to be due to the fact that hydrogen sulfide combines with alkali present in moist surface tissues to form sodium sulfide. Exposure to concentrations higher than 300 ppmv in air can cause death by pulmonar oedema and over 1000 ppmv cause immediate collapse with loss of breathing, even after inhalation of a single breath.

A third major effect of biogenic H_2S consists of the induced corrosion of sewer walls and infrastructures. Biogenic sulfide corrosion is a bacterially mediated process in which hydrogen sulfide gas is subsequently converted to sulfuric acid that attacks concrete and steel within wastewater environments. The hydrogen sulfide gas is biochemically oxidized in the presence of moisture to form sulfuric acid. This process causes critical problems to wastewater managers in terms of repairing and rehabilitation expenses. Sewer assets are under serious threat with an estimated annual asset loss of around $14 billion in the USA alone (Brongers *et al.*, 2002). Sulfide induced concrete corrosion is recognized as a main cause in most cases (US EPA, 1991). For instance, in Australia alone the costs of infrastructure depreciation due to sulfide-induced corrosion are estimated to be in the order of 100 million dollars per year. Full details of this process are presented below.

Methane generation

Hydrogen sulfide is not the only detrimental compound produced in anaerobic sewer systems. Recent field studies showed that a significant amount of methane (CH_4) is formed in sewers, particularly in pressure pipes (Foley *et al.*, 2009). Methanogenesis is carried out by Methanogenic Archaea (MA), a domain distinct from bacteria. Bacteria and archaea differ in cell wall characteristics, membrane lipid composition, in RNA polymerase structure and, therefore, protein synthesis (Gantner *et al.*, 2011). Biogenic formation of methane is a form of anaerobic respiration in which the terminal electron acceptor is not oxygen but carbon compounds of low molecular weight. In contrast to sulfate reducers, methanogens use a limited number of substrates for growth and energy production. Quantitatively, hydrogen, carbon dioxide and acetate are the most important and best-known substrates for methanogens (Muyzer and Stams, 2008). Equations 8.7–8.9 represent the process of methane production.

Ongoing studies in Australia, USA and Spain are addressing this lack of knowledge on the CH_4 formation from sewers. Methane formation in sewers has recently been reported. For instance, Guisasola and co-workers (2008) measured two rising mains in Gold Coast (Australia) where dissolved methane concentrations ranging between 5 and 30 mg/l, equivalent to 20–120 mg COD/l, were measured. Gutierrez and co-workers detected CH_4 production in pressure sewers of la Costa Brava (Spain) (Gutierrez *et al.*, 2012). Measurements at multiple locations of a rising main have shown methane at concentrations between 1.5 and 9 mg COD/l. Similarly to H_2S, methane concentration in sewers has shown to be dependent on the HRT of wastewater and the A/V ratio of the pipe.

Methane production and emission from sewers induce serious concerns as it is a potent greenhouse gas with a global warming potential 21 times that of carbon dioxide over a 100 year horizon. To date, methane production from sewer systems has been largely overlooked in the GHG inventories. The latest report from the Intergovernmental Panel on Climate Change (IPCC) concerning greenhouse emissions did not consider methane production from closed or underground sewer systems (IPCC *et al.*, 2013), despite some previous indications that domestic sewage could be one of the anthropogenic methane sources (Minami and Takata, 1997). Methane is explosive at low concentrations, thus represents a safety risk in confined spaces like sewer manholes because of its low explosion limit (lower explosive level is approximately 5% mix in air) (Spencer *et al.*, 2006). Further, methane production in sewers inevitably consumes the soluble organic compounds (sCOD) including volatile fatty acids (VFAs), which may be required in the receiving wastewater treatment plant for biological nutrient removal.

In anaerobic environments with low redox potentials (< -200 mV), SRB compete with other anaerobes, including fermentative bacteria, proton-reducing acetogenic bacteria, homoacetogens and methanogens, for the available common substrates. In the presence of sulfate in excess, sulfate reducers compete with methanogens for the common substrates hydrogen and acetate and with syntrophic methanogenic communities (Dar *et al.*, 2008). However Guisasola *et al.* (2009) showed that methane and sulfide are simultaneously produced in sewer systems, which implies the coexistence of MA and SRB in sewers biofilms and that these bacteria function simultaneously. The simultaneous functioning of the SRB and MA is related to the spatial arrangement of these bacteria in sewer biofilms. Sewer biofilms are relatively thick (several hundred micrometres;

Methanogenic reactions	ΔGo' (kJ/reaction)	Equation no.
$4H_2 + HCO_3^- + H^+ \rightarrow CH_4 + 3H_2O$	−135.6	8.7
$CO_2 + 4H_2 \rightarrow CH_4 + 2H_2O^-$	−130.7	8.8
$Acetate^- + H_2O \rightarrow CH_4 + HCO_3^-$	−31.0	8.9

Mohanakrishnan *et al.*, 2007) and the Sulfate/Organic Matter ratio (S/COD) shows spatial variation inside the biofilm, being relatively high near the surface in contact with the bulk liquid and close to zero in the inner zone adjacent to the pipe surface (Sun *et al.*, 2014). Fig. 8.5 depicts a schematic view of this hypothesis, which is supported by the sulfide profile measured in sewer biofilms by Mohanakrishnan *et al.* (2007). In contrast, the supply of methanogenesis precursors (VFA) is unlikely to be limiting within the biofilm. For this reason, the lower affinity of MA for these precursors is not a handicap to the growth of methanogens deeper within the biofilm. With sulfate most likely only partially penetrating the biofilm, conceptually two different zones may appear in the biofilm: a sulfate-reducing anaerobic zone (nearer the surface, dominated by SRB) and a deeper anaerobic zone dominated by MA Thus, the extent of methanogenesis in a sewer system is inversely proportional to the sulfate penetration length into the biofilm.

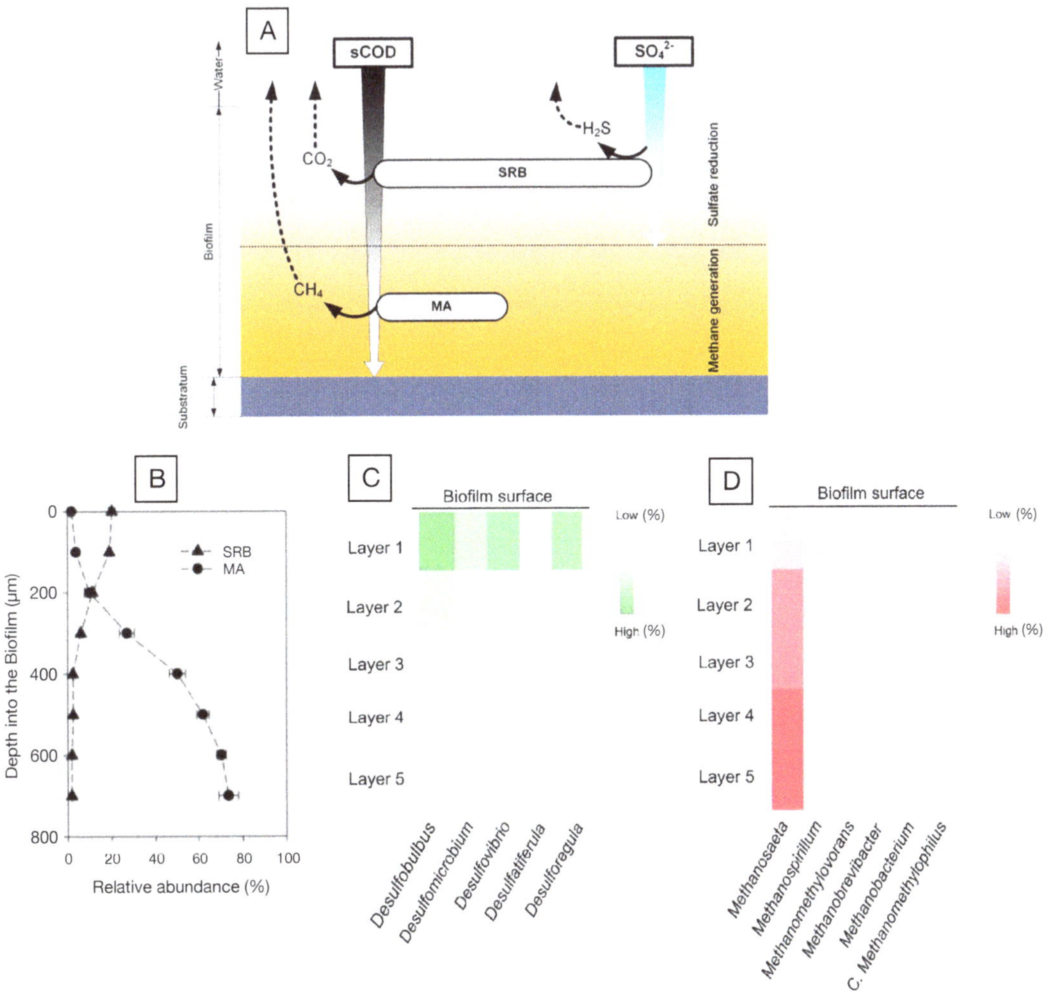

Figure 8.5 (A) Conceptual stratified biofilm model under anaerobic conditions including SRB (sulfate-reducing bacteria) and MA (methanogenic archaea); (B) The SRB and MA proportions of total microorganisms (bacteria and archaea) detected by FISH within the sewer biofilms; (C) Heatmap displaying the abundance and distribution of the predominant SRB genera in different sewer biofilm layers from the biofilm surface to the bottom (Layer 1 to Layer 5). (D) Heatmap displaying the abundance and distribution of the predominant MA genera in different sewer biofilm layers from the biofilm surface to the bottom (layer 1 to layer 5) (Sun *et al.*, 2014).

Fermentation/acetogenic reactions	ΔGo' (kJ/reaction)	Equation no.
$Propionate^- + 3H_2O \rightarrow Acetate^- + HCO_3^- + H^+ + 3H_2$	+76.1	8.10
$Butyrate^- + 2H_2O \rightarrow 2\ Acetate^- + H^+ + 2H_2$	+48.3	8.11
$Lactate^- + 2H_2O \rightarrow Acetate^- + HCO_3^- + H^+ + 2H_2$	−4.2	8.12

In fact, the competition between SRB and MAis not unique to sewer systems. Environmental microbiologists have devoted much attention to such competition in natural anaerobic environments, for example aquatic sediments or paddy rice soils (Abram and Nedwell, 1978; Sørensen *et al.*, 1981; Lovley *et al.*, 1982; Oremland and Polcin, 1982; van Bodegom and Stams, 1999; Abram and Nedwell, 1978; Bodegom and Stams, 1999; Lovley *et al.*, 1982; Oremland and Polcin, 1982; Sørensen *et al.*, 1981) and anaerobic digesters (Bhattacharya *et al.*, 1996; Gupta, 1994; Kalyuzhnyi and Fedorovich, 1998).

Fermentation

Fermentation consists of the partial breakdown of dissolved organic matter that yields organic by-products of low molecular weight. Under anaerobic conditions, degradation of easily biodegradable organic matter is the dominating process. Although fermentation processes do not present direct harmful effects in sewers, its by-products are essential to SRB and MA activities since anaerobic fermentation is the source of H_2 and VFAs. Fermenters create a syntrophic relation with SRB and MA populations in which H_2 and acetate (preferred electron donors) are rapidly scavenged while being produced. Fermentation takes place in the suspended water phase, biofilms and sewer sediments. The number of fermentation reactions occurring in sewers is quite extensive. The more relevant ones are presented in equations 8.10–8.14.

In addition, fermentation also plays a role in the formation of odorous compounds under strictly anaerobic conditions. Therefore there is a strong interaction between the carbon and sulfur cycles in anaerobic sewers that promote the effects described above.

Anoxic sewer microbial processes

Nitrate salts (sodium and calcium nitrate) are sometimes artificially added in to pressure sewers to avoid anaerobic conditions and prevent sulfide formation (Jiang *et al.*, 2009; Mohanakrishnan *et al.*, 2009; Okabe *et al.*, 2007; Zhang *et al.*, 2008). In the last 70 years, nitrate dosing has been used by water industry to control hydrogen sulfide production in sewers (Ganigue *et al.*, 2011). The addition of a thermodynamically favourable electron acceptor to anaerobic sewers aims to shift redox conditions and prevent the sulfate respiration. Anoxic sulfide oxidation by sewer biofilms using nitrate can be described as a two-step process, namely the oxidation of sulfide to elemental sulfur, and oxidation of elemental sulfur to sulfate (equations 8.15 and 8.16, respectively) (Jiang *et al.*, 2009).

Saracevic *et al.* (2006) applied 40 mgN-NO_3^-/l nitrate to a 5.0 km long rising main sewer, and discovered that after a lag time of 3–4 days, the nitrate application reduced sulfide concentrations from 10–20 mg S-S^{2-}/l to below 2–3 mg S-S^{2-}/l. Nitrate concentration of 5 mg N-NO_3^-/l in wastewater were reported to be sufficient to inhibit sulfide production in a 61-km-long gravity sewer (Rodríguez-Gómez *et al.*, 2005).

Nitrate does not have an immediate or long-lasting inhibitory/toxic effect on sulfate reduction on sewer biofilms. Biofilms in a nitrate-receiving laboratory sewer system were found to fully maintain their sulfidogenic activity (sulfide production

Homoacetogenic reactions	ΔGo' (kJ/reaction)	Equation no.
$4H_2 + 2HCO_3^- + H^+ \rightarrow Acetate^- + 4H_2O$	−104.6	8.13
$Lactate^- \rightarrow 1.5\ Acetate^- + 0.5H^+$	−56.5	8.14

Nitrate denitrification reactions	ΔGo' (kJ/reaction)	Equation no.
$5S^{2-} + 2NO_3^- + 12H^+ \rightarrow 5S^0 + N_2 + 6H_2O$	−955	8.15
$5SO + 6NO_3^- + 2H_2O \rightarrow 5SO_4^{2-} + 3N_2 + 4H^+$	−2738	8.16

rate in the absence of nitrate) during several months of nitrate addition (Auguet *et al.*, 2014; Mohanakrishnan *et al.*, 2009). Two main mechanisms have been suggested to control sulfide production by nitrate addition in sewers: anoxic sulfide oxidation and competitive exclusion of SRB. The first involves the growth of a chemolithotrophic sulfide-oxidizing nitrate-reducing community, able to oxidize sulfide to elemental sulfur as a major intermediate coupled to nitrate reduction. The latter triggers the development of a heterotrophic, nitrate reducing bacteria (hNRB) community, competing with SRB for organic electron donors. Jiang *et al.* (2013) proposed a conceptual biofilm model with competitive and synergistic interactions among hNRB, sulfide-oxidizing nitrate-reducing bacteria (soNRB), SRB and MA occurring in anoxic sewer biofilms (Fig. 8.6). Similarly as presented in sections before, microbial stratification within the biofilm plays a major role and H_2S and CH_4 control are related to penetration of nitrate into the biofilm. Sulfate-reducing activity and methanogenesis would persist respectively in the deeper parts of the biofilm where soluble chemical oxygen demand would still be able to penetrate but not nitrate.

These sulfide-oxidizing bacteria (SOB), e.g. *Thiobacillus denitrificans,* are mostly chemoautotrophic and have been extensively studied (Sublette and Sylvester, 1987; Cadenhead and Sublette, 1990; Chazal and Lens, 2000; Yang *et al.*, 2005).

More recently, nitrite (NO_2^-) has also been used for sulfide and methane control in sewer pipes (Jiang *et al.*, 2010; Mohanakrishnan *et al.*, 2008; see equations 8.17 and 8.18 as reaction examples). In addition to the oxidizing capacity as NO_3^-, NO_2^- has a toxic effect on microorganisms thanks to its metabolic inhibitor properties. Nitrite blocks sulfate reduction in SRB by inhibiting the

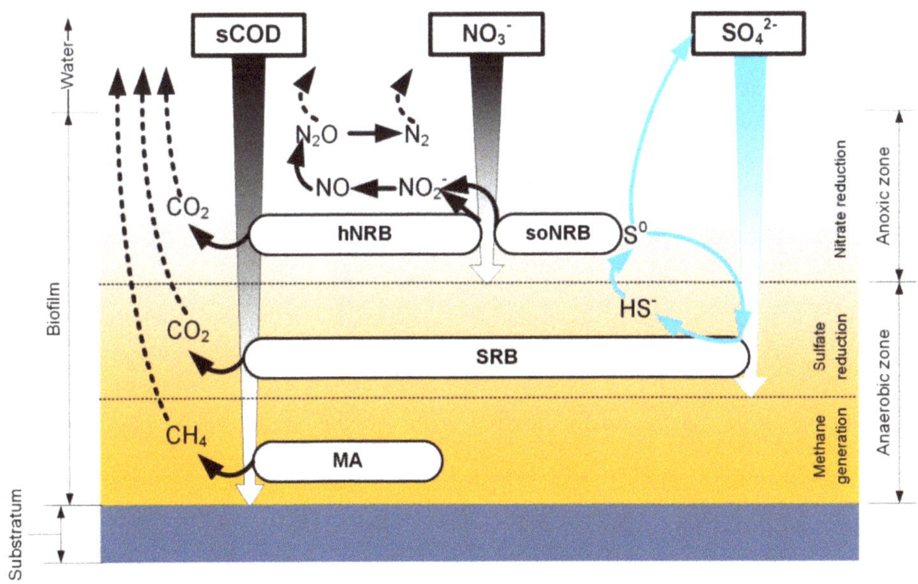

Figure 8.6 Conceptual stratified biofilm model under anoxic conditions: Competitive and synergistic interactions among heterotrophic nitrate reducing bacteria (hNRB), sulfide-oxidizing nitrate-reducing bacteria (soNRB), sulfate-reducing bacteria (SRB), and methanogenic archaea (MA). Blue arrows indicate the sulfur cycling between soNRB and SRB.

Nitrite denitrification reactions	ΔGo' (kJ/reaction)	Equation no.
$3HS^- + 8NO_2^- + 5H^+ \rightarrow 3SO_4^{2-} + 4N_2 + 4H_2O$	−2944	8.17
$3HS^- + 2NO_2^- + 5H^+ \rightarrow 3S^0 + N_2 + 4H_2O$	−917	8.18

dissimilatory sulfite reductase gene that catalyses the conversion of sulfite to sulfide (Hubert *et al.*, 2005; Nemati *et al.*, 2001). Some SRB possess a nitrite reductase gene (which prevents the inhibition of SRB by nitrite), but the function of the enzyme is purely for detoxification, and energy is not generated by the reduction (Greene *et al.*, 2003). Denitrifying bacterial species like *Thiobacillus denitrificans* can oxidize sulfide to elemental sulfur simultaneously reducing nitrogenous species to dinitrogen (Mahmood *et al.*, 2007). Nitrite toxicity produces a long-lasting inhibitory effect for H_2S and CH_4 even after the termination of the nitrite addition Conversely, other oxidants (such as nitrate or oxygen) would need to be present in the bulk at all times and in all sections of the sewer in order to effectively prevent sulfide accumulation.

Jiang and coauthors (2011) revealed that the protonated form of nitrite, HNO_2, free nitrous acid (FNA) is responsible of its biocidal effect. Sewer biofilms in lab reactor exposed to FNA concentrations above 0.2 mgN-NO_2^-/l decreased its viable fraction by 80% only with 6 hours of contact time. The recovery of methane production was about seven times slower compared to the recovery of sulfide production. The biocidal effect was found to be strongly dependent on the FNA concentration rather than the nitrite concentration or the pH level separately (for the pH range of 6.0–7.6). Based on the biocidal effects of FNA, intermittent dosing of nitrite with acid (to form FNA) is potentially a cost-effective strategy to control sulfide and methane production in sewers. Several studies have been confirmed this hypothesis (Jiang *et al.*, 2011a, 2013a).

Aerobic sewer microbial processes

Aerobic microorganisms are commonly found in sewer systems especially in the gravity sections where turbulence and reareation promote conditions optimal to their growth. These organisms

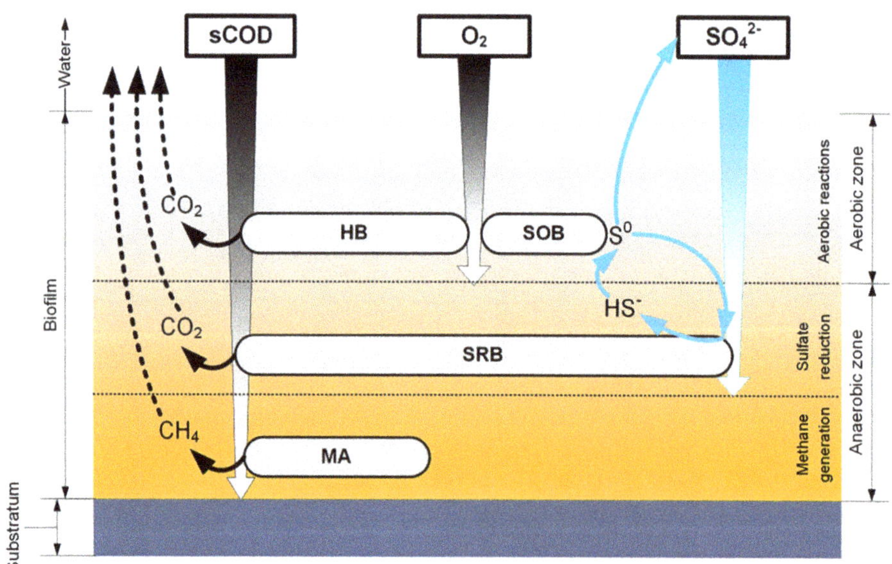

Figure 8.7 Conceptual stratified biofilm model under aerobic conditions: competitive and synergistic interactions among heterotrophic bacteria (HB), sulfide-oxidizing bacteria (SOB), sulfate-reducing bacteria (SRB), and methanogenic archaea (MA). Blue arrows indicate the sulfur cycling between SOB and SRB.

use oxygen as a terminal electron acceptor and consume dissolved organic matter in the sewage in their respiration process (Metcalf and Eddy, 2003). Aerobic processes consist of a major sink of sulfide in sewer systems (Fig. 8.7). The shift between aerobic and anaerobic conditions in wastewater of a sewer system can be very dynamic. Solubility of oxygen depends on the temperature (8.3 mg O_2/l at 25°C, and this increases with the decrease in the temperature) and the respiration rate depends on the availability of degradable organic matter and temperature and may reach values as high as 20–40 mg O_2/l·h (Hvitved-Jacobsen, 2002). Thus the reareation rate in gravity sewers could be insufficient to maintain the oxic conditions. Thus, the gravity sewers with low slopes and in warm climate are likely to develop anaerobic conditions.

Artificial injection of oxygen to sewers is widely used by wastewater authorities to mitigate the sulfide problems (Boon *et al.*, 1998; Ganigue *et al.*, 2011; Hvitved-Jacobsen, 2002; Zhang *et al.*, 2008). A laboratory study carried out in lab-sewer simulating system showed that oxygen is an effective chemical and biological oxidant of sulfide but it did not prevent the sulfide and methane production in the biofilm, which continued in the deeper layers of biofilm irrespective of the oxygen concentration in the bulk (Ganigué and Yuan, 2014; Gutierrez *et al.*, 2008). This is in agreement with the stratification theory proposed by Jiang and co-authors for oxidants addition to sewers (Jiang *et al.*, 2013a). Both studies reported that sulfide and methane accumulation resumed on complete depletion of oxygen at a certain depth of biofilm. Oxygen did not exhibit any toxic effect on sulfate-reducing bacteria (SRB) in the biofilm. With regards to the SRB, it further stimulated its growth and increased its activity in biofilms in downstream sewer sections due to increased availability of sulfate at these locations as the result of oxic conditions upstream. Furthermore the oxygen uptake rate of the system increased with repeated exposure to oxygen, with concomitant increase in the consumption of organic carbon in the wastewater due to the development of an heterotrophic community within the sewer (Ganigué and Yuan, 2014; Gutierrez *et al.*, 2008).

Unsubmerged sewer biofilms

Microbially induced concrete sewer corrosion

Concrete corrosion in sewers is primarily a result of biological processes occurring in the biofilms

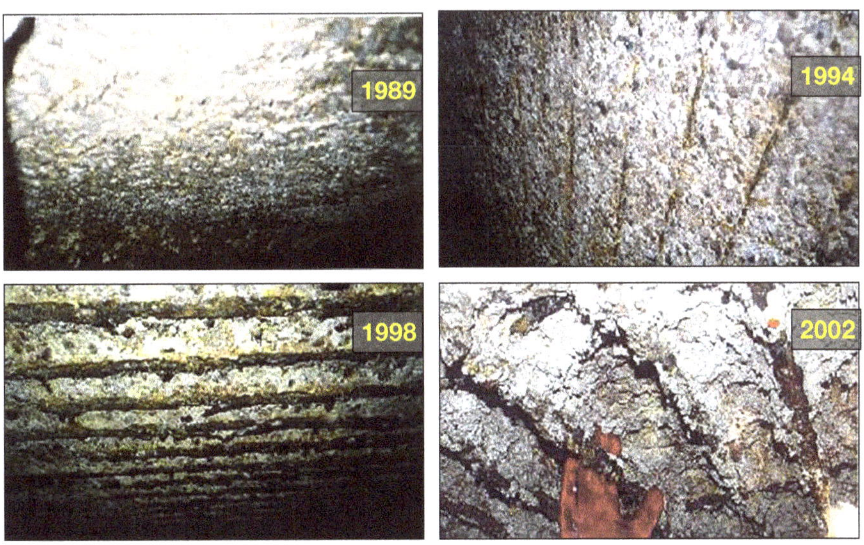

Figure 8.8 Rapid corrosion of a large gravity sewer pipe: 10 km length, 2 m width and 3 m height. The rehabilitation cost of the pipe was 100 Million Australian $. Courtesy of Sydney Water Corporation.

growing on the sewer pipe surface exposed to the gas phase. Due to the significant role played by microorganisms in sewer corrosion, it is termed as microbially induced concrete corrosion (MICC). Corrosion causes loss of concrete mass, cracking of the sewer pipes and ultimately, structural collapse (Fig. 8.8). The rehabilitation and replacement of corrosion damaged sewers involves very high costs. It is estimated that the annual cost of concrete corrosion within the water and wastewater infrastructure is about US$36 billion in USA (Koch *et al.*, 2002). This cost is expected to increase as the ageing infrastructure continues to fail (Sydney *et al.*, 1996; US EPA, 1991).

As shown in Fig. 8.9, sulfide production by sulfate-reducing bacteria and the subsequent emission to the sewer headspace are the primary causes for concrete sewer corrosion. The biological production of sulfuric acid from oxidation of hydrogen sulfide with oxygen (Joseph *et al.*, 2012a; Parker, 1945) causes mass loss of concrete (Islander *et al.*, 1991; Ismail *et al.*, 1993). The corrosion-causing biofilms grow on the surface of corroding concrete (Okabe *et al.*, 2007).

A range of corrosion products is formed during various stages of the corrosion process. The components of uncorroded cement are mainly hydrated calcium silicate ($CaO{\cdot}SiO_2{\cdot}2H_2O$) and portlandite ($Ca(OH)_2$). Abiotic processes that include carbonation and H_2S acidification, result in $CaCO_3$, and $Ca(HS)_2$ and S^0 as the main products (Wei *et al.*, 2014). During active concrete corrosion, biologically produced sulfuric acid leads to the formation of two important corrosion products: gypsum and ettringite, according to the reactions below (equations 8.19–8.22):

$$CaO{\cdot}SiO_2{\cdot}2H_2O + H_2SO_4 \rightarrow CaSO_4 + Si(OH)_4 + H_2O \quad (8.19)$$

$$H_2SO_4 + CaCO_3 \rightarrow CaSO_4 + H_2CO_3 \quad (8.20)$$

$$H_2SO_4 + Ca(OH)_2 \rightarrow CaSO_4{\cdot}2H_2O \text{ (gypsum)} \quad (8.21)$$

$$3CaSO_4{\cdot}2H_2O + 4CaO{\cdot}Al_2O_3{\cdot}13H_2O + 14H_2O \rightarrow (CaO)_3{\cdot}Al_2O_3{\cdot}(CaSO_4)_3{\cdot}32H_2O \text{ (ettringite)} \quad (8.22)$$

Both gypsum and ettringite have significantly higher volumes than intact cement, estimated to range from 124% to 700% (Monteny *et al.*, 2000; Parande *et al.*, 2006). The expansion is believed to cause internal cracking and pitting, which in turn, exposes more surface area for acid attack. However, recent findings revealed that micro-cracking at the corrosion front is more likely caused by the iron rust deposition (Jiang *et al.*, 2014).

Microbial structure and populations of corrosion biofilms

Fresh concrete in sewer pipes immediately after construction is usually immune to biological attack because of its high alkalinity (pH around 12), a result of the formation of calcium hydroxide ($Ca(OH)_2$) as the byproduct of cement

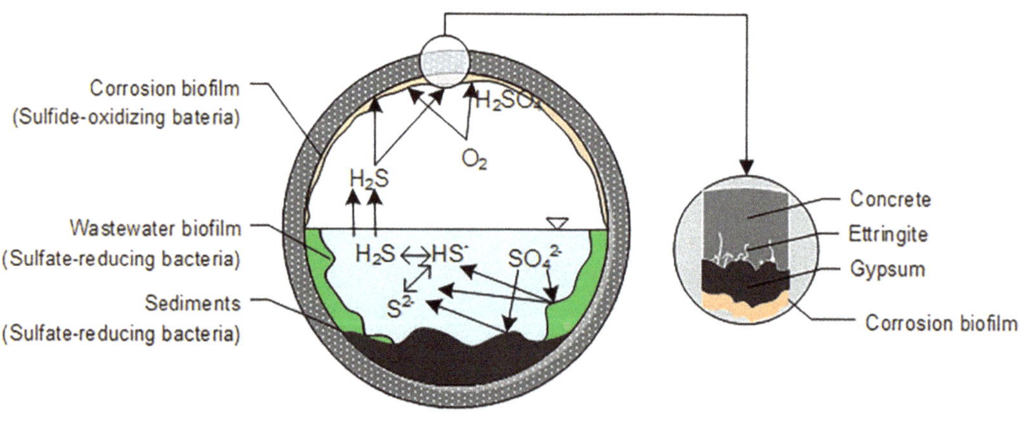

Figure 8.9 Processes involved in the microbially induced concrete corrosion in sewers.

hydration. Anaerobic conditions in wastewater favour the production of hydrogen sulfide and carbon dioxide, which builds up in the sewer air (Guisasola *et al.*, 2009; Lahav *et al.*, 2004; Nielsen *et al.*, 2005). Over time, the pH of the alkaline concrete surface is gradually reduced by the carbonation (equations 8.23–8.24) and neutralization of hydrogen sulfide (equations 8.25–8.27) (Bagreev and Bandosz, 2004, 2005; Joseph *et al.*, 2012b). It was also found that the attachment and colonization of some pioneer microorganisms (mainly heterotrophic, halotolerant, and neutrophilc bacteria) on the concrete surface could have a great impact on the initial pH decrease (Okabe *et al.*, 2007). The initial acidification processes can reduce the pH of concrete sewer from around 12 to 9, this can establish suitable growth conditions for the subsequent emergence and propagation of sulfur and sulfide oxidizing microorganisms on the concrete surface (Joseph *et al.*, 2012a).

$$Ca(OH)_2 + CO_2 \rightarrow CaCO_3 + H_2O \quad (8.23)$$

$$CaO{\cdot}SiO_2{\cdot}2H_2O + CO_2 \rightarrow CaCO_3 + Silica + H_2O \quad (8.24)$$

$$CaCO_3 + H_2S \rightarrow Ca(HCO_3)_2 + Ca(HS)_2 \quad (8.25)$$

$$Ca(OH)_{2+}2H_2S \rightarrow Ca(HS)_2 + 2H_2O \quad (8.26)$$

$$Ca(HS)_2 + O_2 \rightarrow 2S + Ca(OH)_2 \quad (8.27)$$

When the surface pH reaches 9, various microorganisms can colonize on the concrete surface with the availability of moisture and nutrients (i.e. stage 2 of corrosion development shown in Fig. 8.10). Due to the abundance of hydrogen sulfide and oxygen, some neutrophilic sulfide-oxidizing bacteria (SOB) start to colonize and grow on the concrete (Mori *et al.*, 1992). The dominant SOB species of the second stage of corrosion include *Thiothrix* sp., *Thiobacillus plumbophilus*, *Thiomonas* sp., and *Halothiobacillus neapolitanus* (Okabe *et al.*, 2007). These SOB species were probably responsible for the production of sulfuric acid from the gaseous hydrogen sulfide (Vollertsen *et al.*, 2008; Zhang *et al.*, 2008; Zivica and Bajza, 2001). The sulfide-oxidizing bacteria, along with some fungi, algae, and lichens form biofilms (Fig. 8.10), which further reduce down the surface pH of concrete (Domingo *et al.*, 2011; Nica *et al.*, 2000).

In the third stage, at which the pH remains around 2, the acidophilic SOB *Acidithiobacillus thiooxidans* appeared and became the most dominant microbial species (Okabe *et al.*, 2007;

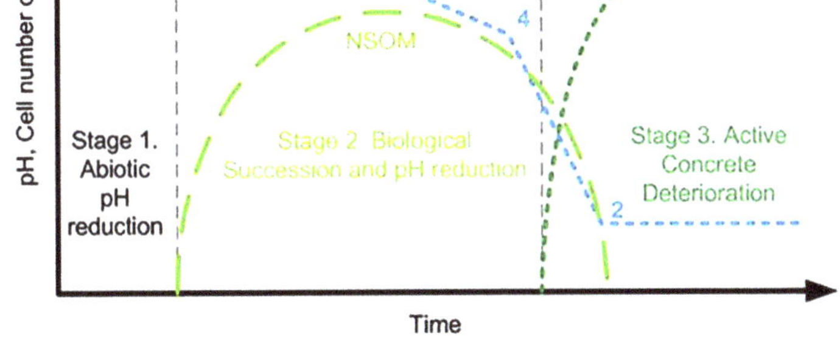

Figure 8.10 The succession of different sulfide-oxidizing bacteria with the changes of surface pH due to the development of concrete corrosion in sewers. Adapted from Islander *et al.* (1991). Neutrophilic sulfide-oxidizing microorganisms (NSOM); acidophilic sulfide-oxidizing microorganism (ASOM).

Table 8.1 Sulfide-oxidizing bacteria (*Thiobacillus*) species involved in the microbially induced concrete corrosion in sewers, and their characteristics (Okabe *et al.*, 2007; Roberts *et al.*, 2002)

Species	Growth pH	Temperature (°C)	Lifestyle	Sulfur substrates
Thiobacillus thioparus	4.5–10	15–42	Autotrophic aerobe	Thiosulfate, sulfide, thiocyanate
Starkeya novella (Thiobacillus novellus)	5–9.2	10–37	Mixotroph	Thiosulfate
Halothiobacillus neapolitanus	4.5–8.5	8–39	Autotroph	Sulfur, sulfide, thiosulfate
Thiomonas intermedia	5–7.5	15–37	Mixotroph	Sulfur, thiosulfate
Thiobacillus plumbophilus	4–6.5	9–41	Autotroph	Sulfide
Thiothrix spp.	Neutral pH	10–30	Mixotroph	Sulfide, thiosulfate
Thiobacillus intermedius (Thiomonas intermedia)	1.7–9	15–37	Mixotroph	Thiosulfate
Acidithiobacillus thiooxidans	0.5–5.5	10–37	Autotroph	Thiosulfate, sulfur
Acidiphilum acidophilum	1.5–6.0	10–35	Heterotroph	Sulfur, thiosulfate

Wei *et al.*, 2014), although there are some conflicting findings about the dominant species in different environments (Cayford *et al.*, 2012; Parker, 1951; Sand and Bock, 1984). The high amount of sulfuric acid produced by corrosion biofilm leads to significant loss of concrete mass at this stage.

The important species like *Thiobacillus* sp. (syn. *Acidithiobacillus* sp.) and other SOB species found in corrosion biofilms are listed in Table 8.1. The first seven species are neutrophilic sulfide-oxidizing microorganisms (NSOM), while the last three are acidophilic sulfide-oxidizing microorganism (ASOM). The NSOM, and probably some fungi, further reduce the concrete pH to 4–5. At the end of stage 2, ASOM starts to grow in the sewer corrosion biofilm due to both the low pH and also the sulfur products produced by NSOM. Sulfur generated by NSOM is an essential substrate for ASOM, which produce significant amounts of sulfuric acid (Jensen *et al.*, 2008; Kelly and Wood, 2000). The optimum growth pH, trophic property (e.g. autotrophic or mixotrophic), and ability to utilize different sulfur compounds (e.g. H_2S, S^0, and $S_2O_3^{2-}$) by SOB determine the order of appearance and the dominance of different SOB species on corroding concrete surfaces in sewer systems (Okabe *et al.*, 2007).

In addition to the important SOB species in the corrosion biofilm, there are many other microorganisms being identified that are able to oxidize sulfur compounds, most of which are heterotrophic bacteria (*Pseudomonas, Streptomyces, Arthrobacter, Bacillus, Flavobacter, Micromonas*, etc.) and fungi which may grow as co-cultures with SOB (Coleman and Gaudet, 1993; Nica *et al.*, 2000). It was suggested that fungus, e.g. *Fusarium* sp., act synergistically with thiobacilli by either oxidizing sulfide to thiosulfate which can then be used by thiobacilli, or excreting extracellular polymeric substances (EPS) that facilitates thiobacilli to attach to elemental sulfur (Cho and Mori, 1995; Gu *et al.*, 1998) (Fig. 8.11). It was found that over 50% of the corrosion biofilm community was heterotrophic bacteria other than SOB (Okabe *et al.*, 2007). The variety of microorganisms, both heterotrophic and autotrophic

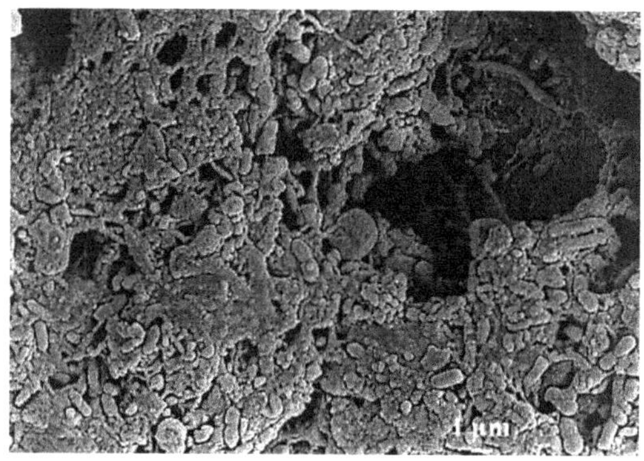

Figure 8.11 Concrete surface (scanning electron microscopy image, 4000×) covered with a layer of biofilms containing many 1–3 mm-sized rods of thiobacilli (Monteny *et al.*, 2000).

bacteria and fungi, isolated from corroding sewers shows the complexity of the microbiological population in the corrosion biofilms.

Factors impacting the corrosion biofilm activities

The H_2S concentration in real sewers varies greatly due to the different hydraulic retention time, flow velocity and wastewater characteristics. In addition to high relative humidity (see below), a H_2S level >2 ppm was suggested to be required for the sulfide oxidation to proceed on concrete sewers (O'Dea, 2007). It is traditionally assumed that corrosion biofilm activity (i.e. biological sulfide oxidation) is directly proportional to H_2S emission rate (De Belie *et al.*, 2004). The well-known Pomeroy model can be used to calculate the deterioration rate of concrete sewer pipes (Pomeroy, 1990).

$$C_r = \frac{11.5k\phi_{sw}}{alk} \qquad (8.28)$$

where C_r = corrosion rate (mm/year); k = factor related to the acid formation, based on climate conditions, 0.8 in moderate climates; f_{sw} = sulfide flux at the air–wall interface $[\text{g } H_2S/(m^2 \cdot h)]$; and alk = alkalinity of the pipe material (g $CaCO_3$/g concrete).

A recent report established a relationship between corrosion rate and controlling factors including H_2S gaseous concentration, relative humidity and temperature for concrete sewers either exposed to air or near the wastewater surface (Jiang *et al.*, 2014). The corrosion rate is proportional to the biofilm activity that is generally formulated as a power function of H_2S concentration (Nielsen *et al.*, 2005).

In sewers, water and nutrients provided by sewage are found to promote the microbial corrosion, especially for the area close to the wastewater level in a sewer pipe (Mori *et al.*, 1992). For the pipe surface further away from wastewater level, the relative humidity of the sewer air and the condensation process on the concrete surface would generate a water film for microbial growth. It was reported that humidity plays a role in surface neutralization at the early stage of sewer concrete corrosion (Joseph *et al.*, 2012a). During a long-term investigation of sewer corrosion, it was found that humidity is important for the corrosion biofilms which are far from the water level in sewer pipes (Jiang *et al.*, 2014). This is because high humidity leads to high moisture content in concrete that facilitates increased biological H_2S oxidation and sulfate production.

Short-term changes of temperature are typical for sewer systems and the interactions between sewer systems and sewage (Vollertsen *et al.*, 1999). One important process for sewer concrete corrosion is the air-water transfer of hydrogen sulfide, which was found to increase with increasing temperature at a constant turbulence level (Yongsiri *et al.*, 2004). It was widely accepted that sulfide oxidation rate, both chemically and biologically, increases with temperature, which can be described with an Arrhenius relationship (Nielsen *et al.*, 2004, 2006). The sulfide oxidation rate was reported to be doubled for a temperature increase of 7–9°C. In addition, sewer systems located in different climates would have been acclimated to different temperatures. The sulfide oxidation rates, and accordingly corrosion rates, would thus be very different for different climatic regions. However, no clear effects of temperature (5–17°C) was found for the hydrogen sulfide oxidation kinetics (Nielsen *et al.*, 2012). This was attributed to the population dynamics of SOB in the corrosion layers. Another study also reported non-significant effects of temperature between 18–35°C on the sulfide oxidation and corrosion rates (Jiang *et al.*, 2014). Further studies are needed to clarify the effects of temperature on the actual corrosion rates in sewers.

Control of concrete sewer corrosion

To alleviate and control the concrete sewer corrosion problems, various gas-phase technologies have been used to reduce or remove H_2S from sewer air. The sewer air treatment technologies include activated carbon adsorption, chemical scrubbing, and biotrickling filters for the biological oxidation of H_2S (Sivret and Stuetz, 2010). In addition to removing H_2S from sewer air, the ventilation also reduce the humidity in the sewer gas phase, which can significantly reduce the sewer concrete corrosion (Jiang *et al.*, 2014).

However, it is difficult to control sulfide in the entire sewer network using chemical dosing or sewer air treatment due to costs and site restrictions. It is thus relevant and essential to make the sewer itself resistant to corrosion by concrete-based technologies that construct new sewers with corrosion-resistant concrete (proactive prevention) or repair, coat and line the corroded concrete surfaces (passive rehabilitation) (Haile *et al.*, 2010; Hewayde *et al.*, 2007). This approach includes applications of admixtures, protective coatings, and acid-resistant cement to prevent chemical attack by sulfuric acid. Antimicrobial coatings and admixtures, such as silver/copper zeolites (Zeomic®) and water-stabilized silicone quaternary ammonium salt (Conshield®) among a few other commercialized products, were also applied to reduce or eliminate the microbial activity (De Muynck *et al.*, 2009; Rivera-Garza *et al.*, 2000; Yamanaka *et al.*, 2002).

Recently, it was reported that *Escherichia coli* DH5*a* biofilm showed great potential to control and minimize microbially induced concrete corrosion by *Thiobacillus neapolitanus* and *Thiobacillus thiooxidans* (Soleimani *et al.*, 2013a). A protective biofilm layer with a depth of 20–40 mm was successfully grown on the surface of cement mortars (Soleimani *et al.*, 2013b). However, due to the highly oligotrophic condition on the sewer surface exposed to air, it is unlikely this technology can be implemented in real gravity sewer.

For new sewers, it is preferable to be constructed with corrosion-resistant concrete because the rehabilitation after corrosion damage is usually difficult and expensive. This can be achieved by using cementitious materials containing the corrosion inhibitors as an admixture (Saraswathy and Song, 2007). Different cements or cements with admixtures have been trialed with limited success. Current admixtures mainly focus on changing the physiochemical properties of concrete, such as reducing permeability or increasing buffering capacity. Concrete with a lower content of tricalcium aluminate (C_3A) was shown to have better resistance to the sulfate attack (Monteny *et al.*, 2000). Polymer additions are also used to modify concrete, with its corrosion resistance to acid attack being improved or worsen for different polymers (Vincke *et al.*, 2001).

Conclusions and perspectives

Biofilms are naturally present in urban sewer systems. Sewer biofilms can grow under different environments (anaerobic/aerobic) leading to the growth of different types of microorganisms. The release of detrimental compounds from sewer biofilms, such as H_2S or CH_4, has been thoroughly described in this chapter. A significant amount of knowledge has been generated in recent years but further investigations are still required to fully understand sewer biofilms and control its impacts on urban water systems.

It is important to have a good understanding of the biofilm-functioning under the specific sewer conditions in order to develop appropriate control strategies. In this regard, new methods for the control of the production of detrimental compounds in sewer biofilms are currently being developed and tested. For instance, free nitrous acid, an emerging chemical, has been shown to be cost-effective and environment-friendly in both lab studies and full scale scenarios (Jiang and Yuan, 2013a,b; Jiang *et al.*, 2011b, 2013b).

Integration of sewers with different subsystems of the urban water cycle is also receiving a most deserved attention. Recent studies have demonstrated that using sulfate-based coagulants in drinking water can have a major impact in sulfate content of sewer downstream, enhancing the development of sewer biofilms (Pikaar *et al.*, 2014).

Moreover further research is required to understand how the sewer biofilms contribute to the in-sewer biotransformation of various chemicals of different interests. Very little is known about the role of biofilms in the transformation of many micro pollutants detected in sewers, including pharmaceuticals and personal care products. A limited number of recent studies have shown sewer biofilms to play an important role in the degradation of illicit drugs and residues, pharmaceuticals and health-related compounds (Jelic *et al.*, 2014; Thai *et al.*, 2013, 2014).

With regards to the microbially induced corrosion problem, microbial communities in corrosion biofilms and their biological processes are still not fully delineated. A better understanding of the microorganisms involved and their metabolisms will lead to more efficient control strategies and optimal sewer management.

Acknowledgements

Oriol Gutierrez would like to acknowledge the project CTM 2011-27163 funded by the Spanish Government Ministerio de Economía y Competitividad, the European Commission project 2010-RG277050 and the support from the Economy and Knowledge Department of the Catalan Government through a Consolidated Research Group (2014 SGR 291)-Catalan Institute for Water Research. Dr Guangming Jiang is the recipient of a Queensland State Government's Early Career Accelerate Fellowship (Australia).

References

Abram, J., and Nedwell, D. (1978). Inhibition of methanogenesis by sulfate reducing bacteria competing for transferred hydrogen. Arch. Microbiol. *117*, 89–92.

Auguet, O., Pijuan, M., Guasch-Balcells, H., Borrego, C., and Gutierrez, O. (2015). Implications of Downstream Nitrate Dosage in anaerobic sewers to control sulfide and methane emissions. Water Res. *68*, 522–532.

Bagreev, A., and Bandosz, T. (2004). Carbonaceous materials for gas phase desulphurization: role of surface. Am. Chem. Soc. Div. Fuel Chem. *49*, 817–821.

Bagreev, A., and Bandosz, T.J. (2005). On the Mechanism of Hydrogen Sulfide Removal from Moist Air on Catalytic Carbonaceous Adsorbents. Ind. Eng. Chem. Res. *44*, 530–538.

De Belie, N., Monteny, J., Beeldens, A., Vincke, E., Van Gemert, D., and Verstraete, W. (2004). Experimental research and prediction of the effect of chemical and biogenic sulfuric acid on different types of commercially produced concrete sewer pipes. Cem. Concr. Res. *34*, 2223–2236.

Bhattacharya, S.K., Uberoi, V., and Dronamraju, M.M. (1996). Interaction between acetate fed sulfate reducers and methanogens. Water Res. *30*, 2239–2246.

Bodegom, P.M. va., and Stams, A.J.M. (1999). Effects of alternative electron acceptors and temperature on methanogenesis in rice paddy soils. Chemosphere *39*, 167–182.

Boon, A.G., Vincent, A.J., and Boon, K.G. (1998). Avoiding the problems of septic sewage. Water Sci. Technol. *37*, 223–231.

Brongers, M.P.H. (2001). Drinking water and sewer systems. In Cost of Corrosion Study. (NACE, USA), pp. 1–11.

Brongers, M., Virmani, P.Y., and Payer, J.H. (2002). Drinking water and sewer systems. In Corrosion Costs and Preventive Strategies in the United States. (United States Department of Transportation Federal Highway Administration Publication No FHWA-RD-01–156), pp. 1–16.

Cayford, B.I., Dennis, P.G., Keller, J., Tyson, G.W., and Bond, P.L. (2012). High-throughput amplicon sequencing reveals distinct communities within a corroding concrete sewer system. Appl. Environ. Microbiol. *78*, 7160–7162.

Cho, K.-S., and Mori, T. (1995). A newly isolated fungus participates in the corrosion of concrete sewer pipes. Water Sci. Technol. *31*, 263–271.

Coleman, R.N., and Gaudet, I.D. (1993). Thiobacillus neopolitanus implicated in the degradation of concrete tanks used for potable water storage. Water Res. *27*, 413–418.

Dar, S.A., Kleerebezem, R., Stams, A.J.M., Kuenen, J.G., and Muyzer, G. (2008). Competition and coexistence of sulfate-reducing bacteria, acetogens and methanogens in a lab-scale anaerobic bioreactor as affected by changing substrate to sulfate ratio. Appl. Microbiol. Biotechnol. *78*, 1045–1055.

Domingo, J.W.S., Revetta, R.P., Iker, B., Gomez-Alvarez, V., Garcia, J., Sullivan, J., and Weast, J. (2011). Molecular survey of concrete sewer biofilm microbial communities. Biofouling *27*, 993–1001.

Foley, J., Yuan, Z., and Lant, P. (2009). Dissolved methane in rising main sewer systems: field measurements and simple model development for estimating greenhouse gas emissions. Water Sci. Technol. *60*, 2963–2971.

Ganigue, R., Gutierrez, O., Rootsey, R., and Yuan, Z. (2011). Chemical dosing for sulfide control in Australia: an industry survey. Water Res. *45*, 6564–6574.

Ganigué, R., and Yuan, Z. (2014). Impact of oxygen injection on CH4 and N2O emissions from rising main sewers. J. Environ. Manage. *144*, 279–285.

Gantner, S., Andersson, A.F., Alonso-Sáez, L., and Bertilsson, S. (2011). Novel primers for 16S rRNA-based archaeal community analyses in environmental samples. J. Microbiol. Meth. *84*, 12–18.

Greene, E.A., Hubert, C., Nemati, M., Jenneman, G.E., and Voordouw, G. (2003). Nitrite reductase activity of sulfate-reducing bacteria prevents their inhibition by nitrate-reducing, sulphide-oxidizing bacteria. Environ. Microbiol. *5*, 607–617.

Gu, J.D., Ford, T.E., Berke, N.S., and Mitchell, R. (1998). Biodeterioration of concrete by the fungus Fusarium. Int. Biodeterior. Biodegradation *41*, 101–109.

Guisasola, A., Sharma, K.R., Keller, J., Yuan, Z., and Jiang, G. (2009). Development of a model for assessing methane formation in rising main sewers. Water Res. *43*, 2874–2884.

Gupta, A. (1994). Methanogenesis and sulfate reduction in chemostats – I. Kinetic studies and experiments. Water Res. *28*, 781–793.

Gutierrez, O., Mohanakrishnan, J., Sharma, K.R., Meyer, R.L., Keller, J., and Yuan, Z. (2008). Evaluation of oxygen injection as a means of controlling sulfide production in a sewer system. Water Res. *42*, 4549–4561.

Gutierrez, O., Park, D., Sharma, K.R., and Yuan, Z. (2009). Effects of long-term pH elevation on the sulfate-reducing and methanogenic activities of anaerobic sewer biofilms. Water Res. *43*, 2549–2557.

Gutierrez, O., Sharma, K.R., Batista, J., Pijuan, M., and Poch, M. (2012). Monitoring and prediction of sulfide and methane emisions from sewer in mediterreanean cities. In META 2012, X Mesa Española de Tratamiento de Aguas.

Gutierrez, O., Sudarjanto, G., Ren, G., Ganigué, R., Jiang, G., and Yuan, Z. (2014). Assessment of pH shock as a method for controlling sulfide and methane formation in pressure main sewer systems. Water Res. *48*, 569–578.

Haile, T., Nakhla, G., Allouche, E., and Vaidya, S. (2010). Evaluation of the bactericidal characteristics of nano-copper oxide or functionalized zeolite coating for bio-corrosion control in concrete sewer pipes. Corros. Sci. *52*, 45–53.

Hewayde, E.H., Nakhla, G.F., Allouche, E.N., and Mohan, P.K. (2007). Beneficial impact of coatings on biological generation of sulfide in concrete sewer pipes. Struct. Infrastruct. Eng. *3*, 267–277.

Hubert, C., Nemati, M., Jenneman, G., and Voordouw, G. (2005). Corrosion risk associated with microbial souring control using nitrate or nitrite. Appl. Microbiol. Biotechnol. *68*, 272–282.

Hvitved-Jacobsen, T. (2002). Sewer systems and processes. In Sewer Processes: Microbial and Chemical Process Engineering of Sewer Networks. Hvitved-Jacobsen, T., Vollertsen, J., and Nielsen, A.H., eds (CRC Press, Taylor and Francis Group LLC, Boca Raton, FL, USA), pp. 1–10.

Islander, R.L., Devinny, J.S., Mansfeld, F., Postyn, A., and Shih, H. (1991). Microbial ecology of crown corrosion in sewers. J. Environ. Eng. *117*, 751–770.

Ismail, N., Nonaka, T., Noda, S., and Mori, T. (1993). Effects of carbonation on microbial corrosion of concretes. J. Constr. Manag. Eng. *20*, 11–138.

Jelic, A., Rodriguez-Mozaz, S., Barcelo, D., and Gutierrez, O. (2014). Impact of in-sewer transformation on 43 pharmaceuticals in a pressurized sewer under anaerobic conditions. Water Res. *68*, 98–108.

Jensen, H.S., Nielsen, A.H., Hvitved-Jacobsen, T., and Vollertsen, J. (2008). Survival of hydrogen sulfide oxidizing bacteria on corroded concrete surfaces of sewer systems. Water Sci. Technol. *57*, 1721–1726.

Jiang, G., and Yuan, Z. (2013a). Synergistic inactivation of anaerobic wastewater biofilm by free nitrous acid and hydrogen peroxide. J. Hazard. Mater. *250–251*, 91–98.

Jiang, G., and Yuan, Z. (2013b). Inactivation kinetics of anaerobic wastewater biofilms by free nitrous acid. Appl. Microbiol. Biotechnol. *98*, 1367–1376.

Jiang, G., Sharma, K.R., Guisasola, A., Keller, J., and Yuan, Z. (2009). Sulfur transformation in rising main sewers receiving nitrate dosage. Water Res. *43*, 4430–4440.

Jiang, G., Gutierrez, O., Sharma, K.R., and Yuan, Z. (2010). Effects of nitrite concentration and exposure time on sulfide and methane production in sewer systems. Water Res. *44*, 4241–4251.

Jiang, G., Gutierrez, O., and Yuan, Z. (2011a). The strong biocidal effect of free nitrous acid on anaerobic sewer biofilms. Water Res. *45*, 3735–3743.

Jiang, G., Gutierrez, O., and Yuan, Z. (2011b). The strong biocidal effect of free nitrous acid on anaerobic sewer biofilms. Water Res. *45*, 3735–3743.

Jiang, G., Sharma, K.R., and Yuan, Z. (2013a). Effects of nitrate dosing on methanogenic activity in a sulfide-producing sewer biofilm reactor. Water Res. *47*, 1783–1792.

Jiang, G., Keating, A., Corrie, S., O'Halloran, K., Nguyen, L., and Yuan, Z. (2013b). Dosing free nitrous acid for sulfide control in sewers: Results of field trials in Australia. Water Res. *47*, 4331–4339.

Jiang, G., Wightman, E., Donose, B.C., Yuan, Z., Bond, P.L., and Keller, J. (2014). The role of iron in sulfide induced corrosion of sewer concrete. Water Res. *49*, 166–174.

Joseph, A.P., Keller, J., Bustamante, H., and Bond, P.L. (2012a). Surface neutralization and H(2)S oxidation at early stages of sewer corrosion: influence of temperature, relative humidity and H(2)S concentration. Water Res. *46*, 4235–4245.

Joseph, A.P., Keller, J., Bustamante, H., and Bond, P.L. (2012b). Surface neutralization and H2S oxidation at early stages of sewer corrosion: Influence of temperature, relative humidity and H2S concentration. Water Res. *46*, 4235–4245.

Kalyuzhnyi, S.V., and Fedorovich, V.V. (1998). Mathematical modelling of competition between sulfate reduction and methanogenesis in anaerobic reactors. Bioresour. Technol. *65*, 227–242.

Kelly, D.P., and Wood, A.P. (2000). Reclassification of some species of Thiobacillus to the newly designated genera Acidithiobacillus gen. nov., Halothiobacillus gen. nov and Thermithiobacillus gen. nov. Int. J. Syst. Evol. Microbiol. *50*, 511–516.

Koch, G.H., Brongers, M.P.H., Thompson, N.G., Virmani, Y.P., and Payer, J. (2002). Corrosion costs and preventive strategies in the United States (NACE International, Houston, TX, USA).

Lahav, O., Lu, Y., Shavit, U., and Loewenthal, R. (2004). Modeling hydrogen sulfide emission rates in gravity sewage collection systems. J. Environ. Eng. *130*, 1382–1389.

Lovley, D.R., Dwyer, D.F., and Klug, M.J. (1982). Kinetic analysis of competition between sulfate reducers and methanogens for hydrogen in sediments. Appl. Environ. Microbiol. *43*, 1373–1379.

Mahmood, Q., Zheng, P., Cai, J., Wu, D., Hu, B., and Li, J. (2007). Anoxic sulfide biooxidation using nitrite as electron acceptor. J. Hazard. Mater. *147*, 249–256.

Metcalf and Eddy, I. (2003). Wastewater Engineering, Treatment and Reuse, 4th edn (McGraw-Hill, New York, NY, USA).

Mohanakrishnan, J., Gutierrez, O., Meyer, R.L., and Yuan, Z. (2008). Nitrite effectively inhibits sulfide and methane production in a laboratory scale sewer reactor. Water Res. *42*, 3961–3971.

Mohanakrishnan, J., Gutierrez, O., Sharma, K.R., Guisasola, A., Werner, U., Meyer, R.L., Keller, J., and Yuan, Z. (2009). Impact of nitrate addition on biofilm properties and activities in rising main sewers. Water Res. *43*, 4225–4237.

Monteny, J., Vincke, E., Beeldens, A., De Belie, N., Taerwe, L., Van Gemert, D., and Verstraete, W. (2000). Chemical, microbiological, and *in situ* test methods for biogenic sulfuric acid corrosion of concrete. Cem. Concr. Res. *30*, 623–634.

Mori, T., Nonaka, T., Tazaki, K., Koga, M., Hikosaka, Y., and Noda, S. (1992). Interactions of nutrients, moisture and pH on microbial corrosion of concrete sewer pipes. Water Res. *26*, 29–37.

De Muynck, W., De Belie, N., and Verstraete, W. (2009). Effectiveness of admixtures, surface treatments and antimicrobial compounds against biogenic sulfuric acid corrosion of concrete. Cem. Concr. Compos. *31*, 163–170.

Muyzer, G., and Stams, A.J.M. (2008). The ecology and biotechnology of sulfate-reducing bacteria. Nat. Rev. Microbiol. *6*, 441–454.

Nemati, M., Mazutinec, T.J., Jenneman, G.E., and Voordouw, G. (2001). Control of biogenic H2S production with nitrite and molybdate. J. Ind. Microbiol. Biotechnol. *26*, 350–355.

Nica, D., Davis, J.L., Kirby, L., Zuo, G., and Roberts, D.J. (2000). Isolation and characterization of microorganisms involved in the biodeterioration of concrete in sewers. Int. Biodeterior. Biodegradation *46*, 61–68.

Nielsen, A.H., Vollertsen, J., and Hvitved-Jacobsen, T. (2004). Chemical sulfide oxidation of wastewater – effects of pH and temperature. Water Sci. Technol. *50*, 185–192.

Nielsen, A.H., Yongsiri, C., Hvitved-Jacobsen, T., and Vollertsen, J. (2005). Simulation of sulfide buildup in wastewater and atmosphere of sewer networks. Water Sci. Technol. *52*, 201–208.

Nielsen, A.H., Vollertsen, J., and Hvitved-Jacobsen, T. (2006). Kinetics and stoichiometry of aerobic sulfide oxidation in wastewater from sewers-effects of pH and temperature. Water Environ. Res. *78*, 275–283.

Nielsen, A.H., Hvitved-Jacobsen, T., and Vollertsen, J. (2012). Effect of sewer headspace air-flow on hydrogen sulfide removal by corroding concrete surfaces. Water Environ. Res. *84*, 265–273.

O'Dea, V. (2007). Understanding biogenic sulfide corrosion. Mater. Perform. *46*, 36–39.

Okabe, S., Odagiri, M., Ito, T., and Satoh, H. (2007). Succession of sulfur-oxidizing bacteria in the microbial community on corroding concrete in sewer systems. Appl. Environ. Microbiol. 73, 971–980.

Oremland, R.S., and Polcin, S. (1982). Methanogenesis and sulfate reduction: competitive and noncompetitive substrates in estuarine sediments. Appl. Environ. Microbiol. *44*, 1270–1276.

Parande, A.K., Ramsamy, P.L., Ethirajan, S., Rao, C.R.K., and Palanisamy, N. (2006). Deterioration of reinforced concrete in sewer environments. Proc. Inst. Civ. Eng. Eng. *159*, 11–20.

Parker, C.D. (1951). Mechanics of corrosion of concrete sewers by hydrogen sulfide. Sewage Ind. Waste. *23*, 1477–1485.

Parker, D. (1945). The corrosion of concrete. 2. The fuction of *Thiobacillus concretivorus* (Nov-spec) in the corrosion of concrete exposed to atmospheres containing hydrogen sulfide. Aust. J. Exp. Biol. Med. Sci. 23, 91–98.

Pikaar, I., Sharma, K.R., Hu, S., Gernjak, W., Keller, J., and Yuan, Z. (2014). Reducing sewer corrosion through integrated urban water management. Science *345*, 812–814.

Pomeroy, R.D. (1990). The Problem of Hydrogen Sulphide in Sewers (Clay Pipe Development Association Limited, London).

Rivera-Garza, M., Olguín, M.T., García-Sosa, I., Alcántara, D., and Rodríguez-Fuentes, G. (2000). Silver supported on natural Mexican zeolite as an antibacterial material. Microporous Mesoporous Mater. *39*, 431–444.

Roberts, D.J., Nica, D., Zuo, G., and Davis, J.L. (2002). Quantifying microbially induced deterioration of concrete: initial studies. Int. Biodeterior. Biodegradation *49*, 227–234.

Rodríguez-Gómez, L.E., Delgado, S., Álvarez, M., and Elmaleh, S. (2005). Inhibition of Sulfide Generation in a Reclaimed Wastewater Pipe by Nitrate Dosage and Denitrification Kinetics. Water Environ. Res. 77, 193–198.

Sand, W., and Bock, E. (1984). Concrete corrosion in the Hamburg Sewer system. Environ. Technol. Lett. *5*, 517–528.

Saraswathy, V., and Song, H.-W. (2007). Improving the durability of concrete by using inhibitors. Build. Environ. *42*, 464–472.

Sivret, E., and Stuetz, R.M. (2010). Sewer odour abatement practices – an Australian survey. Water 37, 77–81.

Soleimani, S., Ormeci, B., and Isgor, O.B. (2013a). Growth and characterization of Escherichia coli DH5α biofilm on concrete surfaces as a protective layer against microbiologically influenced concrete deterioration (MICD). Appl. Microbiol. Biotechnol. 97, 1093–1102.

Soleimani, S., Ormeci, B., and Isgor, O.B. (2013b). Evaluation of *E. coli* biofilm as a protective barrier against microbiologically influenced deterioration of concrete (MICD) under mesophilic temperatures. Water Sci. Technol. *68*, 303–310.

Sørensen, J., Christensen, D., and Jørgensen, B.B. (1981). Volatile fatty acids and hydrogen as substrates for sulfate-reducing bacteria in anaerobic marine sediment. Appl. Environ. Microbiol. *42*, 5–11.

Sun, J., Hu, S., Sharma, K.R., Ni, B.-J., and Yuan, Z. (2014). Stratified microbial structure and activity in sulfide- and methane- producing anaerobic sewer biofilms. Appl. Environ. Microbiol. *80*, 7042–7052.

Sydney, R., Esfandi, E., and Surapaneni, S. (1996). Control concrete sewer corrosion via the crown spray process. Water Environ. Res. *68*, 338–347.

Thai, P.K., O'Briena, J., Jiang, G., Gernjak, W., Yuan, Z., Eaglesham, G., and Mueller, J.F. (2014). Degradability of creatinine under sewer conditions affects its potential to be used as biomarker in sewage epidemiology. Water Res. *55*, 272–279.

Thai, P.K., Jiang, G., Gernjak, W., Yuan, Z., Lai, F.Y., and Mueller, J.F. (2014). Effects of sewer conditions on the degradation of selected illicit drug residues in wastewater. Water Res. *48*, 538–547.

Thauer, R.K., Jungermann, K., Decker, K., and Pi, P.P.H. (1977). Energy conservation in chemotrophic anaerobic bacteria. Bacteriol. Rev. *41*, 809.

US EPA (1991). Hydrogen sulphide corrosion in wastewater collection and treatment system. Technical report (Office of Water WH595, Washington, DC, USA).

Vincke, E., Boon, N., and Verstraete, W. (2001). Analysis of the microbial communities on corroded concrete sewer pipes – a case study. Appl. Microbiol. Biotechnol. *57*, 776–785.

Vollertsen, J., Almeida, M. d C., and Hvitved-Jacobsen, T. (1999). Effects of temperature and dissolved oxygen on hydrolysis of sewer solids. Water Res. *33*, 3119–3126.

Vollertsen, J., Nielsen, A.H., Jensen, H.S., Wium-Andersen, T., and Hvitved-Jacobsen, T. (2008). Corrosion of concrete sewers – the kinetics of hydrogen sulfide oxidation. Sci. Total Environ. *394*, 162–170.

Wei, S., Jiang, Z., Liu, H., Zhou, D., and Sanchez-Silva, M. (2014). Microbiologically induced deterioration of concrete: a review. Braz. J. Microbiol. *44*, 1001–1007.

Water Environ. Res. Foundation (2007). Minimization of odor and corrosion in collection systems: phase I (WERF, Alexandria, VA and IWA Publishing, London).

Yamanaka, T., Aso, I., Togashi, S., Tanigawa, M., Shoji, K., Watanabe, T., Watanabe, N., Maki, K., and Suzuki, H. (2002). Corrosion by bacteria of concrete in sewerage systems and inhibitory effects of formates on their growth. Water Res. *36*, 2636–2642.

Yang, W., Vollertsen, J., and Hvitved-Jacobsen, T. (2005). Anoxic sulfide oxidation in wastewater of sewer networks. Water Sci. Technol. *52*, 191–199.

Yongsiri, C., Vollertsen, J., Rasmussen, M., and Hvitved-Jacobsen, T. (2004). Air-water transfer of hydrogen sulfide: an approach for application in sewer networks. Water Environ. Res. *76*, 81–88.

Zhang, L., De Schryver, P., De Gusseme, B., De Muynck, W., Boon, N., and Verstraete, W. (2008). Chemical and biological technologies for hydrogen sulfide emission control in sewer systems: A review. Water Res. *42*, 1–12.

Zhang, L., Keller, J., and Yuan, Z. (2009). Inhibition of sulfate-reducing and methanogenic activities of anaerobic sewer biofilms by ferric iron dosing. Water Res. *43*, 4123–4132.

Zivica, V., and Bajza, A. (2001). Acidic attack of cement based materials – a review. Part 1. Principle of acidic attack. Constr. Build. Mater. *15*, 331–340.

Part III

New Technologies Using Biofilms

Biofilm Biodegradation Potential

9

Freshta Akbari, Natasha Andrade, Merily Horwat and Birthe V. Kjellerup

Abstract

Persistent organic pollutants (POPs) are present in the environment after decades of industrial activity and have contaminated soils and sediments worldwide. The group of contaminants described as POPs includes toxic compounds such as polychlorinated biphenyls (PCBs), dioxins, chlorinated ethenes and polycyclic aromatic hydrocarbons (PAHs) and brominated flame retardants. Bioremediation of POPs utilizing microbial communities in the biofilm mode of growth has enhanced the removal of POPs from the environments most often converting the organic pollutants to harmless materials. The state-of-the-art for biofilm based solutions for biodegradation of POPs is primarily based on laboratory experiments often performed at optimal conditions. Thus the influence of natural conditions such as nutrient requirement, bioavailability, life style and physico-chemical conditions might vary depending on the POP in question and the environment such as co-contaminants. Field studies of biofilm based solutions are becoming more frequent and most seem promising. Along with these studies knowledge about the mechanisms by which either indigenous or bioaugmented microorganisms forming biofilms enhancing bioremediation is increasingly being expanded.

Biofilms and environmental pollutants

In natural environments, microorganisms mainly exist in biofilms, which can be defined as communities of microorganisms that are set in a self-produced extracellular polymeric substance (EPS) attached to a surface (Costerton, 1999). This mode of life is advantageous for the individual cells compared to a planktonic mode as they are able to communicate and protect each other from the outside environment, while exchanging nutrients and genetic materials among one another and their environment (Costerton, 1999; Davies *et al.*, 1998). Additionally, biofilms are steady structures, since EPS gives the biofilm a dynamic three-dimensional structure consisting of water channels and voids allowing the transport of nutrients and electron acceptors (Ferrera *et al.*, 2004; Gross *et al.*, 2007). Biofilm can be formed of one bacterial species. However, in most cases they are heterogeneous structures containing a multitude of bacterial species as well as fungi, yeast, protozoa, algae, and other microorganisms (Ferrera *et al.*, 2004; Baker *et al.*, 2009).

Biofilms impact natural and industrial systems as well as human health (Hall-Stoodley *et al.*, 2004; Wu *et al.*, 2013). Although biofilms can cause damages to the equipment, impair human health, and contaminate products, they can provide extensive benefits in the treatment of drinking water and wastewater, detoxification, and biodegradation of hazardous contaminants (Bertin *et al.*, 2007; Accinelli *et al.*, 2012). Thus knowledge of their potential benefits to the environment and human health is important and requires a multidisciplinary effort.

Persistent organic pollutants in the environment

Owing to increasing industrial activity the past two centuries, persistent organic pollutants

(POPs) have accumulated in the environment and contaminated many aquatic sites (Lucas *et al.*, 1993; Hu *et al.*, 2014; Kumar *et al.*, 2014). POPs are a group of synthetic compounds such as polychlorinated biphenyls (PCBs), dioxins, chlorinated ethenes and polycyclic aromatic hydrocarbons (PAHs) that were extensively produced during the industrial revolution (Hu *et al.*, 2014; Kumar *et al.*, 2014). Once they are released into the air, water, and sediment, they greatly impact the human health and natural environments due to their high persistency, toxicity, and chemical stability making POPs persist for many decades (Hu *et al.*, 2014; Kumar *et al.*, 2014). After their production stopped, they still were exposed to humans through contaminated food sources such as seafood and dairy products. Their lipophilic nature allows them to persist in the adipose tissues of animals and humans and thus bio-accumulate in the food chain (Kumar *et al.*, 2014). Elimination of POPs from contaminated sites is of utmost importance as they remain in the environment and eventually enter the food chain causing adverse effects in humans (Lucas *et al.*, 1993; Kumar *et al.*, 2014). Commonly applied remediation methods for POPs include dredging and capping (USEPA, 2005). Not only are these techniques expensive, but they can result in high concentrations in the water environment as contaminated sediment is resuspended (Cho *et al.*, 2009). Therefore, less invasive and more effective bioremediation methods are being explored, which will be covered later in this chapter.

Polychlorinated biphenyls (PCBs)

Among the organic pollutants, polychlorinated biphenyls (PCBs) are the most toxic and persistent contaminants. They were applied in multiple industrial products including coolants, transformer oils, flame retardants from the 1930s until their ban in the 1970s (Kimbrough, 1995; Payne *et al.*, 2013). Their low flammability, high stability, and high vaporization temperature made them optimal for industrial purposes (Martinez *et al.*, 2010; Payne *et al.*, 2013). Despite their ban over four decades ago, weathered remains of PCBs are still present in aquatic sediments (Fagervold *et al.*, 2007; Martinez *et al.*, 2010; Payne *et al.*, 2013). PCBs are not easily broken in nature; thus they contaminate soil and sediment, and pose toxicity risks for organisms as they bio-accumulate in the food chain (Vater *et al.*, 1995; Safe *et al.*, 1997; Payne *et al.*, 2013). They are carcinogenic, toxic, and harmful to reproductive and endocrine systems as well as other organs and systems. Therefore, remediation of PCBs is crucial due to their potential carcinogenicity and ability to accumulate to toxic degrees in higher organisms (Kannan *et al.*, 1998).

Current sources are still causing PCBs to enter the environment such as left over industrial installations and accidental synthesis through combustion that enters via atmospheric deposition (Alcock and Behnisch, 1998; Kim, 2004). These new PCB releases can be transported in stormwater and might also be found in waste water treatment plants. Previous research has indicated that bacterial isolates from PCB contaminated sediments can dechlorinate the biphenyl ring structure in both aerobic and anaerobic environments (Kjellerup *et al.*, 2012). However, the performance of native microorganisms is rarely sufficient to degrade PCBs to a required cleanup level due to limited abundance. For these reasons, developing effective and efficient remediation tools is sought after and the application of biofilm based methods would be advantageous.

Dioxins

Polychlorinated dibenzo-*p*-dioxins and dibenzofurans (PCDDs/Fs) are aromatic chemicals that originate as a result of industrial and combustion by-products (Buser *et al.*, 1985; Cormier *et al.*, 2006). In preparing the industrial chemicals, chlorinated phenols are the starting material, mainly in pesticides. As a result of heating commercial chlorinated phenols, dioxins are released into the environment (Buser *et al.*, 1985; Lohman and Seigneur, 2001; Cormier *et al.*, 2006). Dioxins are found around the world in air, soils, sediment, and food products. Upon their release into the environment, a range of developmental, immunological, and neurological toxic risks are exposed to humans and animals (Buser *et al.*, 1985; Cormier *et al.*, 2006). Due to their chemical stability, high dielectric constants, and non-flammability, they tolerate extreme conditions and are a serious threat to natural systems due to accumulation in the food chain (Lohman and Seigneur, 2001).

Chlorinated ethenes

Chlorinated ethenes are among the most widespread organochlorine water pollutants and they are formed from mainly from emission of burning biomass and volcanoes (Lohner *et al.*, 2011; Tobiszewski and Namiesnik, 2012). Chloroethenes are utilized as cleaning materials, paint removers, industrial solvents, and pesticides and due to their often improper disposal, they can be found primarily in groundwater (Lohner *et al.*, 2011;, Tobiszewski and Namiesnik, 2012). One of their properties is low solubility in water, which enables them to form plumes of dense non-aqueous phase liquid giving them the ability to migrate and making their degradation challenging (Tobiszewski and Namiesnik, 2012). TCE is also a volatile compound and can enter the gas phase and penetrate the pores of surrounding soils (USEPA, 1992). Among them, tricholorethenes (TCE) and perchloroethenes (PCE) are commonly found in contaminated sites (Lohner *et al.*, 2011). These compounds can be reductively dechlorinated by natural processes, which can result in the creation of vinyl chloride, which is extremely toxic to humans (USEPA, 1992).

Anaerobically, the bacterial cultures of *Dehalococcoides* class have shown to reductively dechlorinate TCE and PCE (Maymo-Gatell *et al.*, 1999; Krajmalnik-Brown *et al.*, 2007). Table 9.1

Table 9.1 Bacterial species that have been applied in biofilm based bioremediation approaches

Contaminant	Organisms involved	Process	Reference
Polychlorinated biphenyls (PCBs),	*Dehalobium chlororcoceae* DF1	Anaerobic	May *et al.* (2008)
	Burkholderia xenovorans strain LB400	Aerobic	Bartels *et al.* (1999), Chain *et al.* (2006)
Dioxins: Polychlorinated dibenzo-*p*-dioxins and dibenzofurans (PCDDs/Fs)	*Comamonas* sp. strain KD7 and *Trifolium repens*	Aerobic	Wang and Oyaizu (2011)
	Mixed biofilm of *Burkholderia* sp. NK8 and *Pseudomonas aeruginosa* PAO1	Aerobic	Yoshida *et al.* (2009)
	Dehalococcoides mccartyi strain 195	Anaerobic	Liu *et al.* (2013)
Chlorinated ethenes:, tricholorethenes (TCE) and perchloroethenes (PCE)	*Dehalococcoides* genus	Anaerobic	Maymo-Gatell *et al.* (1999), Krajmalnik-Brown *et al.* (2007)
	Dehalobacter restrictus, *Dehalospirilum multivorans*, and *Geobacter lovleyi*	Anaerobic	Mattes *et al.* (2010)
TCE, PCE	Methanogens	Anaerobic co-metabolism	Maymo-Gatell *et al.* (1999)
	Acetogens		
	Sulfate-reducing bacteria		
Polycyclic aromatic hydrocarbons (PAHs)	*Pseudomonas*, *Sphingonomas*, *Mycobacterium*, *Nocardioides*, *Rhodococcus* and *Novosphingobium*	Aerobic	Lyu *et al.* (2014)
	Arthrobacter viscosus	Aerobic	Ferreira *et al.* (2013)
	Rhodococcus erythropolis and *Pseudomonas* spp.	Aerobic	Kurane (1997)
Brominated flame retardants: (1) polybrominated diphenyl ethers (PBDEs), (2) hexabromocyclododecanes, (3) tetrabromobisphenol A, (4) polybrominated biphenyls and (5) others	*Sulfurospirillum multivorans* and *Dehalococcoides* spp.	Anaerobic	He *et al.* (2006)
	Rhodococcus jostii	Aerobic	Robrock *et al.* (2009)
	Burkholderia xenovorans	Aerobic	Lu *et al.* (2008)
	Bacillus cereus		

provides an overview of microorganisms taking part in biodegradation of POPs. These chlorinated ethenes can also be co-metabolically dechlorinated via methanogens, acetogens, and sulfate-reducing bacteria that contain metal cofactors including haems, corrinoids, and cofactor F_{430} (Maymo-Gatell *et al.*, 1999). Degrading chloroethenes primarily convert to less chlorinated compounds and as the process goes further they will be converted to nontoxic ethane products (ETH) that are harmless (Maymo-Gatell *et al.*, 1999; Lohner *et al.*, 2011).

Polycyclic aromatic hydrocarbons (PAHs)

These are compounds with fused benzene rings that are ubiquitously produced in nature (Lyu *et al.*, 2014). PAHs are formed as a result of numerous human activities such as incomplete combustion of coal, gas, oil spills of petroleum, woods, municipal and industrial incinerators (Seo *et al.*, 2009). They have low water solubility and are thermodynamically stable giving rise to detrimental health effects (Johnsen and Karlson, 2004) (Lyu *et al.*, 2014). Naturally, they are biodegraded via multiple methods such as chemical and photo-oxidation, volatilization, and adsorption to sorptive surfaces (Rodriguez and Bishop, 2008; Lyu *et al.*, 2014). Additionally they are remediated through various bacterial communities' activities including *Pseudomonas, Sphingonomas, Mycobacterium, Nocardioides, Rhodococcus,* and *Novosphingobium* (Lyu *et al.*, 2014).

Brominated flame retardants

As civilization progresses, the discovery and use of new chemicals benefits society. Although some chemicals are produced and used with success for purposes that are in general considered good for the majority of the population, these chemicals can have an unexpected consequence after their primary use. This is the case of some flame retardants. Flame retardant chemicals are added to consumer products to reduce the heat release during a fire and ultimately to prevent or delay fires. They work as intended if used properly, however, they can be released from the consumer products and enter the environment. A particular class of flame retardants has become the subject of concern by the general public, environmental advocates and researchers: brominated flame retardants (BFRs).

BRFs are generally mixtures of flame retardant chemicals commonly used in plastics, textiles, electrical equipment, and construction materials (de Wit, 2002). There are five major classes of brominated flame retardants: (1) polybrominated diphenyl ethers (PBDEs), (2) hexabromocyclododecanes, (3) tetrabromobisphenol A, (4) polybrominated biphenyls, and (5) others. The first class, PBDEs, have been under intense scrutiny and, in the US, have gone through an industry voluntary phase-out of production (USEPA, 2009; USEPA, 2014). PBDEs have similar chemical structures as PCBs. Two of the PBDEs (BDE-47 and BDE-99) are now listed as persistent organic pollutants by the Stockholm Convention (UNEP, 2004). BDE-209 is the fully brominated congener and the main component of the deca-BDE commercial formulation and American manufacturers and importers of BDE-209 have announced a commitment to stop production and use by 2013 (USEPA, 2009).

PBDEs are hydrophobic molecules and have the tendency to accumulate in organic phases such as fatty tissue of humans and animals. In the external environment (not including mammalian tissues), PBDEs are ubiquitous and have been reported in house dust and indoor air (de Boer *et al.*, 2003; Harrad *et al.*, 2007; Voorspoels *et al.*, 2007), soils and biosolids (Andrade *et al.*, 2010), humans (Trudel *et al.*, 2011) and all other environmental compartments. The lowly brominated congeners are known to bioaccumulate, biomagnify, to easily adsorb, and to be more bioactive than the BDE-209 (McDonald, 2002; Stapleton *et al.*, 2005). PBDEs have also been shown to display endocrine disrupting potential in rats (Staskal *et al.*, 2005; Lilienthal *et al.*, 2006), but toxicity in humans is not well studied and lacks the availability of data. Most humans are exposed to PBDEs via food ingestion and breathing of air containing these chemicals and workers in the PBDE manufacturing industry are at a higher risk of exposure (Darnerud *et al.*, 2001).

Biodegradation of persistent organic pollutants using biofilms

Remediation of POPs from contaminated sites is a priority due to their ability to enter the food chain and their potent toxic and carcinogenic properties. Various approaches have been taken, but most POP remediation methods are costly and inefficient such as dredging and capping (USEPA, 2005). Dredging is not only expensive, but can also result in increased POP concentrations in the water phase due to re-suspension of contaminated particles into the water phase causing risk for bioaccumulation (Payne *et al.*, 2011; Martins *et al.*, 2012). Capping effectively controls POP equilibrium while keeping them out of the water phase by sequestration with activated carbon and some other substrates (Zimmerman *et al.*, 2004). But hazardous substances still remain in the environment and contaminated sediments can become recalcitrant to microbial degradation because of strong sequestering capacities of soil particles (Hatzinger and Alexander, 1995).

Research has shown that microbial communities can be established that can degrade organic pollutants aerobically and anaerobically (Kjellerup *et al.*, 2012). While *in situ* microbial degradation of POPs would represent a significant improvement in remediation efforts, previous attempts have failed due to POP stability, low bioavailability, and the low abundance and activity of naturally occurring POP-degrading microorganisms. To overcome these issues a solution would be to locate microbial communities onto sorptive surfaces such as activated carbon surfaces and apply these biofilm communities to contaminated sites.

PCBs

A number of organohalide respiring bacterial species have been identified for their ability to transform PCBs and some of these microbial communities have been established as potential agents for PCB bioremediation. Anaerobic bacteria can dechlorinate highly chlorinated PCBs as they use the chlorines as electron acceptor, while aerobic bacteria degrade PCBs with four or less chlorines by breaking the biphenyl ring structure open and give off water and carbon dioxide as their end products (Fagervold *et al.*, 2007; Payne *et al.*, 2011). Enhanced PCB dechlorination has been observed, when granular activated carbon (GAC) was used as a surface for biofilm formation (Edwards and Kjellerup, 2013; Kjellerup *et al.*, 2014; Mercier *et al.*, 2014). Co-localizing PCB-degrading microbes as biofilms and utilizing it as a microbial inoculum delivery system provides a number of benefits and have shown a high efficiency of GAC to quickly adsorb and sequester PCBs from aquatic sediments (Werner *et al.*, 2006; Edwards and Kjellerup, 2013). They also provide additional PCB-degrading microbes adjacent to sequestered PCBs that enhance the degradation capacity. As a result of this close spatial setting, microbes utilize PCB as an electron acceptor and enable subsequent degradation (Edwards and Kjellerup, 2013). Knowing both the aerobic and anaerobic degradation possibilities of PCBs, several studies tested both conditions and their efficiencies compared to one condition alone. A promising study by Payne *et al.* (2013) evaluated the simultaneous presence of DF1 and LB400 with GAC in bioaugmentation efficiency of contaminated sediment. After 3 months, the amount of PCBs had decreased by 80%, while a 25% decrease was observed in non-bioaugmented sediments after 12 months (Payne *et al.*, 2013).

Dioxins

In a similar fashion biodegradation of dioxins has been reported. Wang and Oyaizu (2011) reported that a biofilm of *Comamonas* sp. strain KD7 and *Trifolium repens* increased reduction rates of dioxins in soil samples by 22% after 3 months. Bioremediation efforts have demonstrated that presence of more than one bacterial species increased the degradation efficiency for dioxins. This strategy was evaluated, when a mixed biofilm of *Burkholderia* sp. NK8 and *Pseudomonas aeruginosa* PAO1 was applied that degraded chlorinated benzoates (Yoshida *et al.*, 2009). Studies have also shown that dioxins can be biodegraded by biofilms in unfavourable environments such as River Kymijoki Finland (Liu *et al.*, 2013). Here it was shown that the bioremediation of weathered dioxins present in the sediment biofilms could be enhanced by stimulation with electron donors, co-substrate and/or bioaugmentation with *Dehalococcoides mccartyi* strain 195 (Liu *et al.*, 2013).

Chlorinated ethenes

In water, chlorinated ethenes can be degraded to harmless products. However, this process is not always completed, which results in formation of other toxic byproducts (Maymo-Gatell *et al.*, 1999; Mattes *et al.*, 2010). Chlorinated ethenes can be metabolized via several pathways: anaerobic reductive dechlorination, aerobic co-metabolism and anaerobic/anaerobic oxidation (Mattes *et al.*, 2010; Frascari *et al.*, 2013). Lower oxidized chlorinated ethenes can be degraded by anaerobic dechlorinating bacteria belonging to the *Dehalococcoides* genus, such as *Dehalobacter restrictus, Dehalospirilum multivorans*, and *Geobacter lovleyi* (Mattes *et al.*, 2010). A challenge for biodegradation in the environment is that these microbes are usually present in low numbers and their degradation reaction rates are low, unless biofilm based applications are utilized for instance in biobarriers or biowalls. Biowalls for remediation of TCE can be installed down gradient of the contaminated area and use the natural flow of groundwater to bring the contaminant through the matrix for degradation. The substrate for biofilm growth in biowalls can be plant mulch from pine, hardwood and cypress established as a reactive barrier are benefiting from the presence of indigenous TCE degrading organisms in the system (Lu *et al.*, 2008; Wei and Seo, 2010).

Polycyclic aromatic hydrocarbons (PAHs)

Given the high toxicity of PAHs, many studies have been conducted to evaluate the microbial interactions in PAHs remediation. Enhanced degradation is often reported in situations, where bacteria are associated with surfaces. Plosz *et al.* (2010) showed that PAHs removal occurred faster, when a biofilm delivery system was combined with ozonation (BIOZO system) and applied to landfill leachate. The study utilized a stage moving-bed biofilm reactor (SMBBR), while evaluating the efficiency of ozonation on removal of PAHs between a pre-anoxic zone and an aerobic zone. These results indicate that nitrate was reduced along with PAH degradation occurring in a reduced and anoxic environment. This dual removal of nitrate and PAH highlights the importance of co-metabolism in biofilm based remediation (Heidler and Halden, 2007; Lolas *et al.*, 2012).

Brominated flame retardants (PBDEs)

These substances are considered to be persistent in the environment, but research has shown the potential for biodegradation (Rayne *et al.*, 2003; He *et al.*, 2006; Vonderheide *et al.*, 2008). One anaerobic biodegradation study (Gerecke *et al.*, 2005) showed that BDE-209 can be degraded to nona- and octa-BDEs using biofilms from sewage sludge as the inoculum and the calculated half-life for BDE-209 was 700 days in the laboratory setting. (He *et al.*, 2006) identified two anaerobic bacterial species able to degrade deca-BDE and octa-BDE mixtures to lower brominated congeners: *Sulfurospirillum multivorans* and *Dehalococcoides* species. The anaerobic degradation of PBDEs has also been observed in sediments, where the sorption of PBDEs to sediment particles may play an important role in slowing down the rate of degradation (Tokarz *et al.*, 2008). In a study with BDE-contaminated sediment, the reductive debromination of BDE-209 was observed and formation of lower brominated congeners (from nona- to hexa-BDEs) occurred as a result. However, the half-life in sediment was estimated at well above a decade (Tokarz *et al.*, 2008). In soils, very little biodegradation was observed for a tri-BDE and BDE-209 in contaminated soils brought to the laboratory and incubated at aerobic and anaerobic conditions (Nyholm *et al.*, 2010). In the environment, BDE-209 was found in remote high altitude mountain lakes despite its low volatility (Bartrons *et al.*, 2011). Moreover, analysis of bottom rock and silt biofilms in the lakes revealed that the concentration of PBDEs in silt biofilms was much lower than the PBDEs concentration in rock biofilms, suggesting that PBDE biodegradation was favoured in silt biofilms due to longer anoxic periods (Bartrons *et al.*, 2011).

Aerobic degradation of PBDEs has been observed, though in laboratory conditions. (Robrock *et al.*, 2009) observed that the PCB-degrading bacteria *Rhodococcus jostii* and *Burkholderia xenovorans* were able to aerobically degrade PBDE congeners with 1–5 bromine atoms to lower brominated compounds. In another laboratory study, a metal-resistant bacterial strain, *Bacillus cereus*, was able to use BDE-209 as its carbon and energy source (Lu *et al.*, 2008) and the authors suggested the possibility of using this bacterium for bioremediation of contaminated sites containing co-contaminants, such as other organic pollutants and metals. However, BDE-47 has shown some toxicity to marine bacteria, reducing the abundance of aerobic marine biofilms (Chiu *et al.*, 2012). During laboratory experiments of aerated activated sludge treatment of wastewaters, biodegradation of PBDEs and the debromination of some PBDE congeners was observed. However, PBDE exposure to activated sludge bacteria caused a reduction in the community diversity for a short period of time (Langford *et al.*, 2007). This suggests that creating the right conditions for anaerobic or aerobic degradation of PBDEs may be challenging, but one of the solutions might be to apply biofilms, where favourable conditions for debromination could be established by applying mixed cultures of other environmental species that can survive the toxic exposure to PBDEs. Although significant progress has been made in identifying bacterial species capable of biodegrading PBDEs, research is still needed for improved understanding of biodegradation pathways and to utilize biofilm based degradation as a potential remediation technology for PBDEs.

Applied biofilm based biodegradation methods

Despite the chemical stability and toxicity of POPs, many microorganisms and methods have been proposed that are able to transform these contaminants. Some of the more frequently applied solutions are discussed below.

Biofilm barriers

The lipophilicity and low water solubility of POPs make them persist in the environment and remain primarily in the solid phase. To prevent these contaminants from spreading into larger areas and simultaneously promote biodegradation, biobarriers consisting of biofilms established on support materials such as mulch, saw dust, and peat moss have been utilized (Seo and Bishop, 2008; Seo *et al.*, 2009). In a lab scale mulch biofilm barrier (Seo *et al.*, 2009) reported that after 150 days, a removal efficiency of 97–99% for phenanthrene and 99.9% for pyrene was observed. However, the mulch biobarrier itself could not entirely prevent migration of aqueous and solubilized PAHs that moved through the barrier without being degraded (Seo *et al.*, 2009). In another study investigating PAH bioremediation (Ferreira *et al.*, 2013) evaluated if the bacterium *Arthrobacter viscosus* would function as a permeable reactive biobarrier (PRBB) for treating PAH contaminated groundwater. It was observed that *A. viscosus* formed a functional and effective biobarrier, where benzo[a]anthracene and phenanthrene removals were observed after seven and three days, respectively. The study also showed that *A. viscosus* adhered readily to sepiolite thus providing a growth surface that can be applied as a solution for PAH remediation in groundwater environments (Ferreira *et al.*, 2013).

Studies of biobarriers established for bioremediation of trichloroethylene (TCE) contaminated ground water has shown similar promising results as for PAHs. Instead of using solid surfaces as substratum for biofilm formation a polycolloid releasing substrate was established as a biobarrier system (Liang *et al.*, 2013). The biobarrier consisted of biodegradable substrates such as vegetable oil (slow release), cane molasses (fast release), two types of surfactants to increase the solubility of TCE for microbial degradation (simple green and soya lecithin) that was formulated as an emulsion enhancing the reductive dechlorination taking place by indigenous dehalorespiring bacteria in the groundwater that was fed to the column experiment. The emulsion created anaerobic conditions and was shown to induce more complete removal of TCE due to increased sorption and subsequent microbial degradation. This concept was subsequently tested under field conditions to reduce TCE plume migration (Kuo *et al.*, 2014). In the field study, the polycolloid biobarrier consisting of soybean oil, lactate and surfactants, which was employed via injection wells (5 m intervals) into the TCE contaminated ground water. The monitoring lasted for approximately 100 days and significant reduction of TCE was observed after 35 days (87 to < 0.1 μg/l) simultaneously with the presence of anaerobic conditions (oxygen concentration decreased from 1.6 to < 0.1 mg/l) and reduced conditions due to the oxidation–reduction potential (ORP) going from 124 to −14 mV after 20 days of injection (Kuo *et al.*, 2014). During this time, TCE degradation products were detected in the ground water samples, which support the results showing reductive dechlorination of TCE taking place and reducing the importance of sorption.

A combination of the two above described approaches, solid surface for biofilm development and substrate application, was been taken for the bioremediation of the solvent tetrachloroethylene (PCE) that also is a groundwater contaminant (Kao *et al.*, 2001). The biobarrier was based on deploying a layer of peat, which was meant as the primary electron donor in order to continuously enhance the anaerobic reductive dechlorination of PCE. Experiments performed in continues-flow columns (laboratory scale) for 65 days were used to evaluate the viability of the system for microbial PCE degradation. The results showed up to 98% removal efficiency of PCE in this system due to the continuous supply of electron donor to the indigenous organohalide respiring bacteria from the peat layer, when synthetic PCE contaminated groundwater was tested (Kao *et al.*, 2001). A subsequent study that advanced the solution with peat for PCR bioremediation developed a two-layer biobarrier system that first consisted of organic-releasing layer (sludge cake from wastewater treatment) that was followed by an oxygen releasing layer (calcium peroxide) (Kao *et al.*, 2003). The sludge cake supplied primary substrates (electron donor) for reductive dechlorination of PCE, whereas the oxygen releasing layer enhanced aerobic degradation or co-metabolization of degradation products from the initial anaerobic processes. In this way complete mineralization of PCE and resulting degradation products that often stall the overall PCE remediation could be obtained *in situ*. Similar laboratory based column studies as described above with synthetic PCE contaminated groundwater were performed with this dual layer biobarrier system and showed up to 99% PCE removal efficiency (Kao *et al.*, 2003). If field tests of this system support these results, this dual-layer biobarrier solution will likely be a cost-effective PCE-remediation technology for contaminated groundwater aquifers.

Biobarriers are also deployed for bioremediation of nitrate, atrazine and other less common groundwater contaminants such as nitric acid, uranium and technetium (U(VI), Tc(VII)) (Michalsen *et al.*, 2009; Hunter and Shaner, 2010). Atrazine contaminated groundwater is often containing high levels as nitrate as well, which impedes the biological degradation of the recalcitrant pesticide. Therefore, a dual approach is needed to establish a bioremediation solution that can degrade both contaminants *in situ*. A model of *in situ* biobarriers was established by setting up two reactors in sequence that were inoculated with an atrazine-degrading consortium, where the first reactor was supplying a vegetable-oil-based solution creating denitrifying conditions to remove nitrate and the second reactor was kept aerobic to degrade atrazine. The experiment showed that the initial denitrifying barrier reduced the concentration of nitrate by 98%, making the subsequent aerobic degradation of atrazine possible. Overall, 99.9% of atrazine was removed after 30 weeks of operation (Hunter and Shaner, 2010), where, surprisingly, 30% of the atrazine was removed in the denitrifying reactor despite the oxygen-deprived conditions and the remaining 70% in the aerobic reactor. This dual-biobarrier set-up showed the potential for removal of recalcitrant contaminants such as atrazine that can take place *in situ* if air is injected into the groundwater system or other oxygenation processes such as the calcium peroxide process described above is introduced.

Groundwater contaminated with inorganic compounds such as metals and acids present a different challenge than recalcitrant organic contaminants such as TCE and PCE. The inorganic compounds cannot be biodegraded and instead the effort focuses on bioimmobilization, which is the case for Uranium and Technetium (Michalsen *et al.*, 2009). This was examined in a 21 month long study, where a biobarrier was established in order to neutralize the nitric acid present in the groundwater (pH 4.7–6.9), remove nitrate and immobilize the present radionuclides in the groundwater. Ethanol was added to the groundwater system that also consisted of crushed limestone and sediment and the pH increased to 6.9, while nitrate was removed, 94% of total-U was immobilized and the sediment biomass in the form of biofilm was increased. Similar changes also occurred in the control system showing that the changes were not due to the addition of ethanol by itself. Instead the changes were shown to be caused by microbial activity, which was confirmed be the presence of dissimilatory nitrite reductase genes (nirS, nirK) supporting the increased denitrification. Analysis of the 16S rRNA from sediment samples showed that Beta-proteobacteria (incl. denitrifying bacteria) were dominant near the influent of acidic groundwater. This changed to Gamma- and Alpha-proteobacteria along the flow path where the pH increased and the concentration of nitrate decreased showing a shift in the microbial population (Michalsen *et al.*, 2009). Overall the results showed that this type of biobarrier can be applied for remediation of acidic radionuclide contaminated groundwater containing high concentrations of nitrate.

Activated sludge biofilms

One promising strategy to enhance POP bioremediation by biofilm is to exploit activated sludge's ability to co-metabolize. The ability of activated sludge to increase the rate of PAH bioremediation has been studied by (Kurane, 1997), where inoculation of *Rhodococcus erythropolis* and *Pseudomonas* sp. to activated sludge efficiently removed a group of PAHs. Additionally, Rodriguez and Bishop (2008) determined that a mixed activated sludge biofilm yielded higher biodegradation rates, when applied to a broad range of PAH substrates. This finding hinted that co-metabolism played an important role in degrading multiple PAHs simultaneously rather than the surfactant being the important factor as previously hypothesized (Rodriguez and Bishop, 2008). Furthermore, a concurrent biofilm delivery system with activated sludge demonstrated a successful approach for removal of micropollutants such as diclofenac (Falas *et al.*, 2013). This study emphasized that using biofilm or activated sludge methods separately did not yield significant results compared to the combined hybrid biofilm-activated sludge process.

Anaerobic biofilm processes in activated sludge wastewater treatment systems for instance digestion processes, which are used for biogas production in order to regain energy from the treatment processes, have shown also to be efficient for degradation of POPs such as PCBs (Bertin *et al.*, 2011). A membrane biological reactor (MBR) was fed with sludge that had been spiked with PCBs and run under both mesophilic (35°C) and thermophilic conditions (55°C). The results showed that more than 50% of the PCBs had been removed due to reductive dechlorination under methanogenic conditions in the digestion process. The microbial community consisted of both fermentative eubacteria as well as acetoclastic and hydrogenotrophic methanogens that all persisted exposure to the high PCB concentrations, while still producing methane. In a previous study of the anaerobic digestion process microbial degradation was also observed for PAHs thus showing that several organic contaminants have the potential to be metabolized during the digestion process (Bertin *et al.*, 2007). Based on this it was shown that biofilm based wastewater treatment systems can be multi-functional and produce energy while simultaneous degrading xenobiotic compounds, which afterwards can be used as fertilizer for agricultural purposes without risk from the presence of pathogenic organisms that also was significantly reduced during the process (Bertin *et al.*, 2007, 2011).

The application of a continuous stirred tank reactor system for treatment of naturally contaminated sewage sludge was also investigated for the biodegradation of six priority PCBs under aerobic and anaerobic conditions (Patureau and Trably, 2006). Less extensively chlorinated compounds experienced losses from aerobic and anaerobic processes that were higher than for the extensively chlorinated compounds (all abiotic losses were below 20%). The degradation was increased under methanogenic conditions, where more than 40% was removed independently of the PCB chlorination degree. The heaviest PCB congeners were more efficiently dechlorinated due to higher anaerobic process rates (removal rate approximately 40%), whereas the degradation of lighter chlorinated PCBs was enhanced under aerobic conditions removal rate up to 100% (from 40% before). The results indicated that bioavailability of the PCBs in naturally contaminated sludge impacted the biodegradation significantly under both aerobic and anaerobic conditions despite the fact that the rates were enhanced in this continuous stirred tank reactor system (Patureau and Trably, 2006). In the system, the limiting PCB availability influenced the PCB removal negatively resulting in PCB concentrations that exceeded the current French/European regulations about the presence of PCBs in sewage sludge. Thus the biosolids cannot be applied onto farmland, since the reduction throughout the wastewater system did not reduce the concentration significantly (Patureau and Trably, 2006).

Not all xenobiotics can be degraded during the otherwise efficient activated sludge process such as PCE even though biodegradation was observed in biobarriers as described above. In one study it was shown that PCE (range 5–150 mg/l) and 2-CP (range 25–150 mg/l) decreased the microbial activity significantly measured by decrease in activity of the three key enzymes dehydrogenase, phosphatase and urease as well as a reduction of the microbial diversity (Li *et al.*, 2013). Short term exposures did not cause a shift in the microbial community, but longer term exposure induced a shift from Alpha- and Gamma-proteobacteria to firmicutes, bacteroidetes and synergistetes becoming dominant. *Actinobacteria* were eradiated during the long-term treatment.

Microbial fuel cells

One of the recent approaches is the application of microbial fuel cells (MFCs) that transform chemical energy from organic wastes into electrical energy with the organic waste being the electron donor (Yamamoto *et al.*, 2014). In MFCs, biofilm communities attach to the anode and assist in energy transfer. In a study conducted by (Ki *et al.*, 2008), it was concluded that as the diversity of microbial communities increased in a MFCs combined with activated sludge from a wastewater treatment plant, the electricity production was increased. MFCs can generate electricity from different sources and (Patil *et al.*, 2009) showed that wastewater from a chocolate factory generated higher currents compared to that of a membrane and salt bridge or glucose. The 16S rRNA results showed that while multiple bacterial groups participated in the MFC activity, Beta-proteobacteria was the predominant group, making up 51% of the community (Patil *et al.*, 2009). A symbiotic relationship was seen between the biofilm and the bacterial communities in the electrolyte that enhanced the current density to approximately 100–150 mA/m^2 (Yamamoto *et al.*, 2014; see also Chapter 11). 16S rRNA analysis of the MFC biofilms inoculated from activated sludge revealed the presence of *Dysgonomonas*, *Sporomusa*, and *Desulfovibrio* in anode and cathode biofilms, while *Geobacter* was only present in anode biofilm. Electricity production was in this study based on the organic substrate methanol as electron donor. However, other studies have shown that that POPs can be used in the same manner to be transformed into electrical energy. In a study by (Chun *et al.*, 2013) weathered PCBs were electrically stimulated by MFCs, thus enhancing the bioremediation rate and wastewater was treated simultaneously with electricity production, when a bio-electrochemical treatment system (similar to MFC) was applied (Velvizhi *et al.*, 2014).

For an affective production of electricity by MFCs, organic waste is fed to MFCs, while at the same time microbial communities are developed in the anode part. However, the challenge has been to harness a high density current that can be applied for external purposes and this aspect needs further research regarding identification of efficient current producing microbes, understanding of their ecology, and preparing optimal growth and nutrient conditions for the microbial communities.

Perspectives and conclusion

Recent efforts of bioremediation of POPs showed that biofilm microbial communities have significantly increased the removal rate from contaminated environments. This approach provides carbon and energy sources to the microbes as well as a support system to which they can adsorb, while transforming the contaminants (de Lipthay *et al.*, 2003; Petrie *et al.*, 2003). In return, the biofilm community converts the organic pollutants to harmless materials (de Lipthay *et al.*, 2003; Petrie *et al.*, 2003). Nevertheless, given the different situations and various microbial communities of POPs degraders, more research is necessary to elucidate the nature and biochemical pathways of these degraders.

As mentioned before, POPs are hydrophobic, toxic, and persistent compounds that make them less bioavailable for degradation (Choi *et al.*, 2009). Thus, many biofilm-based biotransformation reactions are added to organic surfaces such as GAC, enabling the POPs to adsorb to these surfaces thus enhancing the efficiency (Payne *et al.*, 2011; Kjellerup *et al.*, 2014). Another bioremediation challenge that the biofilm based approach would help to limit, is the low abundance and availability of indigenous microorganisms that often limits the rate of the bioremediation process (Fagervold *et al.*, 2007; Kjellerup *et al.*, 2008).

Although using microbial communities to enhance POPs bioremediation is a promising strategy, the lack of specific biodegradative pathways (lack of enzymes or cofactors) that are specific for degradation of different POPs might limit the complete mineralization of these compounds. In a study, it was found that *Geobacter* was useful for bioremediation of metals, but a lack of Fe(III) as the electron acceptor slowed the process (O'Neil *et al.*, 2008). Thus the availability of electron donors/acceptors and co-metabolism activities play a significant role in bioremediation (O'Neil *et al.*, 2008; Frascari *et al.*, 2013).

The review of the current state-of-the-art for biodegradation of POPs and the application of biofilm based methods clearly show that most of this knowledge is based on laboratory experiments, where bacterial cultures are grown under controlled conditions. However, the natural conditions, lifestyle, and nutrient requirements of these microorganisms might vary depending on the POP in question and the environment such as co-contaminants and nutrient availability. Each new study adds to the knowledge of POP bioremediation by exploring the mechanism and interactions of these biofilm community members in transforming POPs. Effort should be made to imitate the natural environment, when performing experiments by setting up experiments under relevant environmental conditions for instance as mesocosms in the laboratory or in the environment to study the biodegradation POPs (Edwards and Kjellerup, 2013; Payne *et al.*, 2013). Experiments should account for factors such as system ecology (symbiosis with other organisms), physico-chemical properties of the environmental matrix, co-metabolism of existent microorganisms and sample heterogeneity by setting up a sampling grid and increase the sampling size. To overcome these challenges, applying robust and dense biofilm communities of specific microbes with the addition of electron donors/acceptors has been efficient to enhance the degradation of hazardous POPs contaminated sites. Nevertheless, given the complexity of metabolism pathways and the interaction of microorganisms and their various environments, more research in the field is necessary.

References

Accinelli, C., Sacca, M.L., Mencarelli, M., and Vicari, A. (2012). Application of bioplastic moving bed biofilm carriers for the removal of synthetic pollutants from wastewater. Bioresour. Technol. *120*, 180–186.

Andrade, N.A., McConnell, L.L., Torrents. A., and Ramirez, M.(2010). Persistence of polybrominated diphenyl ethers in agricultural soils after biosolids applications. J. Agric. Food. Chem. *58*, 3077–3084.

Baker, B.J., Tyson, G.W., Goosherst, L., and Banfield, J.F. (2009). Insights into the diversity of eukaryotes in acid mine drainage biofilm communities. Appl. Environ. Microbiol. *75*, 2192–2199.

Bartels, F., Backhaus, S., Moore, E.R.B., Timmis, K.N., and Hofer, B. (1999). Occurrence and expression of glutathiane-S-transferase-encoding bphK genes in *Burkholderia sp* strain LB400 and other biphenyl-utilizing bacteria. Microbiology UK *145*, 2821–2834.

Bartrons, M., Grimalt, J.O., and Catalan, J. (2011). Altitudinal distributions of BDE-209 and other polybromodiphenyl ethers in high mountain lakes. Environ. Pollut. *159*, 1816–1822.

Bertin, L., Capodicasa, S., Occulti, F., Girotti, S., Marchetti, L., and Fava, F. (2007). Microbial processes associated to the decontamination and detoxification of a polluted activated sludge during its anaerobic stabilization. Water Res. *41*, 2407–2416.

Bertin, L., Capodicasa, S., Fedi, S., Zannoni, D., Marchetti, L., and Fava, F. (2011). Biotransformation of a highly chlorinated PCB mixture in an activated sludge collected from a Membrane Biological Reactor (MBR) subjected to anaerobic digestion. J. Hazard. Mater. *186*, 2060–2067.

Buser, H.R., Rappe, C., and Bergqvist, P.A. (1985). Analysis of polychlorinated dibenzofurans, dioxins and related compounds in environmental samples. Environ. Health Perspect. *60*, 293–302.

Chain, P.S., Denef, V.J., and Konstantinidis, K.T.(2006). *Burkholderia xenovorans* LB400 harbors a multi-replicon, 9.73-Mbp genome shaped for versatility. Proc. Natl. Acad. Sci. *103*, 15280–15287.

Chiu, J.M., Po, B.H., Chan, C.Y., Lam, M.H., Qian, P.Y., and Wu, R.S. (2012). Polybrominated diphenyl ethers (PBDEs) alter larval settlement of marine intertidal organisms across three phyla via reducing bacterial abundance on the biofilms. Environ. Sci. Technol. *46*, 7772–7781.

Cho, Y.M., Ghosh, U., and Kennedy, A.J.(2009). Field application of activated carbon amendment for in-situ stabilization of polychlorinated biphenyls in marine sediment. Environ. Sci. Technol. *43*, 3815–3823.

Choi, H., Agarwal, S., and Al-Abed, S.R. (2009). Adsorption and simultaneous dechlorination of PCBs on GAC/Fe/Pd: mechanistic aspects and reactive capping barrier concept. Environ. Sci. Technol. *43*, 488–493.

Chun, C.L., Payne, R.B., Sowers, K.R., and May, H.D. (2013). Electrical stimulation of microbial PCB degradation in sediment. Water Res. *47*, 141–152.

Cormier, S.A., Lomnicki, S., Backes, W., and Dellinger, B. (2006). Origin and health impacts of emissions of toxic by-products and fine particles from combustion and thermal treatment of hazardous wastes and materials. Environ. Health Perspect. *114*, 810–817.

Costerton, J.W. (1999). Introduction to biofilm. Int. J. Antimicrob. Agents. *11*, 217–221.

Darnerud, P.O., Eriksen, G.S., Johannesson, T., Larsen, P.B., and Viluksela, M. (2001).Polybrominated diphenyl ethers: occurrence, dietary exposure, and toxicology. Environ. Health Perspect. *109* (Suppl. 1), 49–68.

de Boer, J., Wester, P.G., van der Horst, A., and Leonards, P.E. (2003). Polybrominated diphenyl ethers in influents, suspended particulate matter, sediments, sewage treatment plant and effluents and biota from the Netherlands. Environ. Pollut. *122*, 63–74.

de Lipthay, J.R., Tuxen, N., Johnsen, K., Hansen, L.H., Albrechtsen, H.J., Bjerg, P.L., and Aamand, J. (2003). In situ exposure to low herbicide concentrations affects microbial population composition and catabolic gene frequency in an aerobic shallow aquifer. Appl. Environ. Microbiol. *69*, 461–467.

de Wit, C.A. (2002). An overview of brominated flame retardants in the environment. Chemosphere *46*, 583–624.

Edwards, S.J., and Kjellerup, B.V. (2013). Application of biofilm-covered activated carbon particles as a microbial inoculum delivery system for enhanced bioaugmentation of PCBs in contaminated sediment. SERDP and ESTCP, Arlington, VA, USA.

Fagervold, S.K., May, H.D., and Sowers, K.R. (2007). Microbial reductive dechlorination of Aroclor 1260 in Baltimore Harbor sediment microcosms is catalyzed by three phylotypes within the *Chloroflexi*. Appl. Environ. Microbiol. 73, 3009–3018.

Falas, P., Longree, P., la Cour Jansen, J., Siegrist, H., Hollender, J., and Joss, A. (2013). Micropollutant removal by attached and suspended growth in a hybrid biofilm-activated sludge process. Water Res. *47*, 4498–4506.

Ferreira, L., Cobas, M., Tavares, T., Sanroman, M.A., and Pazos, M. (2013). Assessment of *Arthrobacter viscosus* as reactive medium for forming permeable reactive biobarrier applied to PAHs remediation. Environ. Sci. Pollut. Res. Int. *20*, 7348–7354.

Ferrera, I., Massana, R., Casamayor, E.O., Balague, V., Sanchez, O., Pedros-Alio, C., and Mas, J. (2004). High-diversity biofilm for the oxidation of sulfide-containing effluents. Appl. Microbiol. Biotechnol. *64*, 726–734.

Frascari, D., Fraraccio, S., Nocentini, M., and Pinelli, D. (2013). Aerobic/anaerobic/aerobic sequenced biodegradation of a mixture of chlorinated ethenes, ethanes and methanes in batch bioreactors. Bioresour. Technol. *128*, 479–486.

Gerecke, A.C., Hartmann, P.C., and Heeb, N.V.(2005). Anaerobic degradation of decabromodiphenyl ether. Environ. Sci. Technol. *39*, 1078–1083.

Gross, R., Hauer, B., Otto, K., and Schmid, A. (2007). Microbial biofilms: new catalysts for maximizing productivity of long-term biotransformations. Biotechnol. Bioeng. *98*, 1123–1134.

Hall-Stoodley, L., Costerton, J.W., and Stoodley, P. (2004). Bacterial biofilms: from the natural environment to infectious diseases. Nat. Rev. Microbiol. *2*, 95–108.

Harrad, S., Robson, M., Hazrati, S., Baxter-Plant, V.S., Deplanche, K., Redwood, M.D., and Macaskie, L.E. (2007). Dehalogenation of polychlorinated biphenyls and polybrominated diphenyl ethers using a hybrid bioinorganic catalyst. J. Environ. Monit. *9*, 314–318.

Hatzinger, P.B., and Alexander, M. (1995). Effect of aging of chemicals in their biodegradability and extractability. Environ. Sci. Tech. *29*, 537–545.

He, J., Robrock, K.R., and Alvarez-Cohen, L. (2006). Microbial reductive debromination of polybrominated diphenyl ethers (PBDEs). Environ. Sci. Technol. *40*, 4429–4434.

Heidler, J., and Halden, R.U. (2007). Mass balance assessment of triclosan removal during conventional sewage treatment. Chemosphere *66*, 362–369.

Hu, J., Zheng, M., Liu, W., Nie, Z., Li, C., Liu, G., and Xiao, K. (2014). Characterization of polychlorinated dibenzo-*p*-dioxins and dibenzofurans, dioxin-like polychlorinated biphenyls, and polychlorinated naphthalenes in the environment surrounding secondary copper and aluminum metallurgical facilities in China. Environ. Pollut. *193*, 6–12.

Hunter, W.J., and Shaner, D.L. (2010), Biological remediation of groundwater containing both nitrate and atrazine. Curr. Microbiol. *60*, 42–46.

Johnsen, A.R., and Karlson, U. (2004). Evaluation of bacterial strategies to promote the bioavailability of polycyclic aromatic hydrocarbons. Appl. Microbiol. Biotechnol. *63*, 452–459.

Kannan, K., Nakata, H., Stafford, R., Masson, G.R., Tanabe, S., and Giesy, J.P. (1998). Bioaccumulation and toxic potential of extremely hydrophobic polychlorinated biphenyl congeners in biota collected at a superfund site contaminated with Aroclor 1268. Environ. Sci. Technol. 32, 1214–1221.

Kao, C.M., Chen, S.C., and Liu, J.K. (2001). Development of a biobarrier for the remediation of PCE-contaminated aquifer. Chemosphere *43*, 1071–1078.

Kao, C.M., Chen, S.C., Wang, J.Y., Chen, Y.L., and Lee, S.Z. (2003). Remediation of PCE-contaminated aquifer by an *in situ* two-layer biobarrier: laboratory batch and column studies. Water Res. *37*, 27–38.

Ki, D., Park, J., Lee, J., and Yoo, K. (2008). Microbial diversity and population dynamics of activated sludge microbial communities participating in electricity generation in microbial fuel cells. Water Sci. Technol. *58*, 2195–2201.

Kimbrough, R.D. (1995). Polychlorinated biphenyls (PCBs) and human health: an update. Crit. Rev. Toxicol. *25*, 133–163.

Kjellerup, B.V., Sun, X., Ghosh, U., May, H.D., and Sowers, K.R. (2008). Site specific microbial communities in three PCB-impacted sediments are associated with different *in situ* dechlorinating activities. Environ. Microbiol. *10*, 1296–1309.

Kjellerup, B.V., Paul, P., Ghosh, U., May, H.D., and Sowers, K.R. (2012). Spatial distribution of PCB dechlorinating bacteria and activities in contaminated soil. Appl. Env. Soil. Sci. 2012.

Kjellerup, B.V., Naff, C., Edwards, S.J., Ghosh, U., Baker, J.E., and Sowers, K.R. (2014). Effects of activated carbon on reductive dechlorination of PCBs by organohalide respiring bacteria indigenous to sediments. Water Res. *52*, 1–10.

Krajmalnik-Brown, R., Sung, Y., Ritalahti, K.M., Saunders, F.M., and Loffler, F.E. (2007), Environmental distribution of the trichloroethene reductive dehalogenase gene (tceA) suggests lateral gene transfer among *Dehalococcoides*. FEMS Microbiol. Ecol. *59*, 206–214.

Kumar, J., Lind, P.M., and Salihovic, S.(2014). Influence of persistent organic pollutants on the complement system in a population-based human sample. Environ. Int. *71*, 94–100.

Kuo, Y.C., Wang, S.Y., Chang, Y.M., Chen, S.H., and Kao, C.M. (2014). Control of trichloroethylene plume migration using a biobarrier system: a field-scale study. Water Sci. Technol. *69*, 2074–2078.

Kurane, R. (1997). Microbial degradation and treatment of polycyclic aromatic hydrocarbons and plasticizers. Ann. N.Y. Acad. Sci. *829*, 118–134.

Langford, K., Scrimshaw, M., and Lester, J. (2007). The impact of process variables on the removal of PBDEs and NPEOs during simulated activated sludge treatment. Arch. Environ. Contam. Toxicol. *53*, 1–7.

Li, H., Shen, T.T., and Wang, X.L.(2013). Biodegradation of perchloroethylene and chlorophenol co-contamination and toxic effect on activated sludge performance. Bioresour. Technol. *137*, 286–293.

Liang, S.H., Kuo, Y.C., Chen, S.H., Chen, C.Y., and Kao, C.M. (2013). Development of a slow polycolloid-releasing substrate (SPRS) biobarrier to remediate TCE-contaminated aquifers. J. Hazard. Mater. *254–255*, 107–115.

Lilienthal, H., Hack, A., Roth-Harer, A., Grande, S.W., and Talsness, C.E. (2006). Effects of developmental exposure to 2,2,4,4,5-pentabromodiphenyl ether (PBDE-99) on sex steroids, sexual development, and sexually dimorphic behavior in rats. Environ. Health. Perspect. *114*, 194–201.

Liu, H., Park, J.W., Fennell, D.E., Rodenburg, L.A., Verta, M., and Haggblom, M.M. (2013). Microbially mediated reductive dechlorination of weathered polychlorinated dibenzofurans in Kymijoki sediment mesocosms. Chemosphere *91*, 212–221.

Lohman, K., and Seigneur, C. (2001). Atmospheric fate and transport of dioxins: local impacts. Chemosphere *45*, 161–171.

Lohner, S.T., Becker, D., Mangold, K.M., and Tiehm, A. (2011). Sequential reductive and oxidative biodegradation of chloroethenes stimulated in a coupled bioelectro-process. Environ. Sci. Technol. *45*, 6491–6497.

Lolas, I.B., Chen, X., Bester, K., and Nielsen, J.L. (2012). Identification of triclosan-degrading bacteria using stable isotope probing, fluorescence *in situ* hybridization and microautoradiography. Microbiology *158*, 2796–2804.

Lu, X., Wilson, J.T., Shen, H., Henry, B.M., and Kampbell, D.H. (2008). Remediation of TCE-contaminated groundwater by a permeable reactive barrier filled with plant mulch (Biowall). J. Environ. Sci. Health, Part A: Tox. Hazard. Subst. Environ. Eng. *43*, 24–35.

Lucas, W.D., and Hettige, H. (1993). Economic development, environmental regulation, and the international migration of toxic industrial pollution, 1960–88. Office of the Vice President, Development Economics, World Bank. Washington, DC, USA.

Lyu, Y., Zheng, W., Zheng, T., and Tian, Y. (2014). Biodegradation of polycyclic aromatic hydrocarbons by *Novosphingobium pentaromativorans* US6–1. PLoS One *9*, e101438.

Martinez, A., Wang, K., and Hornbuckle, K.C. (2010). Fate of PCB congeners in an industrial harbor of Lake Michigan. Environ. Sci. Technol. *44*, 2803–2808.

Martins, M., Costa, P.M., Raimundo, J., Vale, C., Ferreira, A.M., and Costa, M.H. (2012). Impact of remobilized contaminants in *Mytilus edulis* during dredging operations in a harbour area: bioaccumulation and biomarker responses. Ecotoxicol. Environ. Saf. *85*, 96–103.

Mattes, T.E., Alexander, A.K., and Coleman, N.V. (2010). Aerobic biodegradation of the chloroethenes: pathways, enzymes, ecology, and evolution. FEMS Microbiol. Rev. *34*, 445–475.

May, H.D., Miller, G.S., Kjellerup, B.V., and Sowers, K.R. (2008). Dehalorespiration with polychlorinated biphenyls by an anaerobic ultramicrobacterium. Appl. Environ. Microbiol. *74*, 2089–2094.

Maymo-Gatell, X., Anguish, T., and Zinder, S.H. (1999). Reductive dechlorination of chlorinated ethenes and 1,2-dichloroethane by *'Dehalococcoides ethenogenes'* 195. Appl. Environ. Microbiol. *65*, 3108–3113.

McDonald, T.A. (2002). A perspective on the potential health risks of PBDEs. Chemosphere *46*, 745–755.

Mercier, A., Joulian, C., and Michel, C.(2014). Evaluation of three activated carbons for combined adsorption and biodegradation of PCBs in aquatic sediment. Water Res. *59*, 304–315.

Michalsen, M.M., Peacock, A.D., and Smithgal, A.N.(2009). Treatment of nitric acid-, U(VI)-, and Tc(VII)-contaminated groundwater in intermediate-scale physical models of an *in situ* biobarrier. Environ. Sci. Technol. *43*, 1952–1961.

Nyholm, J.R., Lundberg, C., and Andersson, P.L. (2010). Biodegradation kinetics of selected brominated flame retardants in aerobic and anaerobic soil. Environ. Pollut. *158*, 2235–2240.

O'Neil, R.A., Holmes, D.E., and Coppi, M.V.(2008). Gene transcript analysis of assimilatory iron limitation in *Geobacteraceae* during groundwater bioremediation. Environ. Microbiol. *10*, 1218–1230.

Patil, S.A., Surakasi, V.P., Koul, S., Ijmulwar, S., Vivek, A., Shouche, Y.S., and Kapadnis, B.P. (2009). Electricity generation using chocolate industry wastewater and its treatment in activated sludge based microbial fuel cell and analysis of developed microbial community in the anode chamber. Bioresour. Technol. *100*, 5132–5139.

Patureau, D., and Trably, E. (2006). Impact of anaerobic and aerobic processes on polychlorobiphenyl removal in contaminated sewage sludge. Biodegradation *17*, 9–17.

Payne, R.B., May, H.D., and Sowers, K.R. (2011). Enhanced reductive dechlorination of polychlorinated biphenyl impacted sediment by bioaugmentation with a dehalorespiring bacterium. Environ. Sci. Technol. *45*, 8772–8779.

Payne, R.B., Fagervold, S.K., May, H.D., and Sowers, K.R. (2013). Remediation of polychlorinated biphenyl impacted sediment by concurrent bioaugmentation with anaerobic halorespiring and aerobic degrading bacteria. Environ. Sci. Technol. *47*, 3807–3815.

Petrie, L., North, N.N., Dollhopf, S.L., Balkwill, D.L., and Kostka, J.E. (2003). Enumeration and characterization of iron(III)-reducing microbial communities from acidic subsurface sediments contaminated with uranium(VI). Appl. Environ. Microbiol. *69*, 7467–7479.

Plosz, B.G., Vogelsang, C., Macrae, K., Heiaas, H.H., Lopez, A., Liltved, H., and Langford, K.H. (2010). The BIOZO process-a biofilm system combined with ozonation: occurrence of xenobiotic organic micro-pollutants in and removal of polycyclic aromatic hydrocarbons and nitrogen from landfill leachate. Water Sci. Technol. *61*, 3188–3197.

Rayne, S., Ikonomou, M.G., and Whale, M.D. (2003). Anaerobic microbial and photochemical degradation of 4,4'-dibromodiphenyl ether. Water Res. *37*, 551–560.

Robrock, K.R., Coelhan, M., Sedlak, D.L., and Alvarez-Cohen, L. (2009). Aerobic biotransformation of polybrominated diphenyl ethers (PBDEs) by bacterial isolates. Environ. Sci. Technol. *43*, 5705–5711.

Rodriguez, S., and Bishop, P. (2008). Enhancing the Biodegradation of Polycyclic Aromatic Hydrocarbons: Effects of Nonionic Surfactant Addition on Biofilm Function and Structure. J. Env. Eng. *134*, 505–512.

Safe, S., Connor, K., Ramamoorthy, K., Gaido, K., and Maness, S. (1997). Human exposure to endocrine-active chemicals: Hazard assessment problems. Regul. Toxicol. Pharmacol. *26*, 52–58.

Seo, Y., and Bishop, P.L. (2008). The monitoring of biofilm formation in a mulch biowall barrier and its effect on performance. Chemosphere *70*, 480–488.

Seo, Y., Lee, W.H., Sorial, G., and Bishop, P.L. (2009). The application of a mulch biofilm barrier for surfactant enhanced polycyclic aromatic hydrocarbon bioremediation. Environ. Pollut. *157*, 95–101.

Stapleton, H.M., Dodder, N.G., Offenberg, J.H., Schantz, M.M., and Wise, S.A. (2005). Polybrominated diphenyl ethers in house dust and clothes dryer lint. Environ. Sci. Technol. *39*, 925–931.

Staskal, D.F., Diliberto, J.J., Devito, M.J., and Birnbaum, L.S. (2005), Inhibition of human and rat CYP1A2 by TCDD and dioxin-like chemicals. Toxicol. Sci. *84*, 225–231.

Tobiszewski, M., and Namiesnik, J. (2012). Abiotic degradation of chlorinated ethanes and ethenes in water. Environ. Sci. Pollut. Res. Int. *19*, 1994–2006.

Tokarz, J.A., Ahn, M.Y., Leng, J., Filley, T.R., and Nies, L. (2008). Reductive debromination of polybrominated diphenyl ethers in anaerobic sediment and a biomimetic system. Environ. Sci. Technol. *42*, 1157–1164.

Trudel, D., Scheringer, M., von Goetz, N., and Hungerbuhler, K. (2011). Total consumer exposure to polybrominated diphenyl ethers in North America and Europe. Environ. Sci. Technol. *45*, 2391–2397.

UNEP (2004). The Stockholm Convention on Persistent Organic Pollutants. UNEP, Stockholm, Sweden.

USEPA (1992). US EPA 1992: Ground Water Issue. Vol. EPA/540/S-92/002. http://www.epa.gov/superfund/remedytech/tsp/download/tce.pdf USEPA.

USEPA (2005). Contaminated sediment remediation guidance for hazardous waste sites. US Environmental Protection Agency, Washington, DC, USA.

USEPA (2009). DecaBDE Phase-out Initiative. Washington, DC, USA.

USEPA (2014). Polybrominated Diphenyl Ethers (PBDEs) Action Plan Summary. Washington DC, USA.

Vater, S.T., Velazquez, S.F., and Cogliano, V.J. (1995). A case study of cancer data set combinations for PCBs. Regul. Toxicol. Pharmacol. *22*, 2–10.

Velvizhi, G., Goud, R.K., and Venkata Mohan, S. (2014). Anoxic bio-electrochemical system for treatment of complex chemical wastewater with simultaneous bioelectricity generation. Bioresour. Technol. *151*, 214–220.

Vonderheide, A.P., Mueller, K.E., Meija, J., and Welsh, G.L. (2008). Polybrominated diphenyl ethers: causes for concern and knowledge gaps regarding environmental distribution, fate and toxicity. Sci. Total. Environ. *400*, 425–436.

Voorspoels, S., Covaci, A., Jaspers, V.L., Neels, H., and Schepens, P. (2007). Biomagnification of PBDEs in three small terrestrial food chains. Environ. Sci. Technol. *41*, 411–416.

Wang, Y., and Oyaizu, H. (2011). Enhanced remediation of dioxins-spiked soil by a plant–microbe system using a dibenzofuran-degrading *Comamonas* sp. and *Trifolium repens L.* Chemosphere *85*, 1109–1114.

Wei, Z., and Seo, Y. (2010). Trichloroethylene (TCE) adsorption using sustainable organic mulch. J. Hazard. Mater. *181*, 147–153.

Werner, D., Ghosh, U., and Luthy, R.G. (2006). Modeling polychlorinated biphenyl mass transfer after amendment of contaminated sediment with activated carbon. Environ. Sci. Technol. *40*, 4211–4218.

Wu, K., Yau, Y.C., Matukas, L., and Waters, V. (2013). Biofilm compared to conventional antimicrobial susceptibility of *Stenotrophomonas maltophilia* isolates from cystic fibrosis patients. Antimicrob. Agents. Chemother. *57*, 1546–1548.

Yamamoto, S., Suzuki, K., and Araki, Y.(2014). Dynamics of different bacterial communities are capable of generating sustainable electricity from microbial fuel cells with organic waste. Microbes. Environ. *29*, 145–153.

Yoshida, S., Ogawa, N., Fujii, T., and Tsushima, S. (2009). Enhanced biofilm formation and 3-chlorobenzoate degrading activity by the bacterial consortium of *Burkholderia* sp. NK8 and *Pseudomonas aeruginosa* PAO1. J. Appl. Microbiol. *106*, 790–800.

Zimmerman, J.R., Ghosh, U., Millward, R.N., Bridges, T.S., and Luthy, R.G. (2004). Addition of carbon sorbents to reduce PCB and PAH bioavailability in marine sediments: physicochemical tests. Environ. Sci. Technol. *38*, 5458–5464.

Electroactive Biofilms in Water and Air Pollution Treatment

10

Anna Vilajeliu-Pons, Sebastià Puig, Alessandro Carmona-Martínez, Nicolas Bernet, Marta Coma, Federico Aulenta, Jesús Colprim and Maria Dolors Balaguer

Abstract

Biofilms are used in wastewater treatment and in the production of valuable compounds. Bioelectrochemical system (BES) technology represents one practical application of biofilms. In these systems, bioelectrogenic biofilms are a bacterial consortium capable of performing electron transfer to the conductive material on which they are grown. This capacity of these organisms has been used in environmental biotechnology to couple pollutant removal, mainly from water but also from air streams, for the production of energy or valuable products. The following chapter outlines the details of such a consortium, highlighting the mechanisms of extracellular electron transfer and their main applications.

Bioelectrochemical systems

Bioelectrochemical systems (BESs) were described for the first time approximately 100 years ago (Arends and Verstraete, 2012; Schröder, 2011). In a very general sense, BESs represent a group of technologies capable of (1) producing power simultaneously with air and liquid pollution treatment, and (2) recovering and/or producing chemical compounds. The concept of BESs started with Michael C. Potter, who described the electric activity of bacterial and yeast cultures in 1911 (Potter, 1911). After this discovery, it took 20 years until the additional development of the novel concept of BESs. BESs were possible due to Cohen who, like Potter, reported differences in the 'reduction potential' of a bacterial culture after a few days of incubation (Cohen, 1931). Interestingly, at that time, it was strongly believed that such power production was only possible due to the presence of exogenous molecules, such as electron mediators. Since then, BESs have come a long way in their research and development.

In the following decades, the idea of microbial electricity generation did not significantly progress, and the field received little attention until 1980–2000. Since then, BESs have been rapidly developed. Microbial fuel cells (MFCs) were the first BES used to produce electricity (Rabaey *et al.*, 2003). Thereafter, microbial electrolysis cells (MECs) were developed, which required applying a cell potential to overcome the energy barrier to remove pollutants, such as uranium (Gregory and Lovley, 2005) and chlorinated compounds (Aulenta *et al.*, 2010), or to produce high value products (Rabaey and Rozendal, 2010). Both bioelectrochemical cells consist of two electrodes placed in anode and a cathode compartments where oxidation and reduction reactions occur. An external electric circuit joins the anode and cathode electrodes. Unlike in the well-known electrochemical cells, at least one of the two reactions is catalysed by microorganisms in this system.

As a general overview of multiple processes occurring in BESs such as MFCs and MECs, bacteria in the anode compartment break down organic material under anaerobic conditions (Fig. 10.1) from which mainly electrons and protons are harvested. Some of the electrons and protons are used for biomass and CO_2 production (among other gases such as H_2 or CH_4). The electrons and protons migrate to the cathode compartment through an external circuit and an ionic membrane, respectively. In the cathode

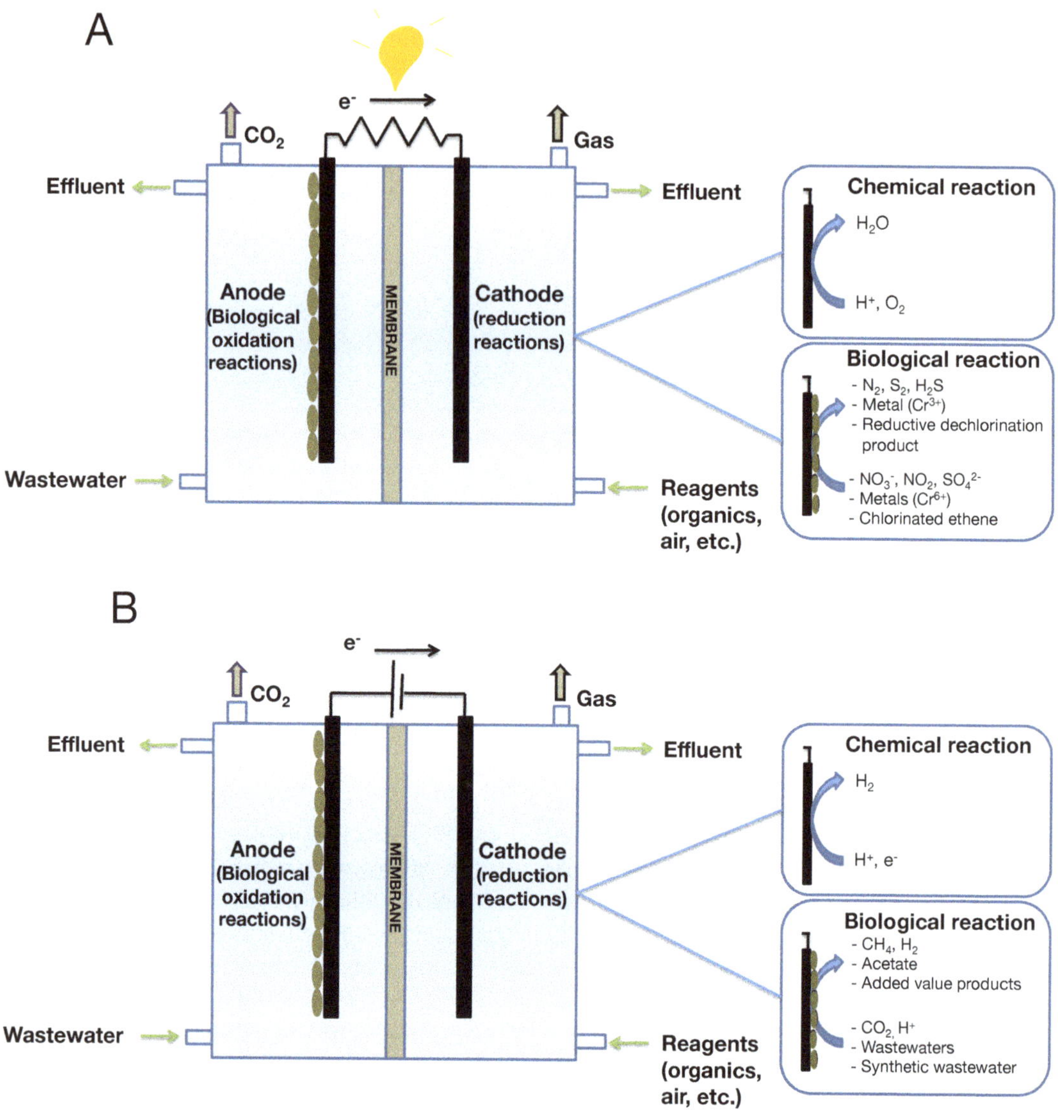

Figure 10.1 General overview of two configurations of bioelectrochemical systems (BESs) performing wastewater treatment. (A) Microbial fuel cell (MFC); (B) microbial electrolysis cell (MEC).

compartment, electrons and protons might be used to electrochemically reduce oxygen to water. They can be biologically captured to reduce different compounds in the so-called 'microbial electrosynthesis' process (Rabaey and Rozendal, 2010). A non-exhaustive list of the reduction reactions that occur at the cathode includes the following: the reduction of nitrate (Gregory *et al.*, 2004) and sulfate (Coma *et al.*, 2013) or the production of H_2 (Oh and Logan, 2005) or other chemical compounds, such as methane (Villano *et al.*, 2011) or acetate (Marshall *et al.*, 2012). The main difference between MFC and MEC is based on the production of electricity (MFC) or the use of electricity (MEC) to allow biological or chemical reactions at the cathode.

Microorganisms adhere to each other, creating a biofilm layer on the electrode surface with extracellular polymeric substances (EPS). This biofilm allows for successful electron transfer through the electrode surface and can comprise either a pure culture (Logan and Regan, 2006) or a mixed culture (complex microbial community) (Zhan *et al.*, 2014). The microorganism–electrode interaction is not yet fully understood and is the subject of active research (Guo *et al.*, 2013). Different materials have been used as both electrode and carrier materials, including carbon, graphite, titanium and stainless steel (Wei *et al.*, 2011; Logan, 2008a). Now, granular graphite is one of the most common and inexpensive materials used because of its natural conductivity properties and affinity for microorganisms (Vilar-Sanz *et al.*, 2013; Arends *et al.*, 2012; Puig *et al.*, 2011a; Logan, 2008; Maignien *et al.*, 2006; Rabaey *et al.*, 2005). However, a different range of conductive materials has been used (Logan *et al.*, 2007), even though they have been modified in order to increase the level of microorganism surface adhesion and the conductivity of the material (Guo *et al.*, 2013).

In the anode compartment, the layer of biofilm contains microorganisms with the ability to transfer electrons extracellularly (Lovley, 2008a). These bacteria, which are commonly called exoelectrogens (Logan and Regan, 2006), have an important role in oxidizing and reducing metals in natural environments. In fact, the microbial electron transfer that some bacteria use in their role as insoluble metal electron acceptors is of special importance in several biogeochemical cycles. Moreover, this phenomenon has been applied to the bioremediation of contaminated sites since the early 1990s (Logan, 2008; Lovley *et al.*, 2004a; Nealson *et al.*, 1991). In the context of the direct conversion of organic wastes into electricity by exoelectrogen bacteria in BESs, it is worth noting that in order to improve bioenergy production, the research on BESs should explore the use of either robust mixed cultures or co-cultures with the ability to break down complex substrates. Such an approach should therefore continue because currently only a few attempts have successfully addressed this aspect (Miceli *et al.*, 2014; Speers *et al.*, 2014).

In the cathode compartment, microorganisms are able to reduce compounds by taking electrons from the electrode surface. These bacteria are called electrotrophs (Lovley, 2011). The role of these microorganisms in such specific systems has been hardly ever studied. Therefore, it is important to gain further fundamental knowledge about these microorganisms for the removal of pollutants from wastewater and for the production of high value products (Patil *et al.*, 2012).

Biofilms in bioelectrochemical systems

The study of MFC biofilms did not take off until 2005 (Rabaey and Verstraete, 2005). Since 2005, knowledge of the role of microorganisms was considered important in order to maximize the energy production and nutrient removal capacity or product production. Thereafter, the number of biofilm studies increased significantly. Some microorganisms responsible for current production, such as *Geobacter sulfurreducens* (Reguera *et al.*, 2005) and *Shewanella oneidensis* (Gorby *et al.*, 2006), were identified. Since then, the mechanisms of extracellular electron transfer between bacteria and electrode materials have been extensively studied by diverse research groups (Marsili and Zhang, 2009; Rosenbaum and Angenent, 2009). Currently, special attention is given to the syntrophic interactions within the biofilm and the obtained end-product because such types of interactions seem more relevant when working on the microbial electrosynthesis of added value products that do not necessarily depend on direct uptake electron transfer (Arends *et al.*, 2013). In the context of BESs, syntrophy is defined as the mutualistic interaction between microorganisms, where the main goal is maximizing the resource utilization (Lovley *et al.*, 2011). Synergy between communities helps degradation because some strains produce metabolites that can be used for other species to complete degradation. Single communities working individually are not capable of removing such substances. This concept is also related to the other communities present in extreme conditions, where microorganisms help each other to survive (McInerney *et al.*, 2008). In BESs, syntrophic interactions have been described on exoelectrogenic and non-exoelectrogenic bacteria in anode biofilms (Parameswaran *et al.*, 2009).

As stated previously, known biofilm structures consist of bacterial cells surrounded by self-produced EPS. Biofilms can be formed by single populations or by mixed communities. Microorganisms in the biofilm show heterogeneity due to their interspersed distribution inside the EPS matrix (Davey and George, 2000). Usually, mixed bacterial cultures are used to inoculate BESs. Then, the biofilm is enriched by applying very specific operational conditions (such as a single substrate or a stable pH/temperature/redox potential) (Harnisch *et al.*, 2011). This enrichment is useful for increasing the biomass of the biofilm and its removal/production capacity. Once dominating bacteria are found in BES studies, isolated or purchased pure cultures can be used to study and acquire more fundamental and specific knowledge about the involved microorganisms and their metabolic pathways. On the other hand, mixed cultures have also been used to treat the complex organic matter that is usually contained in wastewaters due to their ability to adapt to changing conditions, such as those normally encountered in wastewater treatment plants. Consequently, a higher power output has been detected in BES studies based on mixed cultures, which seem to be more robust and resilient (Arends *et al.*, 2011). The maximum power output is not only related to the culture's source. In fact, there are many factors that can modify biofilm formation and behaviour, including the electrode surface, the nutrient availability, the pH (in pure cultures) and the hydrodynamic conditions (Franks *et al.*, 2010).

Exoelectrogenic microorganisms

Description

Lovley (2006) described a new form of microbial respiration in which microorganisms conserve energy for growth while transferring electrons extracellularly to the electrode. Lovley first called these bacteria *exoelectrogens*. Exoelectrogens are also known as anodophiles (Park and Zeikus, 2003), electrochemically active bacteria (EAB) (Chang *et al.*, 2006), anode-respiring bacteria (Torres *et al.*, 2007) and electrogenic microorganisms (Debabov, 2008).

Currently, the study of exoelectrogens has been prompted by their application in MFCs. In this alternative treatment, exoelectrogen microorganisms transform the acetate or complex organic matter from wastewater into electricity (Debabov, 2008; Lovley, 2008b). The most extensively studied microorganism able to produce high current densities and thick biofilms in MFCs is the Gram-negative *Geobacter sulfurreducens* (Bond and Lovley, 2003). Other bacteria with the ability to produce high current densities include *Rhodopseudomonas palustris* DX-1 (Xing *et al.*, 2008), *Thermincola ferriacetica* (Parameswaran *et al.*, 2013), *Geoalkalibacter ferrihydriticus* (Badalamenti *et al.*, 2013) and *Geoalkalibacter subterraneus* (Carmona-Martínez *et al.*, 2013). Such bacterial species have been well characterized, but they are not the only ones with a corroborated electrical activity (Gorby *et al.*, 2006). Different Gram-positive (*Pseudomonas* sp.) and Gram-negative (*Shewanella oneidensis*) bacteria have been identified in bioanodes. Therefore, their presence makes them likely to be responsible for the extracellular electron transfer process at the electrode surface (Arends *et al.*, 2011).

History and identification of exoelectrogens

Potter's discovery (1911) of microbial catalysed electrode reduction revolutionized the world of microbiology. He observed and reported electricity production with *Escherichia coli* and yeast cultures. It was not until the end of the twentieth century that Vargas *et al.* (1998) defined Fe(III) as the first external electron acceptor in MFC systems working with exoelectrogens.

The polymerase chain reaction (PCR) and fluorescence *in situ* hybridization (FISH) techniques were used to identify bacterial populations in the biofilms (Davey and George, 2000). These techniques have since been used to identify exoelectrogens biofilms in BES systems.

Reimers *et al.* (2001) established microbes in seawater batteries, obtaining low power and voltage gradients. One year later, Bond *et al.* (2002) reproduced the work of Reimers by sequencing the microbial community responsible for electricity production. The 16S rRNA sequences obtained were analysed using the most probable number-polymerase chain reaction (MPN-PCR) technique, showing that the *Geobacteraceae* family of Delta-proteobacteria was the most abundant. Finding a bacterial species within the *Geobacteraceae* family (i.e. *Desulfuromonas acetoxidans*) has been of great importance for the BESs field owing to the frequent appearance of these types of bacteria in electroactive biofilms (Yates *et al.*, 2012).

The interest on exoelectrogens has therefore rapidly increased with the identification of anode respiring bacteria in BESs, gradually achieving higher current and power outputs. The complete nucleotide sequences of the *Shewanella oneidensis* (Heidelberg *et al.*, 2002) and *Geobacter sulfurreducens* (Methé *et al.*, 2003) genomes have been described. The utilization of microarray technique (Nielsen *et al.*, 2003), polymerase chain reaction–denaturing gradient gel electrophoresis (PCR-DGGE) (Rabaey *et al.*, 2004) and pyrosequencing (Kiely *et al.*, 2011) have all proven their usefulness in the physiological identification and characterization of the microorganisms responsible for electricity generation. For instance, microarray and genetic analysis have successfully been used for the development of electron transfer mechanisms in *Geobacter sulfurreducens* (Holmes *et al.*, 2006). However, additional cloning, sequencing, and assembly techniques are needed to improve the microbial knowledge about exoelectrogens.

Once the first microorganisms were identified, scientists tried to unravel the mechanisms used to release electrons. The mechanisms used to transfer electrons can be direct (direct contact, conductive biofilm and nanowires) or indirect (mediators). Cylic voltammetry is a technique used to distinguish between the direct and indirect electron transfer processes in mixed or pure cultures (Harnisch and Freguia, 2012). Confocal Resonance Raman Microscopy (CRRM) has also been used to test variations in c-type cytochromes redox state (Virdis *et al.*, 2014).

Geobacter sulfurreducens has been widely studied as a model exoelectrogenic bacterial species that is frequently enriched in electroactive biofilms. Methé *et al.* (2003) interpreted the *G. sulfurreducens* genome, thus facilitating future studies about its electron transport mechanisms. The importance of conductive pili (also called nanowires) was realized when scientists began to understand their utility in electron transport processes (Malvankar *et al.*, 2011; Lovley *et al.*, 2011; Reguera *et al.*, 2005).

Nevin *et al.* (2008) compared pure (*G. sulfurreducens*) and mixed cultures in terms of the power production in different designs of MFCs. Different proportions between anode and cathode compartments have been tested, leading to the conclusion that the MFC design could have more important implications for power production than the use of pure or mixed biofilm cultures. However, other parameters have been evaluated, including biofilm conductivity. Malvankar *et al.* (2012) demonstrated that mixed cultures biofilms had higher electrical conductivity than pure cultures of *Geobacter sulfurreducens*. High conductivity allows for rapid electron transport between electrodes. Recent studies have demonstrated syntrophic interactions between exoelectrogen and non-exoelectrogen microorganisms (Parameswaran *et al.*, 2009) and the synergy in communities to remove complex compounds (Kiely *et al.*, 2011). Kim *et al.* (2011) showed that the proportion of exoelectrogens to non-exoelectrogens was lower at higher organic loading rates.

Finally, the most recent studies have focused on the effects of different parameters on the microbial community. Aelterman *et al.* (2008) studied the influence of anode potential. The microbial community was capable of self-regulating extracellular electron transfer (EET) pathways at different anode potentials and was not affected by this parameter. Moreover, the impact of media composition on the microbial community was studied by Puig *et al.* (2010) and Behera and Ghangrekar (2009). Different organic loading rates (OLRs) and pH values were tested by the authors mentioned above to achieve the maximum power production. It was shown that the microbial community activities improved at OLRs below 1 kg COD/kg VSS·d (where COD was the chemical oxygen demand and VSS was the volatile suspended solid in the anode compartment) with a basic pH of 8–9.5 producing the highest power production.

Applicability of BESs for waste treatment

The first exoelectrogen microorganism applications were studied using synthetic media. The most-used carbon source simulating wastewater was acetate, which is also used in genetic and molecular studies (Lovley *et al.*, 2011; Kiely *et al.*, 2011; Logan and Regan, 2006; Logan *et al.*, 2006; Bond and Lovley, 2003; Bond *et al.*, 2002). Table 10.1 shows a representative but non-exhaustive set of studies regarding exoelectrogen wastewater treatment. In the study by Rabaey *et al.* (2005), microorganisms were able to oxidize 53% of the acetate with a Coulombic Efficiency (CE) of 36%. The term CE correlates with the electrons flowing through the BES, with the process occurring in one of the chambers (i.e. the anode or the cathode). It is the ratio between the electron actually transferred to the electrons potentially transferred to the electrical circuit considering the reaction occurring at the bioelectrode (equation 10.1).

$$CE = \frac{M_S I}{FnQ\ \ C} \cdot 100 \qquad (10.1)$$

Here, M_s is the molecular mass of the substrate (in g/mol), I is the current sourced from the MFC (in A) and F is Faraday's constant (96,485 C/mol). The term n identifies the number of electrons released for each mole of oxidized substrate, Q is the flow (l/day), and ΔC is the substrate concentration change (in mg/l) between the influent and effluent streams.

Thus, the CE from the anode is related to the percentage of organic matter removed by exoelectrogen bacteria in the BES. Several authors have studied the mechanisms involved in order to improve the efficiency of these systems. Recently, Villano *et al.* (2013) built a MEC with the ability to oxidize 94% of acetate in the influent with a CE of 91%.

In addition to acetate, a variety of other substrates have been used in the BES anode. For instance, glucose (Freguia *et al.*, 2008; Rabaey *et al.*, 2005, 2003) and fructose (Liamleam and

Table 10.1 Representative compilation of BESs studies for the treatment of synthetic media and raw wastewaters in terms of their percentage of removal capacity and coulombic efficiency (CE)

Substrate	Load (kg COD/ m^3·day)	Removal capacity (%)	CE (%)	Reference
Acetate	1.10	72	75±7	Rabaey *et al.* (2005)
	2.10	53	36±2	
Acetate	1.08	94	91±2	Villano *et al.* (2013)
Glucose	1.10	84	59±4	Rabaey *et al.* (2005)
Glucose	0.50	85	89±4	Rabaey *et al.* (2003)
	5.00	41	10±2	
Municipal wastewater	10–20	79	<5	He *et al.* (2014)
Municipal wastewater	1.50	78–83	n.d.	Sevda and Sreekrishnan (2014)
Municipal wastewater	7.20	80	n.d.	Puig *et al.* (2011)
Municipal wastewater	9.77	43	n.d.	Puig *et al.* (2011a)
Municipal wastewater	0.10	75	20	Liu and Logan (2004)
Industrial wastewater: bakery	0.05	86	2±1	Velasquez-Orta *et al.* (2011)
Industrial wastewater: brewery		85	2±1	
Industrial wastewater: paper		78	26±6	
Industrial wastewater: dairy		82	2±1	
Industrial wastewater: brewery	6.70	44	7	Wen *et al.* (2010)
Landfill leachate	3.20	13	n.d.	Greenman *et al.* (2009)
Landfill leachate	2.20	10	<2	Puig *et al.* (2011a)
Swine manure	4.50	27	8	Min *et al.* (2005)
Swine manure	2.00	62	n.d.	Lim *et al.* (2012)
Swine manure	1.20	84	<1	Zhuang *et al.* (2012)
	4.90	77	<1	

n.d., no data.

Annachhatre, 2007) have been tested. Interestingly, power outputs with both substances were lower than those of acetate due to previous fermentation processes, which produce hydrogen and acetate and methanogens that compete with the exoelectrogens for hydrogen. In contrast, non-synthetic media (i.e. raw wastewaters) have also been tested. For example, different wastes, including municipal wastewater (He *et al.*, 2014; Sevda and Sreekrishnan, 2014; Puig *et al.*, 2011; Liu and Logan, 2004) and industrial wastewaters from the bakery, brewery, paper, dairy and food industries (Jia *et al.*, 2013; Velasquez-Orta *et al.*, 2011; Wen *et al.*, 2010) have been used as the influent for BESs. Here, the organic loading rate has been generally higher than with the synthetic media based on easily degradable substrates. Despite the higher removal found when testing non-synthetic media, the exoelectrogen activity was lower due to the complex organic matter matrix in the influent, causing a more diversified range of side metabolic reactions that inevitably consume electrons.

Other substrates such as landfill leachate (Puig *et al.*, 2011a; Greenman *et al.*, 2009), swine manure (Lim *et al.*, 2012; Zhuang *et al.*, 2012; Min *et al.*, 2005) and urine (Ieropoulos *et al.*, 2012) have been analysed and found to have similar organic loading rates as wastewaters. The use of complex organic influents has resulted in low CEs and power production. However, such studies have proved the applicability of BESs.

Therefore, higher CEs are obtained when available biodegradable substrates such as acetate are used as the influent. However, other more complex substrates such as glucose or wastewaters have low CEs due to the requirements of earlier electron-consuming steps such as acid genesis (sugars to acetate) or hydrolysis.

Electrotrophic microorganisms

Description

An electrothoph has the microbial capacity to directly accept electrons from the electrode at the cathode compartment in the BES. These microorganisms have not been widely studied thus far. Early studies focused mainly on the anode compartment and on the exoelectrogen microorganisms. Currently, the interest in electrotrophs has increased, as researchers aim to develop biocathodes for the production of different added value molecules (Marshall *et al.*, 2012; Rosenbaum *et al.*, 2011).

Different reduction reactions have been reported in the literature. Some have been useful in the treatment of wastewaters with high nitrate content. For instance, Pous *et al.* (2013) demonstrated the successful conversion of nitrate and nitrite to dinitrogen gas. Recently, with the changing paradigm of converting wastes to new bioresources, other reactions producing high value products are being considered. For example, the conversion of carbon dioxide into methane has been recently demonstrated by Villano *et al.* (2013). Additionally, Marshall *et al.* (2013) showed that the carbon chain of certain organic acids commonly produced as the dead-end products of fermentation could be elongated. Rozendal *et al.* in 2008 provided a proof of concept of protons being reduced to hydrogen by means of a process that is microbiologically catalysed by an electroactive biofilm (Arends *et al.*, 2011). Hence, the variability of electron acceptors in cathode microbial reactions has demonstrated that the electrotroph capacity is a promising route to remove pollutants and/or to produce high-value products.

History and identification of electrotrophic metabolism

The study of these microorganisms was not developed until recently. Only a few studies have focused on the characterization of electrotrophic microorganisms. Wrighton *et al.* (2010) deeply analysed the denitrifying microbial communities of two different BESs and examined their phylogenetic affiliation and community structures. The study performed by Vilar-Sanz *et al.* (2013) about denitrifying communities focused on the functional genes of denitrification pathways.

Once the microorganisms were identified, efforts became more focused on the identification of the electron transfer mechanisms utilized by such electrotrophic microorganisms (Rosenbaum *et al.*, 2011; Lovley, 2011) and the relationship between members of the communities. Marshall *et al.* (2013) focused on the physical (scanning electron microscope) and molecular (RNA, DNA) analysis of microbiomes (synergic communities) for electrosynthesis purposes. On the other hand, Ross *et al.* (2011) studied *Shewanella oneidensis* strain MR-1 to learn about the complex oxidation/reduction reactions occurring at the electrodes. This genus uses the Mtr respiratory pathway to catalyse electron flow from cytoplasmic oxidative reactions to electrodes. It has the ability to drive microbial reductive metabolism. This result suggested the possibility of obtaining valuable fuels and chemicals from that pathway.

Electrotrophic process applications

Electrotrophic microorganisms have been used in bioremediation and in the production/recovery of chemical compounds. A great variety of environmental pollutants have been treated by taking advantage of electrotrophic microorganisms in BES. For example, due to the intensive agriculture and livestock activities, the release of nitrate from water has become a serious problem, requiring large amounts of organic matter for its removal through conventional processes (denitrification). In this scenario, electrotrophs might provide a great benefit because they can reduce nitrates and/or nitrites from contaminated groundwater and wastewater to dinitrogen gas (Pous *et al.*, 2013; Kondaveeti and Min, 2013; Vilar-Sanz *et al.*, 2013; Puig *et al.*, 2012, 2011b; Wrighton *et al.*, 2010; Virdis *et al.*, 2008). Electron transport increases in highly conductive media but is limited in media with low conductivity (Pous *et al.*, 2013). Despite this, the possibility of treating groundwater without affecting the drinking water quality by BESs has been reported. Biocathodes can also reduce other types of inorganic contaminants, such as sulfates (Coma *et al.*, 2013), different metals such as uranium (Anderson *et al.*, 2003) and chromium (VI) (Tandukar *et al.*, 2009), and chlorinated compounds (Aulenta *et al.*, 2009). Table 10.2 provides a compilation of representative reducible pollutant studies using electrotrophic treatments.

Electrotrophic activity is not only linked to the removal of contaminants. Recent studies have suggested a promising future for these bacteria to produce (or recover) high value products from waste (Table 10.3). Different organic and inorganic products such as methane, hydrogen, acetate, ethanol, 1–3 propanediol and succinate have been obtained from the electrotroph activity (Logan and Rabaey, 2012). Hydrogen is the main product of water electrolysis and occurs easily in electrochemical cells. However, the presence of microorganisms in the cathode catalyses this reaction, such that production can be obtained at higher redox potentials.

Methane is another compound studied in BESs by different scientists and is of interest due to its high energetic value. Recently, the interest in this type of methane production route has gained increased attention. It has been demonstrated that methane production is feasible using synthetic and raw wastewaters and CO_2 carbon based substrates. For example, methane and acetate were

Table 10.2 Summary of literature studies about waste streams treated by electrotrophic bacteria and their percentage of removal capacity and Coulombic Efficiency (CE) with different substrates (S)

Substrate	Source (kg S/m^3·day)	Removal capacity (%)	CE (%)	Reference
Nitrate	0.06	64	60–80	Pous *et al.* (2013)
Nitrite	0.28	77	41±17	Vilar-Sanz *et al.* (2013)
Nitrate	0.37	15	85±11	
Nitrate	0.05	42	73±18	Puig *et al.* (2012)
Nitrate	0.50	35	48±11	Puig *et al.* (2011b)
Sulfates	0.21	<1	n.d.	Coma *et al.* (2013)
Chromium	0.01	99	n.d.	Tandukar *et al.* (2009)
Uranium	6.10^{-6}	70	n.d.	Anderson *et al.* (2003)

n.d., no data.

Table 10.3 Summary of literature studies about different high value ending products production of electrotrophic bacteria from different initial products and cathode potentials

Initial product (electron acceptor)	Final product	Potential (mV versus SHE)	Reference
H^+	Hydrogen	–1000	Batlle-Vilanova *et al.* (2014)
Synthetic wastewater	CH_4	<–1000	Xafenias and Mapelli (2014)
	Acetate		
Wastewater	CH_4	–590 to –900	Marshall *et al.* (2013)
CO_2	H_2		
H^+	Fatty acids		
CO_2	CH_4, H_2	–1047 to –1147	Jiang *et al.* (2013)
H^+	Acetate, CH_4, H_2	<–1147	
CO_2	CH_4, H_2	–590	Marshall *et al.* (2012)
H^+			
CO_2	Multicarbon organic compounds	–400	Nevin *et al.* (2011)
CO_2	Multicarbon organic compounds	–400	Nevin *et al.* (2010)
H^+	Hydrogen	–700	Rozendal *et al.* (2008)

n.d., no data.

the predominant products at a cathode potential of −590 mV versus SHE (Marshall *et al.*, 2012). Another interesting example of the production of added-value molecules was recently described by Xafenias and Mapelli (2014), where the production of multicarbon organic compounds such as acetate was reported with significant efficiencies of approximately 60% when the cathode potential was poised below −1000 mV versus SHE (standard hydrogen electrode). This might be due to a synergistic reaction with the hydrogenotrophic methanogens, as this potential hydrogen could be produced.

Hydrogen production is one of the most well-studied reactions in biocathode compartments. The first attempts to use MECs to couple organic wastewater and hydrogen production were completed by Bruce Logan's group with small-scale prototypes. In 2009, Wagner *et al.* made a laboratory scale MEC for the treatment of swine wastewater in batch mode. They produced 0.9–1 m^3 H_2/m^3·day while removing up to 72% of the COD. The gas produced was up to 77% hydrogen but also contained up to 13% methane. Lalaurette *et al.* (2009) coupled a dark-fermentation reactor to a MEC to convert maize stover lignocellulose or cellobiose into hydrogen. Hydrogen yields approximately 1 l H_2/g COD and 1 l H_2/l·day were obtained. Cusick *et al.* (2010) evaluated the capacities for treatment and energy recovery of an MFC and MEC applied to winery and domestic wastewater. They concluded that energy recovery and organic removal from wastewater are more effective with MFCs than with MECs but that hydrogen production from wastewater-fed MECs can be cost-effective.

Cusick *et al.* (2011) published the first pilot-scale study carried out with a 1000 L-MEC treating winery wastewater in California. The reactor contained 144 electrode pairs in 24 modules. The development of an electroactive biofilm required approximately 60 days, which is longer than the period typically needed for laboratory-scale MECs. This was mainly due to the lack of volatile fatty acids (VFA) in the feeding, the low pH and the low and unstable temperature. Controlling these different parameters made it possible to enhance the performance in terms of current production, with a maximum current of 7.4 A/m^3 at the end of the test after 100 days of operation at a cathodic potential of 0.9 V. The maximum gas production was 0.19 ± 0.04 l/l·day. However, in the absence of a membrane, the hydrogen was directly converted to methane (86 ± 6% of product gas) by hydrogenotrophic methanogens.

In 2011, a 120 L-MEC was set up by Tom Curtis' group (Newcastle University, UK) in a municipal wastewater treatment plant in northern England. It was fed with raw wastewater at ambient temperatures. The organic loading rate was 0.14 kg COD/m^3·day, which is just below the typical loading rates for activated sludge. The energetic cost was 2.3 kJ/g DOD, which is below the values for activated sludge (2.5–7.2 kJ/g COD) (Heidrich *et al.*, 2013). The reactor maintained performance for 12 months (including over winter with influent temperatures as low as 6°C) achieving 70% electrical energy recovery and producing more than 1 litre of almost 100% pure hydrogen gas per day (Cotterill *et al.*, 2013). This work represents the first proof-of-concept demonstration of the feasibility of domestic wastewater treatment in a large-scale MEC at an ambient temperature.

Due to these results, a new prototype was developed and installed in another wastewater treatment plant in the north of England. The process was modified to reduce the hydrogen losses and to optimize the fluxes to prevent the bypass of the anode biofilm and sludge accumulation. The anode surface area was increased by 90% for the same volume compared to the previous prototype (Cotterill *et al.*, 2013). As in the previous pilot-reactor, the MEC was made of removable cassettes placed in a tank, following a design focused on improving the service ability of the reactor, minimizing hydrogen losses and improving flow distribution. Each cassette consisted of a carbon felt anode and a stainless steel wire wool cathode, with a battery separator instead of the expensive Nafion membrane generally used in laboratory-scale systems (Cotterill *et al.*, 2013).

Escapa *et al.* (2012) studied the potential inclusion of a MEC in an existing domestic wastewater treatment plant from an economic point of view. They estimated that for a full-scale MEC operating at a current density of 5 A/m^2 anode and an energy consumption of 0.9 kWh/kg COD, an anodic chamber cost of 1220 €/m^3 is the target purchase cost for which a break-even point can be reached after 7 years.

The main challenges for the development of MEC technology are enhancing the hydrogen-production rate and lowering the energy input (Liu *et al.*, 2010). This includes increasing the performance of the anodic biofilms in terms of current density, the development of efficient cathode electrode materials and novel MEC architecture to overcome the high cathode overpotential and the large internal resistance caused by the neutral pH conditions. In addition to the domestic wastewater, the development of cost-efficient pretreatment should favour the combined use of dark fermentation and MEC applied to high-strength wastewaters to significantly enhance the overall hydrogen-production rate and yield (Liu *et al.*, 2010).

Microbial extracellular electron transfer (EET) mechanisms

Certain microorganisms that are considered strict anaerobes and that are able to reduce metallic oxides such as iron are also able to transfer electrons via extracellular electron transfer (EET) to an electrode material (exoelectrogen). However, such metal oxide reduction ability does not necessarily confer EET ability (Richter *et al.*, 2007). Due to the original focus on the improvement of MFC performance, EET mechanisms have long remained unknown, although they represent a critical step in understanding the power generation phenomenon in BESs. Fortunately, advances in electrochemical techniques have clarified part of these EET mechanisms (Manohar *et al.*, 2008) through the size of the polarization curves (Logan, 2008a). Cyclic voltammetry (Harnisch and Freguia, 2012), electrochemical impedance spectroscopy (Dominguez-Benetton *et al.*, 2012) and surface-enhanced Resonance Raman Spectroscopy (Millo, 2012) are all techniques used to understand the EET mechanisms and to allow for BES characterization (Franks and Nevin, 2010).

Recent biocathode studies have shown that electrochemically active bioanodes may be turned into biocathodes when the environmental and operation conditions change (Rozendal *et al.*, 2008). According to Rozendal *et al.* (2008), similar EET mechanisms as those occurring in bioanodes could occur in biocathodes. However, this hypothesis remains to be experimentally tested. One of the main differences between anodic and cathodic EET mechanisms is that the redox active components could operate at higher redox potentials (Arends *et al.*, 2011).

Due to the knowledge generated by these techniques, scientists have been able to distinguish between different EET mechanisms, such as direct or indirect extracellular electron transfer. Interestingly, in mixed microbial biofilms, both mechanisms are believed to occur simultaneously in order to maximize the microbial benefits (Logan *et al.*, 2006).

Direct electron transfer mechanisms

Direct electron transfer (DET) is defined as the transport of electrons from the cofactor of a redox active enzyme (oxidoreductase) or a redox protein in the bacterial cell membrane to the electrode surface in the absence of redox mediators (Logan *et al.*, 2006). Thus, the main advantage of direct electron transport consists of the absence of diffusion limitations between microorganisms and the electrode (Rabaey and Rozendal, 2010).

Initial investigations of microbial DET were based on pure cultures. The microorganisms selected for this type of studies were *Geobacter sulfurreducens* (Gregory *et al.*, 2004; Bond and Lovley, 2003) and *Shewanella oneidensis* (Gorby *et al.*, 2006) because they are well known for being dissimilatory metal reducing organisms in nature (Lovley *et al.*, 2004; Nealson *et al.*, 1991). Studies on EET have focused on Gram-negative bacteria because they have shown more efficient EET capacities in terms of their achieved currents (1 A/m^2) and greater EET diversity in both direct and mediated mechanisms (Arends *et al.*, 2011). However, *Thermincola ferriacetica* has recently been reported to produce current values of 8 A/m^2, indicating that Gram-positive bacteria can also efficiently conduct electrons via DET (Parameswaran *et al.*, 2013).

Microorganisms use three different direct electron transfer mechanisms, including the following: (1) direct contact between the membrane and the electrode, (2) conductive biofilms, and (3) conductive pili called nanowires.

Single-cell direct contact

Exoelectrogen microorganisms can attach to the electrode surface. *Geobacter sulfurreducens* is a clear example of this transfer mechanism. It has been demonstrated by spectroelectrochemical studies that electron transfer occur via c-type cytochromes displayed on the outer cell surface (Millo *et al.*, 2011; Busalmen *et al.*, 2008). Furthermore, *G. sulfurreducens* grows in layers, which allows for close contact between cells (Bond and Lovley, 2003) (Fig. 10.2).

On the other hand, *Shewanella oneidensis* is a well-known example of another anode respiring

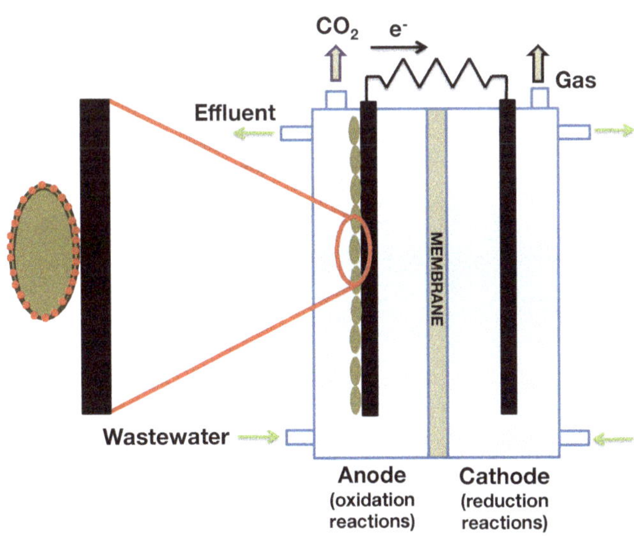

Figure 10.2 Scheme of direct electron transport mechanism. Small red circles represent c -type cytochromes. A colour characteristic of electroactive biofilms due to their high content of haems (Jensen *et al.*, 2010).

bacterium. Interestingly, it adheres to its substrate five times stronger in anaerobic conditions than in aerobic conditions, indicating that direct contact is a good fixation strategy for EET in adverse conditions. Therefore, based on such findings, EET studies between electrodes and microbes were initially accomplished under anaerobic conditions (Kim *et al.*, 1999).

Conductive biofilm

As mentioned above, exoelectrogen microorganisms within conductive biofilms attach to each other by EPS (Fig. 10.3). This strategy results in the formation of thick and conductive biofilm layers that allow for high current productions in BES ($1 A/m^2$). For instance, it has been recently demonstrated that in pure-culture biofilms of *Geobacter sulfurreducens*, the microorganisms are able to release cytochromes to the matrix (Lovley *et al.*, 2011), thereby increasing the conductivity of the biofilm structure.

Conductive pili (Nanowires)

An additional mechanism of microbial DET in biofilms involves self-produced conductive hair-like filaments. These filaments are produced by some microorganisms, attach to the cell wall and are known as nanowires (Malvankar and Lovley, 2014, 2012). Such nanowires are involved in long-range EET due to the high content of c-type cytochromes in their structure. Additionally, the nanowires are shaped as thin single strands that connect the microorganism to the solid electrode (Fig. 10.4). Therefore, in studies where scanning

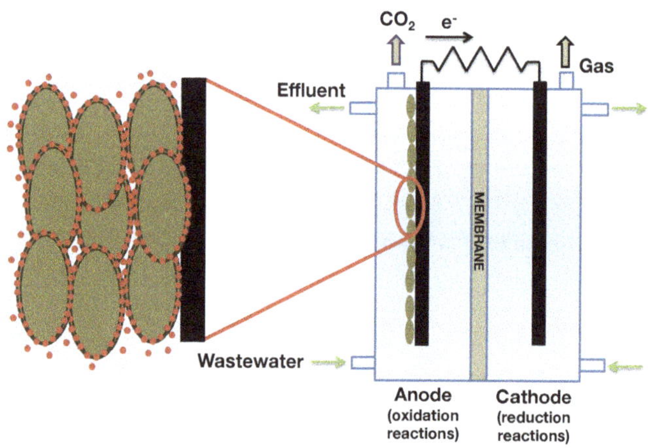

Figure 10.3 Scheme of direct electron transport mechanism of conductive biofilms. Blue matrix: EPS.

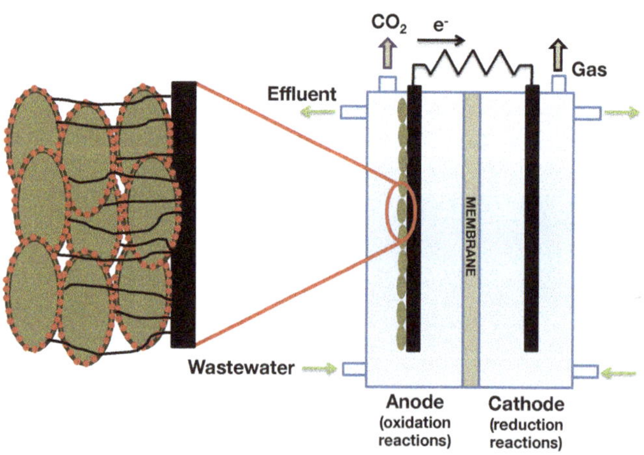

Figure 10.4 Scheme of direct electron transport mechanism by nanowires. Black lines: conductive pili.

electron microscopy is used for the inspection of electroactive biofilms, other web-like extracellular material resulting from the sample preparation could be wrongly identified as nanowires (Badalamenti *et al.*, 2013).

Nanowires have been proposed as a possible EET mechanism that is used by a variety of different microorganisms. For example, *Geobacteraceae* and *Shewanellaceae* species have been studied as nanowire producers by Reguera *et al.* (2006) and Gorby *et al.* (2006), respectively. Later, the work of Malvankar and Lovley (2014) has revealed that the function of the nanowire in these two organisms is different. While the nanowires in *Geobacter sulfurreducens* have metal conductivity and transfer electrons, the electrons seem to 'jump' between cytochromes located on the non-conductive pili in *Shewanella oneidensis*. This characteristic seems to confer *Geobacter sulfurreducens* a better ability to transfer electrons than *Shewanella oneidensis*. Moreover, *G. sulfurreducens* is able to produce a much thicker and uniform biofilm layer (> 50 mm) than *S. oneidensis*, which further supports experimental observations of *G. sulfurreducens* as a prominent electroactive bacteria, as both strains have been compared for current/power production in identical experimental set-ups (Call and Logan, 2011).

Indirect electron transfer mechanisms

Indirect electron transfer (IET) is defined as the transport of electrons from the bacterial cell to the electrode surface or from the cell to another bacterial cell through redox mediators (Rabaey and Rozendal, 2010) (Fig. 10.5).

These mediators could be chemically added or could be self-produced by the microorganisms as secondary metabolites (Arends *et al.*, 2011) (Table 10.4). Pyocyanin is one of the endogenous chemical mediators produced by *Pseudomonas aeruginosa* described in BES (Rabaey *et al.*, 2004).

A variety of chemicals have also been used to facilitate IET. These mediators include neutral red (Park *et al.*, 1999), anthraquinone-2–6, disulfonate (AQDS) (Holmes *et al.*, 2004), thionin, potassium ferricyanide (Bond *et al.*, 2002), methyl viologen, and others (Rabaey and Verstraete, 2005). The main problems of the chemical additions are related to the high cost of the mediators, their possible toxic effects and their wash-out effects.

Within the context of anodic biofilms, it is worth noting that the study of DET mechanisms has gained more attention over IET mechanisms. This has resulted in the partial abandon of IET by the scientific community. However, recent preliminary studies on microbial electrosynthesis of acetate at the cathode and the absence of evident biofilms suggest that an electron uptake based on IET mechanism might be responsible for the observed conversion of CO_2 into acetate (Arends *et al.*, 2013). Therefore, it is expected that the interest in IET will soon increase rapidly.

As in the case of anode EET mechanisms, cathode IET requires mediators to carry out the

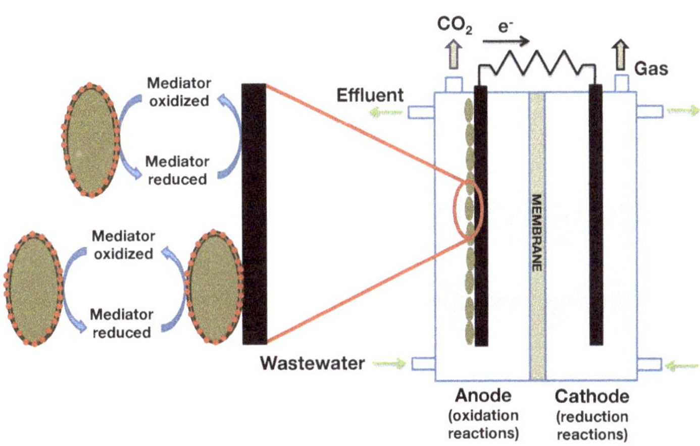

Figure 10.5 Scheme of indirect electron transport mechanism.

Table 10.4 Representative biogenic production of redox mediators

Microorganism	Mediator molecule	Reference
Sphingomonas xenophaga	4-Amino-1,2-naphthoquinone	Keck *et al.* (2002)
Pseudomonas aeruginosa	Phenazine-1-carboxylic acid	Price-Whelan *et al.* (2006)
Pseudomonas chlororaphis	Phenazine-1-carboxamide	van Rij *et al.* (2004)
Shewanella oneidensis	Flavin mononucleotide	von Canstein *et al.* (2008)
Shewanella algae	Melanin	Turick *et al.* (2002)
Bacillus pyocyaneus	Pyocyanine	Friedheim and Michaelis (1931)
Propionibacterium freundenreichii	2-Amino-3-carboxy-1,4-naphthoquinone	Hernandez and Newman (2001)
Shewanella alga	Cyanocobalamin	Workman *et al.* (1997)
Acetobacterium woodii	Hydroxycobalamin	Hashsham and Freedman (1999)
Pseudomonas stutzeri	Pyridine-2,6-bis	Lewis *et al.* (2001)
Methanosarcina thermophila	Porphorinogen-type molecules	Koons *et al.* (2001)
Shewanella oneidensis	1,4-Dihydroxy-2-naphthoate derivative	Ward *et al.* (2004)

More detailed information can be found in the following references: Hernandez and Newman, 2001; Li *et al.*, 2009; Marsili and Zhang, 2009; Schröder, 2007; Watanabe *et al.*, 2009.

reduction of certain final electron acceptors (see above). Although little information is available at present, it can be assumed that the mediators responsible for the cathode IET could be artificially added or biologically self-produced as has been described for the anode compartments. One recent study demonstrated the use of methyl viologen as a mediator for the electrochemically assisted microbial dechlorination of trichloroethene (TCE) and cis -dichloroethene (cis -DCE) (Aulenta *et al.*, 2009). Other mediators have been applied to the cathodes, such as anthraquinone-2,6-disulfonate (Thrash *et al.*, 2007) and neutral red (Park and Zeikus, 1999) for the reduction of perchlorate and fumarate, respectively. Taking into account all available information, it is expected that the cathodic IET mechanism will be very similar to the anodic IET (Thrash and Coates, 2008).

Microbial intracellular electron transport (ICET) mechanisms

The internal microbial mechanism to transport and release electrons inside the membranes of microorganisms is currently being researched. The study of these pathways is usually performed with pure cultures. One of the most studied species for ICET is *Geobacter sulfureducens*, which is studied due to its enrichment in multiple BES studies (Yates *et al.*, 2012) and due to its demonstrated ability to produce high current densities (Chen *et al.*, 2012). Despite these efforts, the mechanism of electron transfer remains unclear as unknown pathways exist in the process.

The mechanism that allows electron transfer from electron donor (reductant) to the electron acceptor (oxidant) occurs inside the lipid-membranes and involves the oxidoreductase enzyme (Hartshorne *et al.*, 2009). First, electrons are generated by the NADH dehydrogenases located in the inner mitochondrial membrane (Fig. 10.6). This enzyme catalysis the conversion of NADH to NAD^+ and thereby releases protons to the periplasm zone. NAD^+ is reduced through a route that recycles it back to the active form via the Krebs cycle.

Electrons are transferred to the cytochrome b-c by the coenzyme ubiquinone. Coenzyme ubiquinone, also called Q10, has various functions related to its redox capacity, including electron transport between cytochromes. Once the electrons are in this cytochrome, more protons are released into the intermembrane space. Thereafter, electrons from the cytochrome b-c are transferred to the c-type cytochrome. The C-type cytochrome is mobile, connecting the inner membrane, the periplasm and the outer membrane.

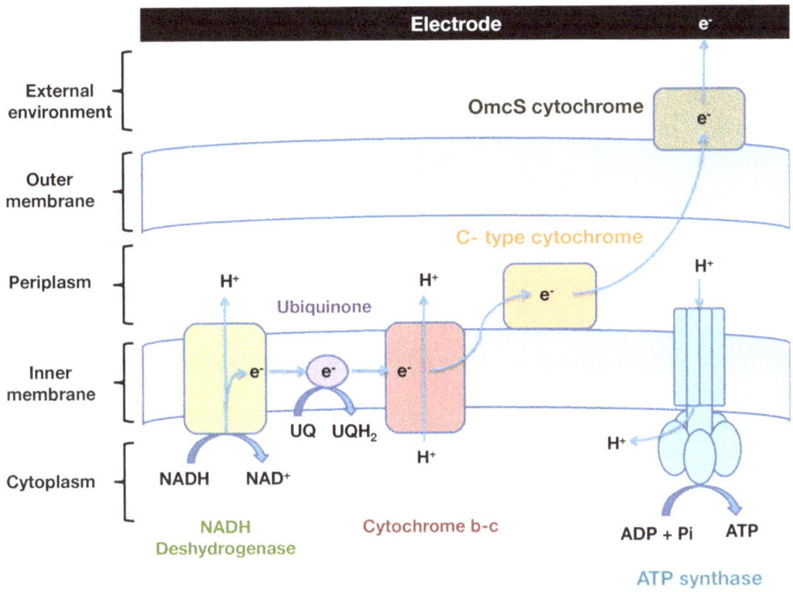

Figure 10.6 Electron transport chain to move electrons directly from the mitochondrial membrane to the external membrane until the electrode.

This cytochrome is responsible for the electron transfer between both membranes (Lovley *et al.*, 2004). Then, electrons are transferred through different cytochromes until they reach the OmcS cytochrome, which is in contact with the surface of the cell (Lovley, 2008b).

During electron transfer, the gradient of protons between the cytoplasm and the periplasm is used to generate energy in the form of adenosine triphosphate (ATP). The accumulation of protons is generated by proton pumping from the different cytochromes, increasing the pH of the periplasm. ATP synthase pumps these protons across the inner membrane and thereby produces ATP from adenosine diphosphate (ADP) and inorganic phosphate. This ability has been intensively studied with the aim of increasing the power production of MFCs. Different genetic engineering modifications have been applied to *Geobacter sulfurreducens* (Lovley *et al.*, 2011): cell abilities to produce more cytochromes, which increase the amount of electron transfer within the membrane and therefore increase the nanowire expression levels. These effects in turn enhance the contact cell–cell and/or cell–electrode contact, but not the current production. To improve the current production, Malvankar *et al.* (2011) increased the biofilm conductivity, observing a positive correlation. These findings highlight the necessity of a better understanding of the processes occurring inside the bacteria.

It is believed that the ability of *Geobacter sulfurreducens* to create very thick biofilms (> 50 mm) through the direct contact mechanism can be substituted by electron transport through conductive pili (Fig. 10.4). Such an additional DET mechanism is actively being discussed by the scientific community. However, recent studies on microbial nanowires suggest that these pili-like structures, which are located along the membranes on the external environment of the cell, are covered by OmcS cytochromes (Leang *et al.*, 2010; Lovley, 2006). Interestingly, this nanowire DET pathway allows the microorganisms to transfer electrons between them without intermediates, enabling biofilm communication (Logan and Regan, 2006).

The mechanisms of electron transfer through the membrane in electrotroph bacteria remain largely unknown. Scientists suspect that the process of moving electrons is similar to the exoelectrogenic process. Ross *et al.* (2011) studied *Shewanella oneidensis* strain MR-1 and demonstrated that the same pathway can be used for oxidation and reduction reactions. The reduction of these compounds consumes protons in the cytoplasm, generating a proton gradient across the inner membrane (Lovley, 2011). Therefore, a proton gradient is necessary to generate energy by ATP synthase, as is the case for exoelectrogens.

Conclusions

This chapter provides a comprehensive description about the microbial aspects of bioelectrochemical systems. The potential of BES technology relies on its use as an alternative, cost-effective technology for the removal of contaminants (organic matter, carbon dioxide, nitrate, metals, chlorinated compounds, etc.) and the production of high value products (hydrogen, caustic soda, methane, volatile fatty acids, alcohols, etc.).

The rapid advances in electric and microbial techniques over the last decade have allowed for the improvement of such systems. Despite this, it remains necessary to gain fundamental and applied knowledge about the microbiological field to better understand the complex metabolic routes of the microorganisms involved. Several opportunities will be available once the wastewater treatment capacity and high-value production rate of BESs is improved.

Acknowledgement

This research was financially supported by the Spanish Government through the BEST ENERGY project (CTQ2011-23632).

References

Aelterman, P., Freguia, S., Keller, J., Verstraete, W., and Rabaey, K. (2008). The anode potential regulates bacterial activity in microbial fuel cells. Appl. Microbiol. Biotechnol. *78*, 409–418.

Aelterman, P., Rabaey, K., Clauwaert, P., and Verstraete, W. (2006). Microbial fuel cells for wastewater treatment. Water Sci. Technol. *54*, 1–9.

Anderson, R.T., Vrionis, H.A., Ortiz-bernad, I., Resch, C.T., Long, P.E., Dayvault, R., Marutzky, S., Metzler, D.R., Peacock, A., White, D.C., Lowe, M., Lovley, D.R., and Karp, K. (2003). Stimulating the In Situ activity of *Geobacter* species to remove uranium from the groundwater of a uranium-contaminated aquifer. Appl. Environ. Microbiol. *69*, 5884–5891.

Arends, J.B.A., Blondeel, E., Tennison, S.R., Boon, N., and Verstraete, W. (2012). Suitability of granular carbon as an anode material for sediment microbial fuel cells. J. Soils Sediments *12*, 1197–1206.

Arends, J.B.A., Boon, N., and Rabaey, K. (2013). Microbial Electrosynthesis on Stainless Steel Electrodes by a Homoacetogenic Community: Are Carbon Electrodes Needed?, in: Microbial Fuel Cells, 4th International Conference, Abstracts. pp. 34–35.

Arends, J.B.A., Desloover, J., Puig, S., and Verstraete, W. (2011). Principles and technology of Microbial Fuel Cells (Wiley-VCH Books, West Sussex, UK).

Arends, J.B.A., and Verstraete, W. (2012). 100 years of microbial electricity production: three concepts for the future. Microb. Biotechnol. *5*, 333–346.

Aulenta, F., Canosa, A., De Roma, L., Reale, P., Panero, S., Rossetti, S., and Majone, M. (2009). Influence of mediator immobilization on the electrochemically assisted microbial dechlorination of trichloroethene (TCE) and cis -dichloroethene (cis -DCE). J. Chem. Technol. Biotechnol. *84*, 864–870.

Aulenta, F., Reale, P., Canosa, A., Rossetti, S., Panero, S., and Majone, M. (2010). Characterization of an electro-active biocathode capable of dechlorinating trichloroethene and cis-dichloroethene to ethene. Biosens. Bioelectron. *25*, 1796–1802.

Badalamenti, J.P., Krajmalnik-brown, R., and Torres, C.I. (2013). Generation of high current densities by pure cultures of Anode- Respiring *Geoalkalibacter* spp. under alkaline and saline conditions in Microbial Electrochemical Cells. Am. Soc. Microbiol. *4*, e00144–e00213.

Batlle-Vilanova, P., Puig, S., Gonzalez-Olmos, R., Vilajeliu-Pons, A., Bañeras, L., Balaguer, M.D., and Colprim, J. (2014). Assessment of biotic and abiotic graphite cathodes for hydrogen production in microbial electrolysis cells. Int. J. Hydrogen Energy *39*, 1297–1305.

Behera, M., and Ghangrekar, M.M. (2009). Performance of microbial fuel cell in response to change in sludge loading rate at different anodic feed pH. Bioresour. Technol. *100*, 5114–5121.

Bond, D.R., and Lovley, D.R. (2003). Electricity production by *Geobacter sulfurreducens* attached to electrodes. Appl. Environ. Microbiol. *69*, 1548–1555.

Bond, D.R., Holmes, D.E., Tender, L.M., and Lovley, D.R. (2002). Electrode-reducing microorganisms that harvest energy from marine sediments. Science *295*, 483–485.

Busalmen, J.P., Esteve-Núñez, A., Berná, A., and Feliu, J.M. (2008). C-type cytochromes wire electricity-producing bacteria to electrodes. Angew. Chem. *47*, 4874–4877.

Call, D.F., and Logan, B.E. (2011). Lactate oxidation coupled to iron or electrode reduction by *Geobacter sulfurreducens* PCA. Appl. Environ. Microbiol. *77*, 8791–8794.

Carmona-Martínez, A.A., Pierra, M., Trably, E., and Bernet, N. (2013). High current density via direct electron transfer by the halophilic anode respiring bacterium *Geoalkalibacter subterraneus*. Phys. Chem. Chem. Phys. *15*, 19699–19707.

Chang, I.S., Moon, H., Bretschger, O., Jang, J.K., Park, H. Il, Nealson, K.H., and Kim, B.H. (2006). Electrochemically active bacteria (EAB) and mediator-less microbial fuel cells. J. Microbiol. Biotechnol. *16*, 163–177.

Chen, S., He, G., Liu, Q., Harnisch, F., Zhou, Y., Chen, Y., Hanif, M., Wang, S., Peng, X., Hou, H., and Schröder, U. (2012). Layered corrugated electrode macrostructures boost microbial bioelectrocatalysis. Energy Environ. Sci. 5, 9769–9774.

Cohen, B. (1931). J. Bacteriol. *21*, 18–19.

Coma, M., Puig, S., Pous, N., Balaguer, M.D., and Colprim, J. (2013). Biocatalysed sulphate removal in a BES cathode. Bioresour. Technol. *130*, 218–223.

Cotterill, S., Dolfing, J., Jones, C., and Curtis, T.P. (2013). Plant-scale development of a Microbial Electrolysis Cell, in: Microbial Fuel Cells, 4th International Conference, Abstracts. pp. 218–223.

Cusick, R.D., Bryan, B., Parker, D.S., Merrill, M.D., Mehanna, M., Kiely, P.D., Liu, G.,and Logan, B.E. (2011). Performance of a pilot-scale continuous flow microbial electrolysis cell fed winery wastewater. Appl. Microbiol. Biotechnol. *89*, 2053–2063.

Cusick, R.D., Kiely, P.D., and Logan, B.E. (2010). A monetary comparison of energy recovered from microbial fuel cells and microbial electrolysis cells fed winery or domestic wastewaters. Int. J. Hydrogen Energy *35*, 8855–8861.

Davey, M.E., and George, A.O. (2000). Microbial Biofilms: from Ecology to Molecular Genetics. Microbiol. Mol. Biol. Rev. *64*, 847–867.

Debabov, V.G. (2008). Electricity from microorganisms. Microbiology 77, 123–131.

Dominguez-Benetton, X., Sevda, S., Vanbroekhoven, K., and Pant, D. (2012). The accurate use of impedance analysis for the study of microbial electrochemical systems. Chem. Soc. Rev. *41*, 7228–7246.

Escapa, A., Gómez, X., Tartakovsky, B., and Morán, A. (2012). Estimating microbial electrolysis cell (MEC) investment costs in wastewater treatment plants: Case study. Int. J. Hydrogen Energy *37*, 18641–18653.

Franks, A.E., Malvankar, N., and Nevin, K.P. (2010). Bacterial biofilms: the powerhouse of a microbial fuel cell. Biofuels *1*, 589–604.

Franks, A.E., and Nevin, K.P. (2010). Microbial Fuel Cells, A Current Review. Energies 3, 899–919.

Freguia, S., Rabaey, K., Yuan, Z., and Keller, J. (2008). Syntrophic processes drive the conversion of glucose in Microbial Fuel Cell anodes. Environ. Sci. Technol. *42*, 7937–7943.

Friedheim, E., and Michaelis, L. (1931). Potentiometric study of piocyanine. J. Biol. Chem. *91*, 355–368.

Gorby, Y.A., Yanina, S., McLean, J.S., Rosso, K.M., Moyles, D., Dohnalkova, A., Beveridge, T.J., Chang, I.S., Kim, B.H., Kim, K.S., *et al.* (2006). Electrically conductive bacterial nanowires produced by *Shewanella oneidensis* strain MR-1 and other microorganisms. Proc. Natl. Acad. Sci. U.S.A. *103*, 11358–11363.

Greenman, J., Gálvez, A., Giusti, L., and Ieropoulos, I. (2009). Electricity from landfill leachate using microbial fuel cells: Comparison with a biological aerated filter. Enzyme Microb. Technol. *44*, 112–119.

Gregory, K.B., Bond, D.R., and Lovley, D.R. (2004). Graphite electrodes as electron donors for anaerobic respiration. Environ. Microbiol. *6*, 596–604.

Gregory, K.B., and Lovley, D.R. (2005). Remediation and recovery of uranium from contaminated subsurface environments with electrodes. Environ. Sci. Technol. *39*, 8943–8947.

Guo, K., Freguia, S., Dennis, P.G., Chen, X., Donose, B.C., Keller, J., Gooding, J.J., and Rabaey, K. (2013). Effects of surface charge and hydrophobicity on anodic biofilm formation, community composition, and current generation in Bioelectrochemical Systems. Environ. Sci. Technol. *47*, 7563–7570.

Harnisch, F., and Freguia, S. (2012). A basic tutorial on cyclic voltammetry for the investigation of electroactive microbial biofilms. Chem. Asian J. 7, 466–475.

Harnisch, F., Koch, C., Patil, S.A., Hübschmann, T., Müller, S., and Schröder, U. (2011). Revealing the electrochemically driven selection in natural community derived microbial biofilms using flow-cytometry. Energy Environ. Sci. *4*, 1265–1272.

Hartshorne, R.S., Reardon, C.L., Ross, D., Nuester, J., Clarke, T.A., Gates, A.J., Mills, P.C., Fredrickson, J.K., Zachara, J.M., Shi, L., *et al.* (2009). Characterization of an electron conduit between bacteria and the extracellular environment. Proc. Natl. Acad. Sci. U.S.A. *106*, 22169–22174.

Hashsham, S.A., and Freedman, D.L. (1999). Enhanced biotransformation of carbon tetrachloride by *Acetobacterium woodii* upon addition of hydroxocobalamin and fructose. Appl. Environ. Microbiol. *65*, 4537–4542.

He, W., Feng, Y., Liu, J., Wang, X., Qu, Y., and Ren, N. (2014). A horizontal plug flow and stackable pilot microbial fuel cell for municipal wastewater treatment. Bioresour. Technol. *156*, 132–138.

Heidrich, E.S., Dolfing, J., Scott, K., Edwards, S.R., Jones, C., and Curtis, T.P. (2013). Production of hydrogen from domestic wastewater in a pilot-scale microbial electrolysis cell. Appl. Microbiol. Biotechnol. *97*, 6979–6989.

Hernandez, M.E., and Newman, D.K. (2001). Cellular and Molecular Life Sciences Extracellular electron transfer. Cell. Mol. Life Sci. *58*, 1562–1571.

Holmes, D.E., Bond, D.R., and Lovley, D.R. (2004). Electron Transfer by *Desulfobulbus propionicus* to Fe (III) and Graphite Electrodes. Appl. Environ. Microbiol. *70*, 1234–1238.

Holmes, D.E., Chaudhuri, S.K., Nevin, K.P., Mehta, T., Methé, B.A., Liu, A., Ward, J.E., Woodard, T.L., Webster, J., and Lovley, D.R. (2006). Microarray and genetic analysis of electron transfer to electrodes in *Geobacter sulfurreducens*. Environ. Microbiol. *8*, 1805–1815.

Ieropoulos, I., Greenman, J., and Melhuish, C. (2012). Urine utilisation by microbial fuel cells; energy fuel for the future. Phys. Chem. Chem. Phys. *14*, 94–98.

Jia, J., Tang, Y., Liu, B., Wu, D., Ren, N., and Xing, D. (2013). Electricity generation from food wastes and microbial community structure in microbial fuel cells. Bioresour. Technol. *144*, 94–99.

Jiang, Y., Su, M., Zhang, Y., Zhan, G., Tao, Y., and Li, D. (2013). Bioelectrochemical systems for simultaneously production of methane and acetate from carbon dioxide at relatively high rate. Int. J. Hydrogen Energy *38*, 3497–3502.

Keck, A., Rau, J., Reemtsma, T., Mattes, R., Stolz, A., and Klein, J. (2002). Identification of quinoide redox mediators that are formed during the degradation of Naphthalene-2-Sulfonate by *Sphingomonas xenophaga* BN6. Appl. Environ. Microbiol. *68*, 4341–4349.

Kiely, P.D., Regan, J.M., and Logan, B.E. (2011). The electric picnic: synergistic requirements for exoelectrogenic microbial communities. Curr. Opin. Biotechnol. *22*, 378–385.

Kim, B.H., Ikeda, T., Park, H.S., Kim, H.J., Hyun, M.S., Kano, K., Takagi, K., and Tatsumi, H. (1999). Electrochemical activity of an Fe (III)-reducing bacterium, *Shewanella putrefaciens* IR-1, in the presence of alternative electron acceptors *200*, 475–478.

Kim, J.R., Rodríguez, J., Hawkes, F.R., Dinsdale, R.M., Guwy, A.J., and Premier, G.C. (2011). Increasing power recovery and organic removal efficiency using extended longitudinal tubular microbial fuel cell (MFC) reactors. Energy Environ. Sci. *4*, 459–464.

Koons, B.W., Baeseman, J.L., and Novak, P.J. (2001). Investigation of cell exudates active in carbon tetrachloride and chloroform degradation. Biotechnol. Bioeng. *74*, 12–17.

Lalaurette, E., Thammannagowda, S., Mohagheghi, A., Maness, P.-C., and Logan, B.E. (2009). Hydrogen production from cellulose in a two-stage process combining fermentation and electrohydrogenesis. Int. J. Hydrogen Energy *34*, 6201–6210.

Leang, C., Qian, X., Mester, T., and Lovley, D.R. (2010). Alignment of the c-type cytochrome OmcS along pili of *Geobacter sulfurreducens*. Appl. Environ. Microbiol. *76*, 4080–4084.

Lewis, T.A., Paszczynski, A., Gordon-Wylie, S.W., Jeedigunta, S., Lee, C.H., and Crawford, R.L. (2001). Carbon tetrachloride dechlorination by the bacterial transition metal chelator pyridine-2,6-bis(thiocarboxylic acid). Environ. Sci. Technol. *35*, 552–559.

Li, X.M., Zhou, S.G., Li, F.B., Wu, C.Y., Zhuang, L., Xu, W., and Liu, L. (2009). Fe(III) oxide reduction and carbon tetrachloride dechlorination by a newly isolated *Klebsiella pneumoniae* strain L17. J. Appl. Microbiol. *106*, 130–139.

Liamleam, W., and Annachhatre, A.P. (2007). Electron donors for biological sulfate reduction. Biotechnol. Adv. *25*, 452–463.

Lim, S.J., Park, W., Kim, T.-H., and Shin, I.H. (2012). Swine wastewater treatment using a unique sequence of ion exchange membranes and bioelectrochemical system. Bioresour. Technol. *118*, 163–169.

Liu, H., Hu, H., Chignell, J., and Fan, Y. (2010). Microbial electrolysis: novel technology for hydrogen production from biomass. Biofuels *1*, 129–142.

Liu, H., and Logan, B.E. (2004). Electricity generation using an air-cathode single chamber microbial fuel cell in the presence and absence of a proton exchange membrane. Environ. Sci. Technol. *38*, 4040–4046.

Logan, B.E. (2008a). Power generation, in: Microbial Fuel Cells, 4th International Conference, Abstracts. pp. 44–61.

Logan, B.E. (2008b). Microbial Fuel Cells (New Jesey, USA: John Wiley and Sons, Inc.).

Logan, B.E., and Rabaey, K. (2012). Conversion of wastes into bioelectricity and chemicals by using microbial electrochemical technologies. Science *337*, 686–690.

Logan, B.E., and Regan, J.M. (2006). Electricity-producing bacterial communities in microbial fuel cells. Trends Microbiol. *14*, 512–518.

Logan, B.E., Hamelers, B., Rozendal, R., Schröder, U., Keller, J., Freguia, S., Aelterman, P., Verstraete, W., and Rabaey, K. (2006). Microbial fuel cells: methodology and technology. Environ. Sci. Technol. *40*, 5181–5192.

Logan, B., Cheng, S., Watson, V., and Estadt, G. (2007). Graphite fiber brush anodes for increased power production in air-cathode microbial fuel cells. Environ. Sci. Technol. *41*, 3341–3346.

Lovley, D.R. (2006). Bug juice: harvesting electricity with microorganisms. Nat. Rev. Microbiol. *4*, 497–508.

Lovley, D.R. (2008a). Extracellular electron transfer: wires, capacitors, iron lungs, and more. Geobiology *6*, 225–231.

Lovley, D.R. (2008b). The microbe electric: conversion of organic matter to electricity. Curr. Opin. Biotechnol. *19*, 564–571.

Lovley, D.R. (2011). Powering microbes with electricity: direct electron transfer from electrodes to microbes. Environ. Microbiol. Rep. *3*, 27–35.

Lovley, D.R., Holmes, D.E., and Nevin, K.P. (2004). Dissimilatory Fe (III) and Mn (IV) reduction. Adv. Microb. Physiol. *49*, 219–286.

Lovley, D.R., Ueki, T., Zhang, T., Malvankar, N.S., Shrestha, P.M., Flanagan, K.A., Aklujkar, M., Butler, J.E., Giloteaux, L., Rotaru, A.-E., *et al.* (2011). *Geobacter*: the microbe electric's physiology, ecology, and practical applications. Adv. Microb. Physiol. *59*, 1–100.

Malvankar, N.S., and Lovley, D.R. (2012). Microbial nanowires: a new paradigm for biological electron transfer and bioelectronics. ChemSusChem *5*, 1039–1046.

Malvankar, N.S., and Lovley, D.R. (2014). Microbial nanowires for bioenergy applications. Curr. Opin. Biotechnol. *27C*, 88–95.

Malvankar, N.S., Vargas, M., Nevin, K.P., Franks, A.E., Leang, C., Kim, B.-C., Inoue, K., Mester, T., Covalla, S.F., Johnson, J.P., Rotello, V.M., Tuominen, M.T., and Lovley, D.R. (2011). Tunable metallic-like conductivity in microbial nanowire networks. Nat. Nanotechnol. *6*, 573–579.

Malvankar, N.S., Lau, J., Nevin, K.P., Franks, A.E., Tuominen, M.T., and Lovley, D.R. (2012). Electrical conductivity in a mixed-species biofilm. Appl. Environ. Microbiol. *78*, 5967–5971.

Manohar, A.K., Bretschger, O., Nealson, K.H., and Mansfeld, F. (2008). The use of electrochemical impedance spectroscopy (EIS) for the evaluation of the electrochemical properties of bioelectrochemical systems. Bioelectrochemistry *72*, 149–154.

Marshall, C.W., LaBelle, E.V., and May, H.D. (2013). Production of fuels and chemicals from waste by microbiomes. Curr. Opin. Biotechnol. *24*, 391–397.

Marshall, C.W., Ross, D.E., Fichot, E.B., Norman, R.S., and May, H.D. (2012). Electrosynthesis of commodity chemicals by an autotrophic microbial community. Appl. Environ. Microbiol. *78*, 8412–8420.

Marsili, E., and Zhang, X. (2009). Shuttling via soluble compounds. In Bioelectrochemical Systems: From Extracellular Electron Transfer to Biotechnological Application, Rabaey, K., Angenent, L., Schroder, U., and Keller, K., eds. (IWA Publishing, London, UK), pp. 59–75.

McInerney, M.J., Struchtemeyer, C.G., Sieber, J., Mouttaki, H., Stams, A.J.M., Schink, B., Rohlin, L., and Gunsalus, R.P. (2008). Physiology, ecology, phylogeny, and genomics of microorganisms capable of syntrophic metabolism. Ann. N.Y. Acad. Sci. *1125*, 58–72.

Methé, B.A., Nelson, K.E., Eisen, J.A., Paulsen, I.T., Nelson, W., Heidelberg, J.F., Wu, D., Wu, M., Ward, N., Beanan, M.J., *et al.* (2003). Genome of *Geobacter sulfurreducens*: metal reduction in subsurface environments. Science *302*, 1967–1969.

Miceli, J.F., Garcia-Peña, I., Parameswaran, P., Torres, C.I., and Krajmalnik-Brown, R. (2014). Combining microbial cultures for efficient production of electricity from butyrate in a microbial electrochemical cell. Bioresour. Technol. *169C*, 169–174.

Millo, D. (2012). Spectroelectrochemical analyses of electroactive microbial biofilms. Biochem. Soc. Trans. *40*, 1284–1290.

Millo, D., Harnisch, F., Patil, S.A., Ly, H.K., Schröder, U., and Hildebrandt, P. (2011). In situ spectroelectrochemical investigation of electrocatalytic microbial biofilms by surface-enhanced resonance Raman spectroscopy. Angew. Chem. Int. Ed. Engl. *50*, 2625–2627.

Min, B., Kim, J., Oh, S., Regan, J.M., and Logan, B.E. (2005). Electricity generation from swine wastewater using microbial fuel cells. Water Res. *39*, 4961–4968.

Nealson, K.H., Myers, C.R., and Wimpee, B.B. (1991). Isolation and identification of manganese-reducing bacteria and estimates of microbial Mn (IV)-reducing potential in the Black Sea. Deep Sea Res. Part A. Oceanogr. Res. Pap. *38*, S907–S920.

Nevin, K.P., Hensley, S. a, Franks, A.E., Summers, Z.M., Ou, J., Woodard, T.L., Snoeyenbos-West, O.L., and Lovley, D.R. (2011). Electrosynthesis of organic compounds from carbon dioxide is catalyzed by a diversity of acetogenic microorganisms. Appl. Environ. Microbiol. *77*, 2882–2886.

Nevin, K.P., Richter, H., Covalla, S.F., Johnson, J.P., Woodard, T.L., Orloff, A.L., Jia, H., Zhang, M., and Lovley, D.R. (2008). Power output and columbic efficiencies from biofilms of *Geobacter sulfurreducens* comparable to mixed community microbial fuel cells. Environ. Microbiol. *10*, 2505–2514.

Nevin, K.P., Woodard, T.L., and Franks, A.E. (2010). Microbial Electrosynthesis: Feeding microbes electricity to convert carbon dioxide and water to multicarbon extracellular organic compounds. Am. Soc. Microbiol. *1*, e00103–e00110.

Nielsen, J.L., Wagner, M., and Nielsen, P.H. (2003). Use of microautoradiography to study *in situ* physiology of bacteria in biofilms. Environ. Sci. Biotechnol. *2*, 261–268.

Oh, S.E., and Logan, B.E. (2005). Hydrogen and electricity production from a food processing wastewater using fermentation and microbial fuel cell technologies. Water Res. *39*, 4673–4682.

Parameswaran, P., Bry, T., Popat, S.C., Lusk, B.G., Rittmann, B.E., and Torres, C.I. (2013). Kinetic, electrochemical, and microscopic characterization of the thermophilic, anode-respiring bacterium *Thermincola ferriacetica*. Environ. Sci. Technol. *47*, 4934–4940.

Parameswaran, P., Torres, C.I., Lee, H.-S., Krajmalnik-Brown, R., and Rittmann, B.E. (2009). Syntrophic interactions among anode respiring bacteria (ARB) and Non-ARB in a biofilm anode: electron balances. Biotechnol. Bioeng. *103*, 513–523.

Park, D.H., Laivenieks, M., Guettler, M.V., Jain, M.K., and Zeikus, J.G. (1999). Microbial utilization of electrically reduced neutral red as the sole electron donor for growth and metabolite production. Appl. Environ. Microbiol. *65*, 2912–2917.

Park, D.H., and Zeikus, J.G. (1999). Utilization of electrically reduced neutral red by *Actinobacillus succinogenes*: Physiological function of neutral red in membrane-driven fumarate reduction and energy conservation. J. Bacteriol. *181*, 2403–2411.

Park, D.H., and Zeikus, J.G. (2003). Improved fuel cell and electrode designs for producing electricity from microbial degradation. Biotechnol. Bioeng. *81*, 348–355.

Patil, S.A., Hägerhäll, C., and Gorton, L. (2012). Electron transfer mechanisms between microorganisms and electrodes in bioelectrochemical systems. Bioanal. Rev. *4*, 159–192.

Potter, M.C. (1911). Electrical effects accompanying the decomposition of organic compounds. Proc. R. Soc. London. Ser. B. *84*, 260–276.

Pous, N., Puig, S., Coma, M., Balaguer, M.D., and Colprim, J. (2013). Bioremediation of nitrate-polluted groundwater in a microbial fuel cell. J. Chem. Technol. Biotechnol. *88*, 1690–1696.

Price-Whelan, A., Dietrich, L., and Newman, D. (2006). Rethinking 'secondary' metabolism: physiological roles for phenazine antibiotics. Nat. Chem. Biol. *2*, 221–225.

Puig, S., Coma, M., Desloover, J., Boon, N., Colprim, J., and Balaguer, M.D. (2012). Autotrophic denitrification in microbial fuel cells treating low ionic strength waters. Environ. Sci. Technol. *46*, 2309–2315.

Puig, S., Serra, M., Coma, M., Balaguer, M.D., and Colprim, J. (2011). Simultaneous domestic wastewater treatment and renewable energy production using microbial fuel cells (MFCs). Water Sci. Technol. *64*, 904–910.

Puig, S., Serra, M., Coma, M., Cabré, M., Balaguer, M.D., and Colprim, J. (2010). Effect of pH on nutrient dynamics and electricity production using microbial fuel cells. Bioresour. Technol. *101*, 9594–9599.

Puig, S., Serra, M., Coma, M., Cabré, M., Balaguer, M.D., and Colprim, J. (2011a). Microbial fuel cell application in landfill leachate treatment. J. Hazard. Mater. *185*, 763–767.

Puig, S., Serra, M., Vilar-Sanz, A., Cabré, M., Bañeras, L., Colprim, J., and Balaguer, M.D. (2011b). Autotrophic nitrite removal in the cathode of microbial fuel cells. Bioresour. Technol. *102*, 4462–4467.

Rabaey, K., Boon, N., Siciliano, S.D., Verstraete, W., and Verhaege, M. (2004). Biofuel cells select for microbial consortia that self-mediate electron transfer. Appl. Environ. Microbiol. *70*, 5373–5383.

Rabaey, K., Clauwaert, P., Aelterman, P., and Verstraete, W. (2005). Tubular microbial fuel cells for efficient electricity generation. Environ. Sci. Technol. *39*, 8077–8082.

Rabaey, K., Lissens, G., Siciliano, S.D., and Verstraete, W. (2003). A microbial fuel cell capable of converting glucose to electricity at high rate and efficiency. Biotechnol. Lett. *25*, 1531–1535.

Rabaey, K., and Rozendal, R.A. (2010). Microbial electrosynthesis – revisiting the electrical route for microbial production. Nat. Rev. Microbiol. *8*, 706–716.

Rabaey, K., Van der Sompel, K., Maignien, L., Boon, N., Aelterman, P., Clauwaert, P., De Schamphelaire, L., Pham, H.T., Vermeulen, J., Verhaege, M., Lens, P., and Verstraete, W. (2006). Microbial Fuel Cells for sulfide removal. Environ. Sci. Technol. *40*, 5218–5224.

Rabaey, K., and Verstraete, W. (2005). Microbial fuel cells: novel biotechnology for energy generation. Trends Biotechnol. *23*, 291–298.

Reguera, G., McCarthy, K.D., Mehta, T., Nicoll, J.S., Tuominen, M.T., and Lovley, D.R. (2005). Extracellular electron transfer via microbial nanowires. Nature *435*, 1098–1101.

Reguera, G., Nevin, K.P., Nicoll, J.S., Covalla, S.F., Woodard, T.L., and Lovley, D.R. (2006). Biofilm and nanowire production leads to increased current in *Geobacter sulfurreducens* fuel cells. Appl. Environ. Microbiol. *72*, 7345–7348.

Reimers, C.E., Tender, L.M., Fertig, S., and Wang, W. (2001). Harvesting energy from the marine sediment–water interface. Environ. Sci. Technol. *35*, 192–195.

Richter, H., Lanthier, M., Nevin, K.P., and Lovley, D.R. (2007). Lack of electricity production by *Pelobacter carbinolicus* indicates that the capacity for Fe(III) oxide reduction does not necessarily confer electron transfer ability to fuel cell anodes. Appl. Environ. Microbiol. *73*, 5347–5353.

Rosenbaum, M., and Angenent, L.T. (2009). Genetically modified microorganisms for bioelectrochemical systems. In Bioelectrochemical Systems: From Extracellular Electron Transfer to Biotechnological Application, Rabaey, K., Angenent, L., Schroder, U., and Keller, K., eds. (IWA Publishing, London, UK), pp. 101–113.

Rosenbaum, M., Aulenta, F., Villano, M., and Angenent, L.T. (2011). Cathodes as electron donors for microbial metabolism: which extracellular electron transfer mechanisms are involved? Bioresour. Technol. *102*, 324–333.

Ross, D.E., Flynn, J.M., Baron, D.B., Gralnick, J.A., and Bond, D.R. (2011). Towards electrosynthesis in *shewanella*: energetics of reversing the mtr pathway for reductive metabolism. PLoS One *6*, e16649–e16658.

Rozendal, R.A., Jeremiasse, A.W., Hamelers, H.V.M., and Buisman, C.J.N. (2008). Hydrogen production with a Microbial Biocathode. Environ. Sci. Technol. *42*, 629–634.

Schröder, U. (2007). Anodic electron transfer mechanisms in microbial fuel cells and their energy efficiency. Phys. Chem. Chem. Phys. *9*, 2619–2629.

Schröder, U. (2011). Discover the possibilities: microbial bioelectrochemical systems and the revival of a 100-year old discovery. J. Solid State Electrochem. *15*, 1481–1486.

Sevda, S., and Sreekrishnan, T.R. (2014). Removal of organic matters and nitrogenous pollutants simultaneously from two different wastewaters using biocathode microbial fuel cell. J. Environ. Sci. Health. A. Tox. Hazard. Subst. Environ. Eng. *49*, 1265–1275.

Speers, A.M., Young, J.M., and Reguera, G. (2014). Fermentation of glycerol into ethanol in a microbial electrolysis cell driven by a customized consortium. Environ. Sci. Technol. *48*, 6350–6358.

Tandukar, M., Huber, S.J., Onodera, T., and Pavlostathis, S.G. (2009). Biological chromium (VI) reduction in the cathode of a microbial fuel cell. Environ. Sci. Technol. *43*, 8159–8165.

Thrash, J.C., and Coates, J.D. (2008). Review: direct and indirect electrical stimulation of microbial metabolism. Environ. Sci. Technol. *42*, 3921–3931.

Thrash, J.C., Van Trump, J.I., Weber, K. a, Miller, E., Achenbach, L. a, Coates, J.D. (2007). Electrochemical stimulation of microbial perchlorate reduction. Environ. Sci. Technol. *41*, 1740–1746.

Torres, C.I., Marcus, A.K., and Rittmann, B.E. (2007). Kinetics of consumption of fermentation products by anode-respiring bacteria. Appl. Microbiol. Biotechnol. *77*, 689–697.

Turick, C.E., Tisa, L.S., and Caccavo, F. (2002). Melanin production and use as a soluble electron shuttle for Fe (III) oxide reduction and as a terminal electron acceptor by *Shewanella algae* BrY. Appl. Environ. Microbiol. *68*, 2436–2444.

Van Rij, E.T., Wesselink, M., Chin-A-Woeng, T.F.C., Bloemberg, G.V., and Lugtenberg, B.J.J. (2004). Influence of environmental conditions on the production of phenazine-1-carboxamide by *Pseudomonas chlororaphis* PCL1391. Mol. Plant. Microbe. Interact. *17*, 557–566.

Vargas, M., Kashefi, K., Blunt-Harris, E.L., and Lovley, D.R. (1998). Fe (III) reduction on early Earth. Lett. to Nat. *395*, 65–68.

Velasquez-Orta, S.B., Head, I.M., Curtis, T.P., and Scott, K. (2011). Factors affecting current production in microbial fuel cells using different industrial wastewaters. Bioresour. Technol. *102*, 5105–5112.

Vilar-Sanz, A., Puig, S., García-Lledó, A., Trias, R., Balaguer, M.D., Colprim, J., and Bañeras, L. (2013). Denitrifying bacterial communities affect current production and nitrous oxide accumulation in a microbial fuel cell. PLoS One *8*, e63460–63471.

Villano, M., Monaco, G., Aulenta, F., and Majone, M. (2011). Electrochemically assisted methane production in a biofilm reactor. J. Power Sources *196*, 9467–9472.

Villano, M., Scardala, S., Aulenta, F., and Majone, M. (2013). Carbon and nitrogen removal and enhanced methane production in a microbial electrolysis cell. Bioresour. Technol. *130*, 366–371.

Virdis, B., Millo, D., Donose, B.C., and Batstone, D.J. (2014). Real-time measurements of the redox states of c-type cytochromes in electroactive biofilms: a confocal resonance Raman Microscopy study. PLoS One *9*, e89918–89928.

Virdis, B., Rabaey, K., Yuan, Z., and Keller, J. (2008). Microbial fuel cells for simultaneous carbon and nitrogen removal. Water Res. *42*, 3013–3024.

Von Canstein, H., Ogawa, J., Shimizu, S., and Lloyd, J.R. (2008). Secretion of flavins by Shewanella species and their role in extracellular electron transfer. Appl. Environ. Microbiol. *74*, 615–623.

Wagner, R.C., Regan, J.M., Oh, S.-E., Zuo, Y., and Logan, B.E. (2009). Hydrogen and methane production from swine wastewater using microbial electrolysis cells. Water Res. *43*, 1480–1488.

Ward, M.J., Fu, Q.S., Rhoads, K.R., Yeung, C.H.J., Spormann, A.M., and Criddle, C.S. (2004). A derivative of the menaquinone precursor 1,4-dihydroxy-2-naphthoate is involved in the reductive transformation of carbon tetrachloride by aerobically grown Shewanella oneidensis MR-1. Appl. Microbiol. Biotechnol. *63*, 571–577.

Watanabe, K., Manefield, M., Lee, M., and Kouzuma, A. (2009). Electron shuttles in biotechnology. Curr. Opin. Biotechnol. *20*, 633–641.

Wei, J., Liang, P., and Huang, X. (2011). Recent progress in electrodes for microbial fuel cells. Bioresour. Technol. *102*, 9335–9344.

Wen, Q., Wu, Y., Zhao, L., and Sun, Q. (2010). Production of electricity from the treatment of continuous brewery wastewater using a microbial fuel cell. Fuel *89*, 1381–1385.

Workman, D.J., Woods, S.L., Gorby, Y.A., Fredrickson, J.K., and Truex, M.J. (1997). Microbial reduction of vitamin B 12 by *Shewanella alga* Strain BRY with subsequent transformation of carbon tetrachloride. Environ. Sci. Technol. *31*, 2292–2297.

Wrighton, K.C., Virdis, B., Clauwaert, P., Read, S.T., Daly, R.A., Boon, N., Piceno, Y., Andersen, G.L., Coates, J.D., and Rabaey, K. (2010). Bacterial community structure corresponds to performance during cathodic nitrate reduction. ISME J. *4*, 1443–1455.

Xafenias, N., and Mapelli, V. (2014). Performance and bacterial enrichment of bioelectrochemical systems during methane and acetate production. Int. J. Hydrogen Energy *3*, 1–12.

Xing, D., Zuo, Y., Cheng, S., Regan, J.M., and Logan, B.E. (2008). Electricity generation by Rhodopseudomonas palustris DX-1. Environ. Sci. Technol. *42*, 4146–4151.

Yates, M.D., Kiely, P.D., Call, D.F., Rismani-Yazdi, H., Bibby, K., Peccia, J., Regan, J.M., and Logan, B.E. (2012). Convergent development of anodic bacterial communities in microbial fuel cells. ISME J. *6*, 2002–2013.

Zhan, G., Zhang, L., Tao, Y., Wang, Y., Zhu, X., and Li, D. (2014). Anodic ammonia oxidation to nitrogen gas catalyzed by mixed biofilms in bioelectrochemical systems. Electrochim. Acta *135*, 345–350.

Zhuang, L., Zheng, Y., Zhou, S., Yuan, Y., Yuan, H., and Chen, Y. (2012). Scalable microbial fuel cell (MFC) stack for continuous real wastewater treatment. Bioresour. Technol. *106*, 82–88.

Biofilms for One-stage Autotrophic Nitrogen Removal

11

José M. Carvajal-Arroyo, Tiago Rogeiro Vitor Akaboci, Maël Ruscalleda, Jesús Colprim, Emilie Courtens and Siegfried E. Vlaeminck

Abstract

About 20 years after the discovery of microbial anoxic ammonium oxidation (anammox), the autotrophic nitrogen removal through partial nitritation–anammox (PNA) for ammoniacal wastewater treatment has become a mature technology. The application of these slow growing anoxic ammonium-oxidizing bacteria (AnAOB) requires engineered systems with efficient biomass retention. In the last decade, several one-stage PNA technologies have been developed that promote the growth of AnAOB in biofilms along with aerobic ammonium-oxidizing bacteria (AerAOB). Such biofilms grow on the surface of a carrier material or in mm-scale bio-aggregates (granules). Thanks to the easy retention of biofilm carriers or good settleability of granules, long sludge retention times can be maintained. Additionally, diffusional oxygen transfer limitation within the biofilm allows for the creation of aerobic and anoxic microniches where AerAOB and AnAOB, respectively, can thrive. This chapter describes and discusses the engineering and ecological characteristics of the different technologies developed so far, including rotating biological contactors (RBC), moving bed biofilm reactors (MBBR), membrane-aerated biofilm reactors (MABR) and granular systems. Moreover, the recent literature on operation parameters that influence the greenhouse gas emissions (i.e. N_2O) during PNA are described. Finally the future trends in the biofilm-PNA applications to new effluents, with special attention to mainstream sewage treatment, are discussed.

Introduction

The discovery of the anoxic ammonium-oxidizing (anammox) bacteria in the early 1990s enabled for the resource-efficient treatment of ammonium rich wastewaters with low biodegradable organic matter content, through completely autotrophic microbial nitrogen transformations. Autotrophic nitrogen removal occurs through two sequential conversion steps. First, about half of the ammonium in the wastewater is converted to nitrite (partial nitritation). This reaction is performed by aerobic ammonium-oxidizing bacteria (AerAOB). Subsequently, the residual ammonium is oxidized by anoxic ammonium-oxidizing bacteria (AnAOB) using the nitrite formed in the first step as a terminal electron acceptor, and yielding nitrogen gas as the main product as well as some nitrate. Although nitrite oxidizing bacteria (NOB) naturally occur in one-stage PNA biofilms, nitratation is minimized to avoid nitrate production which would turn into N removal inefficiency. The combination of these two microbial conversions for completely autotrophic nitrogen removal, in a one-stage process, accepts different terminologies, like oxygen limited autotrophic nitrification-denitrification (OLAND) (Kuai and Verstraete, 1998), completely autotrophic nitrogen removal over nitrite (CANON) (Third *et al.*, 2001), or single-stage nitrogen removal using anammox and partial nitritation (SNAP) (Furukawa *et al.*, 2005). The industry has adopted the name of deammonification. The complete process of PNA, can be carried out in two separate reactors (aerobic and anoxic), or in a single stage, with nitritation and anammox reactions separated in time or in space (within the biofilm).

AerAOB are chemolithoautotrophic bacteria, falling predominantly within the taxonomic group of Beta-proteobacteria. *Nitrosomonas* is the dominant genus of AerAOB in one-stage PNA reactors (Vlaeminck *et al.*, 2012), among which *N. europaea* is the best known species. In the AerAOB metabolic pathway, NH_3 oxidation is catalysed by an ammonia monooxygenase (AMO), yielding hydroxylamine (NH_2OH) as a reaction intermediate. NH_2OH is further oxidized to nitrite by hydroxylamine oxidoreductase (HAO). The oxidation of NH_2OH yields four electrons. Two electrons are returned to AMO to allow for NH_3 oxidation, and the remaining two electrons are used in cell metabolism, including fixation of inorganic carbon (Arp and Stein, 2003) (Fig. 11.1A).

AnAOB are chemolithoautotrophic bacteria belonging to the phylum Planctomycetes. So far, five different genera (*Brocadia, Kuenenia, Anammoxoglobus, Jettenia* and *Scalindua*) have been described (Kartal *et al.*, 2007; Kuypers *et al.*, 2003; Quan *et al.*, 2008; Schmid *et al.*, 2000; Strous *et al.*, 1999a). The first four genera, have been found in freshwater natural ecosystems, and bioreactors (except *Anammoxoglobus*), while *Scalindua* spp. thrive in saline water such as sea water, or highly saline anammox bioreactors (Kartal *et al.*, 2006). In the anammox reaction, nitrite is initially reduced to nitric oxide (NO), by the enzyme nitrite oxidase (NirS). Hydrazine synthase (HZS) combines NO and ammonium to generate hydrazine, which is finally oxidized to dinitrogen gas by hydrazine dehydrogenase (HDH). Carbon fixation in AnAOB is supported by nitrite oxidation to nitrate, which is catalysed by a nitrate reductase (NAR) (Fig. 11.1B) (Kartal *et al.*, 2011).

The stoichiometry of the one-stage PNA (Vlaeminck *et al.*, 2012) can be obtained by combination of the two independent reactions of nitritation (Barnes and Bliss, 1983) and anammox (Strous *et al.*, 1998) were AerAOB cells are accounted as $CH_{1.4}O_{0.4}N_{0.2}$ and AnAOB are accounted for as $CH_2O_{0.5}N_{0.15}$ (equation 11.1).

$$NH_4^+ + 0.792O_2 + 0.080HCO_3^- \rightarrow 0.435N_2 + 0.111NO_3^- + 1.029H^+ + 0.052CH_{1.4}O_{0.4}N_{0.2} + 0.028CH_2O_{0.5}N_{0.15} + 1.460H_2O \quad (11.1)$$

When compared with the conventional nitrification/denitrification, one-stage PNA provides significant advantages. Firstly, only about half of the ammonium needs to be oxidized to nitrite, which translates in up to 57% savings in aeration requirements. Secondly, the lithoautotrophicity of PNA eliminates the need for addition of organic carbon for 100%. Finally, PNA generates ca. 80% less biomass, which significantly reduces the sludge management costs.

Efficient retention of biomass is a major requirement for autotrophic nitrogen removal. AnAOB are slow growing bacteria (Strous *et al.*, 1998). For example, the cellular doubling times

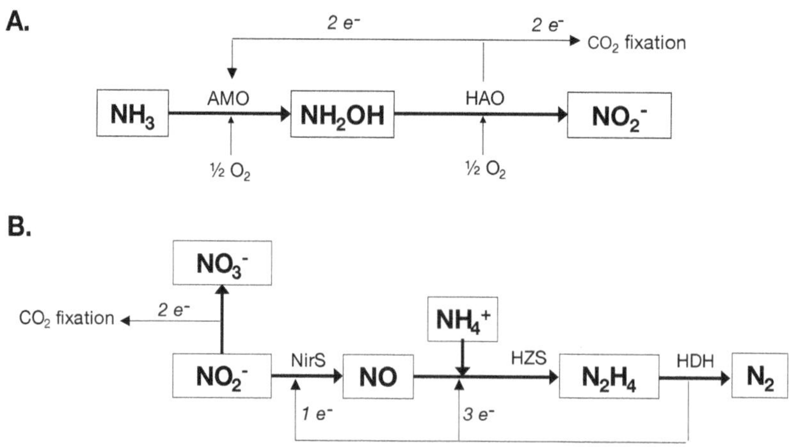

Figure 11.1 A: Metabolic pathway of nitritation by AerAOB. B: Metabolic pathway of anammox by AnAOB.

measured in full scale installations range from 11 to 27 days (Joss *et al.*, 2009; van der Star *et al.*, 2007), although the shortest observed doubling time for AnAOB is 3.3 days, in a laboratory-scale membrane anammox reactor (Lotti *et al.*, 2014). Therefore, the retention time of the biomass in the reactor must exceed those numbers. In general, 30–45 days of sludge retention time (SRT) is recommended in reactors operated at 30–35°C (Vlaeminck *et al.*, 2012). This can be achieved by promoting bacterial growth in biofilms, either on synthetic carriers that are easily retained, or in dense bio-aggregates or granules with fast settling properties. In fact, AnAOB tend to grow in aggregates (flocs or biofilm), and only under strictly anoxic conditions AnAOB are able to grow as free cells (Lotti *et al.*, 2014). The cell density seems to play an important role in AnAOB growth, since AnAOB activity can only be detected above a certain threshold (Strous *et al.*, 1999a). Moreover, De Clippeleir and colleagues (2011) observed higher specific AnAOB activity at high cell densities than in low density cultures, and proved the stimulating role of AerAOB-produced quorum sensing molecules in a PNA biofilm.

Growth of PNA biomass in biofilms offers several advantages aside from improved retention of the microorganisms. Because biofilms have higher cell densities, compared to suspended growth systems, a higher concentration of biocatalyst (microorganisms) can be achieved in the reactor, which translates in a significantly smaller reactor size, and consequently lower capital costs. As a result, full scale plants based on biofilms accept higher volumetric nitrogen loading rates than suspended growth installations (Lackner *et al.*, 2014).

Biofilms for one-stage PNA

The diffusive transport and conversion of the substrates (molecular oxygen and nitrogen species) in one-stage PNA biofilms allows for the formation of microniches with optimal conditions for the growth of both groups of bacteria (Fig. 11.2). In the aerobic part of the biofilm, AerAOB consume the oxygen and produce nitrite. The oxygen penetration depth in PNA depends on ammonium concentration, bulk dissolved oxygen (DO) level, and size of the biofilm, in the range of 0.06–0.02 mm for granules sized 0.79–2.55 mm diameter (Volcke *et al.*, 2012). The depletion of oxygen by AerAOB allows for the growth of AnAOB in the anoxic part of the biofilm, -which would otherwise be inhibited by oxygen (Strous *et al.*, 1999b) – where they benefit from the *in situ* production of nitrite, their terminal electron acceptor. DO inhibition of AnAOB has been further described in the literature (Egli *et al.*, 2001) and a 50% inhibitory concentration (IC_{50}) of 2.2 mg O_2 l^{-1} was calculated from batch tests (Carvajal-Arroyo *et al.*, 2013). However, AnAOB inhibition by DO is reversible, which allows the separation of nitritation and anammox reaction not only in space, but also in time, by alternating imposition of aerobic and anoxic conditions in the reactor.

The immediate uptake of nitrite by AnAOB in PNA biofilms represents an additional advantage over two-stage systems. Maintaining low nitrite concentrations is crucial for a stable and well performing process, since nitrite is a potential inhibitor for both groups of bacteria. While for AerAOB, the undissociated form HNO_2 (free nitrous acid, FNA) is responsible for the bactericidal effect in the range 0.2–2.8 mg FNA-N/l (Anthonisen *et al.*, 1976), ionized nitrite causes toxicity to AnAOB (Puyol *et al.*, 2014). The inhibitory effect of nitrite on AnAOB is strongly exacerbated when the electron donating substrate ammonium is not available (Carvajal-Arroyo *et al.*, 2014). This situation may occur if nitritation and anammox rates are not well balanced.

Aerobic nitrate generation by NOB has to be avoided as it lowers the N removal efficiency, in case this is not compensated by denitrifying activity (Fig. 11.2). Additionally, NOB compete for oxygen with AerAOB. Several strategies are used to reduce NOB growth in PNA reactors. Firstly, the prevention of nitrite accumulation helps maintaining low NOB activity as AnAOB have higher affinity for nitrite than NOB (Lackner *et al.*, 2008). Besides, the temperature plays another important role in allowing AerAOB to outcompete of NOB. AerAOB grow faster than NOB at high temperatures. Although some authors fix the breakthrough temperature in the range 15–18°C (Hellinga *et al.*, 1998; Wyffels *et al.*,

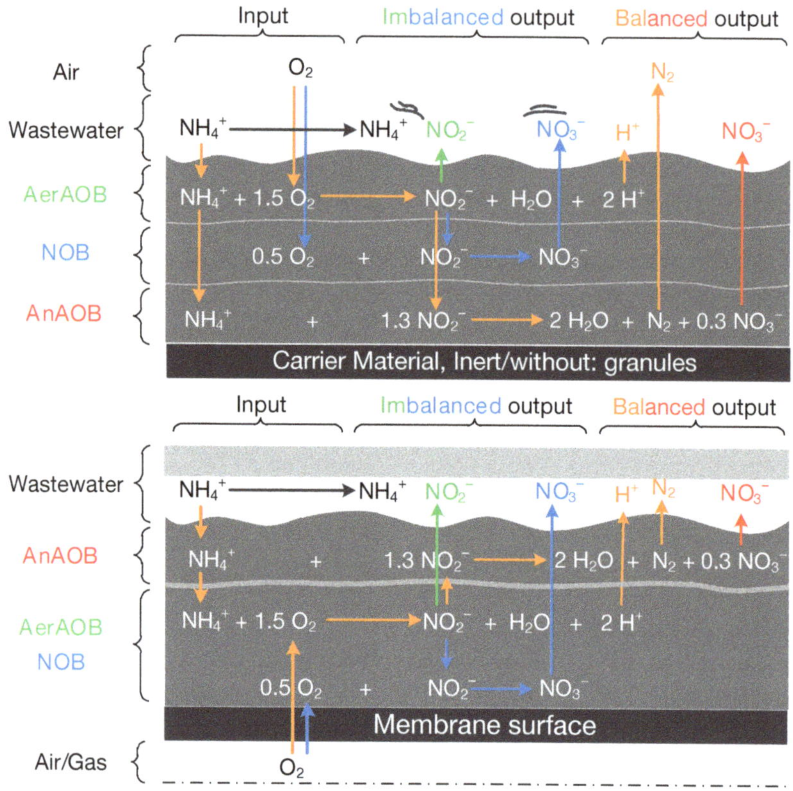

Figure 11.2 Scheme of PNA biofilms with co-diffusion (upper diagram) and counter-diffusion (lower diagram) of ammonium and oxygen. Yellow arrows: microbial transformations leading to balanced performance of PNA biofilms (maximum theoretical N removal, according to stoichiometry in equation 1). Red arrows: NO_3^- production inherent to a balanced PNA biofilm, limiting the N removal efficiency. Green arrows: Excessive nitritation leading to imbalanced PNA outcome. Blue arrows: nitratation (by NOB) leading to imbalanced PNA system (N removal efficiency lower than maximum theoretical). In counter-diffusive systems (MABR) AerAOB and NOB share space in the aerobic zone of the biofilm (Pellicer-Nacher *et al.*, 2014).

2003), other reports show NOB growth greater than AerAOB growth up to 28°C (Knowles *et al.*, 1965). Although adjusting the sludge age would allow for selective washout of NOB, biofilm reactors do not allow for such strategy, since the high SRT maintained in PNA-biofilm reactors enables growth of bacteria with very different growth rates (Fux *et al.*, 2004). Finally, in systems treating high strength effluents, depending on the feeding strategy, the concentrations of free ammonia (FA) can be high enough (>5 mg N/l) to cause inhibition of NOB, giving advantage to AerAOB growth. However, in more diluted effluents FA inhibition of NOB is not likely and other methods need to be applied (see 'Future trends', below).

The first full-scale installations were two-stage systems, in which PNA occurred in separate compartments. Currently however, the combination of both processes in a single reactor is largely preferred. By 2014, around 100 full scale plants were in operation, among which only four facilities operated with two stage systems (Lackner *et al.*, 2014). Biofilm based reactors for one-stage PNA, have been applied in full scale for the treatment of a wide variety of high strength ammoniacal wastewaters. The most popular application of one-stage PNA reactors is the treatment of reject water from sludge digesters. Although, other ammonium rich effluents are currently treated through one-stage PNA, including landfill leachate, food industry effluents (winery, distillery, starch production, sweetener production, potato and meat processing), monosodium glutamate production, or digested black water (Lackner *et al.*, 2014).

Several biofilm reactor configurations that are currently in use at full-scale are rotating biological contactors (RBC) (Hippen *et al.*, 1997), moving bed biofilm reactors (MBBR) (Rosenwinkel and Cornelius, 2005) and granular systems (Abma *et al.*, 2010) (Table 11.1). Additionally membrane-aerated biofilm reactors (MABR) (Gong *et al.*, 2007) have been evaluated at laboratory scale. A detailed description of these technologies can be found in the following sections.

Reactor configurations

Rotating biological contactor (RBC)

A reactor configuration suitable for the application of one-stage autotrophic nitrogen removal is the rotating biological contactor (RBC). First observations of nitrogen losses in the absence of biodegradable organic carbon for denitrification, occurred in RBC treating landfill leachate in Germany (Hippen *et al.*, 1997), Switzerland (Siegrist *et al.*, 1998), and UK (Schmid *et al.*, 2003). AnAOB were later identified as responsible for the N losses.

RBCs have been widely used for wastewater treatment, and their application to autotrophic nitrogen removal has an extensive track record (Patwardhan, 2003; Pynaert *et al.*, 2002). A RBC consists of a series parallel discs mounted on a horizontal shaft that is partially submerged in the wastewater (Fig. 11.3). The biofilm grows on the surface of the discs, and oxygenation is facilitated by continuous rotation of the shaft, alternately exposing the biomass to the air and to the wastewater. Another type of RBC consists of rotating cages, instead of discs, that enclose biofilm carriers, similar to those used in MBBR reactors (Mathure *et al.*, 2005). Although the RBC technology has proved robust and its operation is cost-effective, the nature of the design leaves little flexibility in terms of control strategies.

In RBC, aeration occurs through three different mechanisms (Kim and Molof, 1982): (i) oxygen transfer to the liquid film on the surface of the biofilm during the air exposure period; (ii) direct oxygen uptake by the bacteria exposed to the air; and (iii) oxygen transfer from the air to the bulk liquid in the reactor, which is enhanced by the turbulence caused by the rotation of the discs. The final oxygen transfer coefficient (K_La) is influenced by the rotational speed and the immersion level of the discs. Several works have reported an enhancement of the oxygen transfer rates due to the presence of oxygen consumming bacteria (Kim and Molof, 1982; Paolini, 1986; Courtens *et al.*, 2013), and introduce an 'enhancement factor' in the oxygenation mass balance (equation 11.2):

$$O_{UR} = E \times \alpha K_L a \times (C_{sat} - C_{bulk}) \qquad (11.2)$$

Table 11.1 Qualitative comparison of one-stage PNA reactor configurations based on biofilms

	Reactor configuration			
	RBC	MBBR	MABR	Granules
Investment costs	High	Medium	High	Low
Operational costs	Low	Medium/high	Medium/high	Medium
Area requirement	High	Low	Low	Low
Aeration	Passive	Active	Active*	Active
Ease of DO control	Medium	Medium/high	Low	High
Inoculation feasibility	Low/medium	Medium/high	Low/medium	High
Risk for mechanical failure	High	Low	High	Low
Operational flexibility	Low	Medium	Low	Medium/high
Full-scale installations§	6	9	0	18

*Aeration in MABR occurs by oxygen diffusion through the pores of the membrane without generating bubbles, and it is controlled through the gas pressure in the lumen of the membrane fibres.
§Data from Lackner *et al.* (2014).
Adapted from Vlaeminck *et al.* (2012).

Figure 11.3 Lab-Scale OLAND-rotating biological contactor at the Laboratory of Microbial Ecology and Technology (Ghent University), started up in 2000 (Pynaert *et al.*, 2002). Picture from 2014.

where O_{UR} is the oxygen uptake rate, E is the enhancement factor, $\alpha K_L a$ is the oxygen transfer coefficient in wastewater, C_{sat} is the saturation DO and C_{bulk} is the DO in the bulk liquid.

Overall, lower immersion level and faster rotational speed lead to better oxygenation, although excessive rotation velocity can cause biomass sloughing (Cortez *et al.*, 2008). As in other technologies for autotrophic nitrogen removal, the management of the oxygen budget in RBC affects the balance between nitritation and anammox rates, and plays a role in the suppression of nitratation activity. Courtens *et al.* (2013) showed that high immersion level (80%) effectively suppressed NOB activity. Two mechanisms were identified to contribute to the minimization of aerobic nitrate generation. Firstly, lower oxygen availability caused accumulation of higher FA concentration, inducing NOB inhibition. Secondly, the exposure to anoxic conditions during longer periods, probably caused lag phases in NOB activity.

As in other biofilm technologies, AnAOB grow in the anoxic zones of the biofilm. In a study carried out in a RBC treating landfill leachate in Kölliken, Switzerland, *K. stuttgartiensis* dominated the AnAOB community in the biofilm (Egli *et al.*, 2003). In another RBC treating synthetic wastewater, *K. stuttgartiensis* shared the anoxic areas of the biofilm with *B. anammoxidans* (Pynaert *et al.*, 2003). In the work of Egli and colleagues, AerAOB affiliated to *N. europaea* and *N. eutropha* could be found in a dense layer in the outer part of the biofilm, whereas in a laboratory scale RBC, AerAOB were evenly distributed in the biofilm, including putatively anoxic areas (Pynaert *et al.*, 2003). Along with AerAOB and AnAOB, a great diversity of heterotrophic Planctomycetes (*Pirellula*, *Gemmata*, *Isosphaera*, and *Planctomyces*) were found to coexist. The presence of heterotrophic bacteria in a reactor fed with a medium devoid of organic carbon, was explained by the decay of biomass in the biofilm (Pynaert *et al.*, 2003). Indeed, *Pirellula* has been demonstrated capable of denitrification (Fuerst, 1995), which would further improve total N removal. Unlike the genera *Kuenenia* and *Brocadia*, *Scalindua* is commonly associated with samples from sediments and oxygen minimum zones in the ocean (Jetten *et al.*, 2003).

Although the application of RBC for one-stage autotrophic nitrogen removal has been widely explored at the laboratory scale, there are only two references of RBC intentionally designed to perform PNA at full scale. In Sneek, The Netherlands, a 6-m^3 PNA RBC (cage type) is operated to treat digested black water, serving 464 population equivalents. In this facility, the rotation speed is controlled (1–4 rpm) to achieve the DO set point (0.60–0.65 mg O_2/l), and the pH is maintained in the range 7.0–7.5 by NaOH dosage. Another RBC was built by Advanced Wastewater Solutions (AWWS) in Hulst, The Netherlands, to treat the effluent from a fertilizer production industry. In this case, the DO is controlled by variation of the disc immersion level and rotational velocity. The feeding control is performed on the basis of online measurement of ammonium concentration in the effluent, and the pH is adjusted by acid/base dosage.

Moving-bed biofilm reactor (MBBR)

MBBR have been used for municipal and industrial wastewater treatment, such as pharmaceutical, paper industry, refinery and poultry processing, among others (Barwal and Chaudhary, 2014). MBBR technology utilizes mobile carrier materials to support biomass growth in biofilms. The carriers are kept in suspension inside the reactor, occupying the whole working reactor volume, by mechanical stirring or internal recirculation, together with active aeration in aerobic reactors. In contrast with other biofilm processes, the MBBR dispensed the need for sludge recirculation and backwashing, as well as problems related to clogging, common in fixed biofilm reactors (Ødegaard, 2006).

After the discovery of the anammox process, the application of MBBR configuration to one-stage PNA gained attention due to its high biomass retention efficiency. Since both nitritation and anammox pathways can occur simultaneously in the biofilm, there is no limitation for the anoxic/aerobic reaction times. In MBBR the biofilm is based on a co-diffusion scheme for oxygen and ammonium. AerAOB preferably settle in the outer layer (aerobic layer) while AnAOB grow in the inner layer (anoxic layer), as showed in Fig. 11.2 (Helmer *et al.*, 1999). According to microbial community analysis in a MBBR performing PNA, the biofilm developed is dominated by *Nitrosomonas europaea* and *N. eutropha*, for AerAOB, and Candidatus *Brocadia fulgida*, Candidatus *Anammoxoglobus propionius* and Candidatus *Kuenenia stuttgartiensis* for AnAOB (Almstrand *et al.*, 2014; Gilbert *et al.*, 2014b; Helmer *et al.*, 2002).

Most MBBR performing PNA have been started-up utilizing biofilm carriers developed previously (Christensson *et al.*, 2013), corresponding mainly for commercial application. The development of biofilm carrier from conventional activated sludge can be used as initial start-up strategy, but generally it takes long time for the biofilm to develop (Mehrdad *et al.*, 2014). A successful start-up strategy consists of an initial inoculation with nitrifying biomass, which is subjected to oxygen limitation in order to promote accumulation of nitrite. Subsequently, anammox biomass is seeded enabling for N removal (Daverey *et al.*, 2013).

Biofilm thickness and density are dependent on the hydrodynamics and biochemical reactions, and it influences the nitrogen removal efficiency in MBBR. A biofilm with thickness less than 0.2 mm might not support AnAOB growth due to the difficulty to obtain an anoxic layer under aerobic conditions (Hao *et al.*, 2002). However, no changes in nitrogen removal rate were obtained when the biofilm thickness increased from 0.27 mm to 0.77 mm during operation of a PNA- MBBR (Cema *et al.*, 2011). It took about 30 days for AerAOB enrichment and attachment on carrier, under low turbulence, and about 150 days for AnAOB growth in the biofilm (Mehrdad *et al.*, 2014). The biofilm in the carrier is protected from shearing forces and, therefore, the biofilm thickness is only limited by the carrier itself. The structure and activity of the microbial community in the biofilm developed in carriers, and consequently, the nitrogen removal efficiency are influenced by the operational conditions. Real-time DO control is the most common strategy, since it allows for the management of the nitrite level in the reactor. Ammonium, nitrite and nitrate are monitored in the influent and effluent with the help of online sensors. The DO set point can be increased or decreased, if the ratio $NO_3^-{}_{produced}/NH_4^+{}_{removed}$ is lower or higher than 11%, respectively. This strategy minimizes the growth of NOB in the reactor (Christensson *et al.*, 2013). Conductivity and pH can also be used for process performance assessment (Szatkowska *et al.*, 2007). The pH values decrease during partial nitritation due to consumption of alkalinity and to a lower extent, increase due to anammox reaction. In turn, conductivity is depleted by removal of the main ions (ammonium and hydrogen carbonate).

A significant number of publications show the applications for MBBR for nitrogen removal in one-stage reactor and several full-scale MBBR are currently in operation with several commercial applications in the market (Lackner *et al.*, 2014).

Granular systems

Compared to other biofilm reactors, granular biomass requires neither a carrier nor chemicals for microbial attachment. Granular biomass was first applied for anaerobic digestion of low strength wastewaters (Lettinga *et al.*, 1980) although later on, its main application shifted to highly loaded wastewaters, including agroindustrial wastewaters. Years later, granulation was shown feasible also in aerobic reactors for sewage treatment (Mishima and Nakamura, 1991).

In the last years, some studies demonstrated the feasibility of enriching aerobic nitrifying biomass in granular systems (Campos *et al.*, 2000). As AnAOB also grow in granules (Abma *et al.*, 2007), autotrophic nitrogen removal is carried out by combination of partial nitritation and anammox processes in granules. Similar to other biofilm systems, the microbial community is stratified in an outer aerobic layer – dominated by AerAOB – surrounding the anoxic zone – with AnAOB – in the core of the granule (Fig. 11.4).

Granular nitritation-anammox processes have been mainly carried out in sequencing batch reactor (SBR), but also in airlift and up-flow column reactors (Vlaeminck *et al.*, 2009; Winkler *et al.*, 2012; Zhang *et al.*, 2012; Wang *et al.*, 2014).

Different strategies are applied to start up autotrophic nitrogen removal in reactors with granular biomass, depending on the source of inoculum. One of them consists in inoculating the reactor with AnAOB sludge and enriching the reactor by feeding ammonium and nitrite. Once stable nitrogen removal is obtained, nitrifying biomass can be added in the reactor, changing influent to contain only ammonium as nitrogen source. Simultaneously, oxygen must be provided for ammonium oxidation into nitrite, necessary for anammox reaction (Zhang *et al.*, 2012). Other strategies consist in obtaining first stable partial nitritation from nitrifying biomass, under oxygen limitation, and then proceed with AnAOB inoculation (Vázquez-Padín *et al.*, 2009). According to Vázquez-Padín (2009), the second start-up strategy seems to be more suitable for two reasons: (i) AnAOB activity in the first strategy can decrease after nitrifying biomass inoculation, and (ii) less AnAOB sludge is necessary to start up the process. In case of neither nitrifying nor anammox biomass are available, cultivating AerAOB and AnAOB from activated sludge can be also utilized (Hu *et al.*, 2013), requiring a long start-up period due to slow microbial growth rate.

Microbial balance between AerAOB and AnAOB is also relevant for autotrophic nitrogen removal in granular systems. This balance is determined by granule size and substrate concentration in the bulk liquid, which will influence substrate mass transfer into the granule, and consequently the activities in each granule layer. Under DO limitation (~5 µm O_2), Nielsen and colleagues (2005) verified that large type aggregates (>500 µm), accounted for 68% of anammox potential, with AerAOB located at outer granule

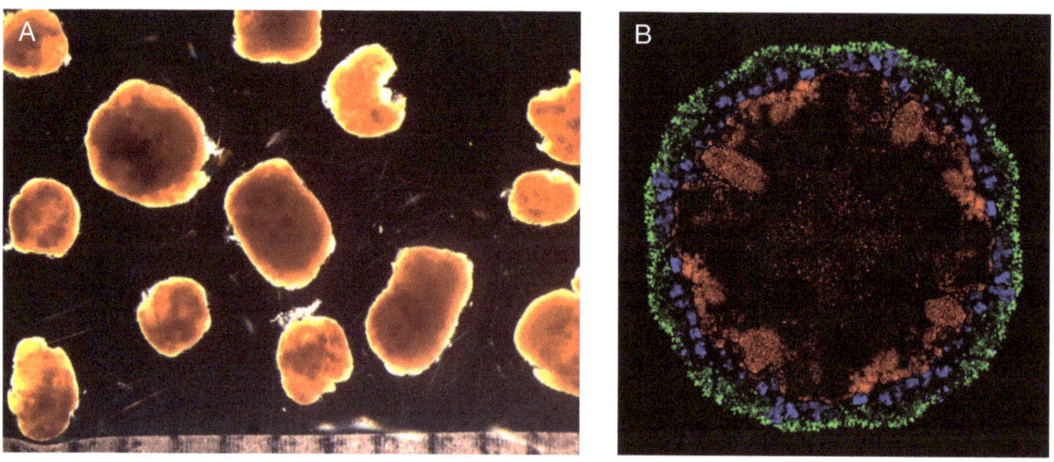

Figure 11.4 (A) Micrograph of a granular biofilm. (B) Fluorescence *in situ* hybridization (FISH) micrograph of a PNA granule, showing stratification of AerAOB (green, probes Nso1225 and Nso190), NOB (*Nitrospira* spp.) (blue, probe Ntspa662) and AnAOB (red, probe Amx820). Image extrapolated from Figure 1 in Vlaeminck *et al.* (2010).

layer and AnAOB at inner granule, whereas 65% of the nitrification potential was found in the smaller aggregates (<500 μm), primarily composed by AerAOB. Therefore, hydrodynamic conditions and sludge age, both of which affect the size distribution of the aggregates (or the biofilm depth), are key to obtain efficient nitrogen removal, minimizing both ammonium and nitrite in the effluent of the reactors. Accumulation of nitrite provides and advantage for NOB growth, and can result in excessive accumulation of nitrate.

Although granulation mechanisms are still not so clear, some factors have been pointed out to trigger biomass aggregation into anaerobic and aerobic granules. Hydrodynamics can be considered one of the most relevant among these factors. Granulation pathways in one-stage partial nitritation and anammox process were investigated by Vlaeminck and colleagues (2010), who hypothesized that granules replicate by division and budding, driven by bacterial growth and/or decay based on species-specific physiology and by hydrodynamic shear and mixing. Internal decay and subsequent shear on collisions of a weakened granule, can lead to granule division at a certain size. These new, small aggregates can grow into a new granule.

Short hydraulic retention time (HRT) allows the formation of larger and denser granules (Jin *et al.*, 2008), by selecting bigger granules that have good settling velocity. This parameter can be used in reactors in continuous operation, such as airlift and up-flow column reactor, and in batch mode operation, e.g. SBR. On the other hand, ammonium conversion can be optimized by adjusting the HRT (Vázquez-Padín *et al.*, 2010). For SBR reactor, settling time also could be a parameter to wash-out the particles with smaller diameter (De Clippeleir *et al.*, 2009). Applying this conditions, flocs and smaller aggregates composed majorly by AerAOB leave the system, avoiding an excess of nitritation rate.

The ecology of granular biomass can be influenced by organic carbon, due to competition of heterotrophic denitrifying bacteria with AerAOB and AnAOB for oxygen and nitrite, respectively. Based on this, PNA has been extensively applied as an alternative to denitrify wastewater with low chemical oxygen demand (COD):N ratio content (around 0.5). Nonetheless, growth of heterotrophic microorganisms in presence of organic matter can cause an increase in sludge production, which translates in a shorter SRT of the autotrophic biomass, impacting the stability of the autotrophic process. Moreover, heterotrophic growth negatively affects the settleability of the biomass, contributing further to biomass loss (Jenni *et al.*, 2014).

Membrane-aerated biofilm reactors

The application of membrane-aerated biofilm reactors (MABR) to one-stage autotrophic nitrogen removal has been explored at laboratory scale. In MABR, a hydrophobic membrane is used to provide oxygenation, and it also serves as carrier material for biofilm growth, while ammonium is present in the bulk liquid. Therefore, the electron donor (ammonium) and the terminal electron acceptor (oxygen) flow in counter-diffusion through the biofilm (Fig 11.2). In MABR, the oxygenation occurs by diffusion of the oxygen across the pores of the membrane, without producing bubbles. This avoids biofilm detachment from the carrier (the surface of the membrane), and additionally, it minimizes stripping of NH_3, NO and N_2O. As a consequence of the counter-diffusive feeding of ammonium and oxygen, the stratification pattern in MABR is different from other biofilms formed in systems where both ammonium and oxygen are present in the bulk liquid. AerAOB grow in the interface biofilm-membrane, were DO concentrations are highest. Then, a DO gradient is created as oxygen is consumed for ammonium oxidation, creating anoxic conditions near the biofilm–liquid interface, where AnAOB grow (Fig. 11.5)

The feasibility of one-stage autotrophic nitrogen removal in a MABR was first proved by (Gong *et al.*, 2007). In this study, the surface of the membrane was covered by a non-woven fabric as carrier material and a maximum N removal efficiency of 89% was achieved. This is the maximum achievable N removal efficiency according to PNA stoichiometry, therefore some endogenous heterotrophic denitrification may have occurred, which is expectable in a thick biofilm supported in such a carrier. In a modelling study, Terada and colleagues (2007) revealed the importance of the

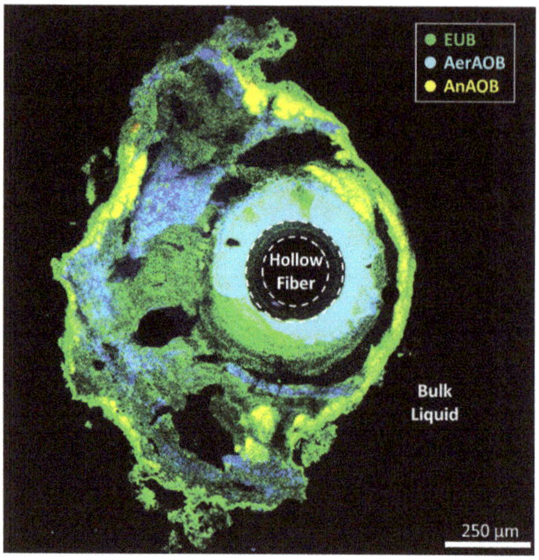

Figure 11.5 Fluorescence *in situ* hybridization (FISH) micrograph of cross-sectional stratification of bacteria in the biofilm of a MBBR for PNA reported in Pellicer-Nacher *et al.*, 2014. Eubacteria (EUB mix) stained with FLUO (green), AerAOB (Nso190-Nmo218-Cluster 6a192) stained with Cy3 (blue) and AnAOB (AMX820) stained with Cy5 (red).

biofilm depth for high N removal efficiency in MABR. While thick biofilms (450–1400 µm) provided superior performance of counter-diffusive systems, greater thicknesses caused the removal rates to decrease, likely due to ammonium mass transfer limitation. Additionally, denitrification fuelled by biomass decay may improve total N removal. However, excessive COD in the influent can compromise the stability of the reactor, and influents with COD/N greater than 2 may not be treatable in a one-stage autotrophic MABR (Lackner *et al.*, 2008).

A major drawback of counter-diffusion of the main substrates in MABR is the difficulty to control NOB growth. The exact DO level that NOB experience is hard to control. Furthermore, DO concentration is highest at the surface of the membrane. Here, the concentration of ammonium is the lowest, and therefore inhibition of NOB by free ammonia is unlikely. Different strategies have been evaluated to suppress NOB activity in MABR. For example, the application of intermittent aeration proved successful in limiting nitrite uptake by NOB. The exposure to transient anoxic conditions caused a lag phase in NOB activity, giving AerAOB an advantage to compete for oxygen, and lowering nitratation from 85%, during continuous aeration, to 9% of total nitrite consumption, during sequential aeration (Pellicer-Nàcher *et al.*, 2010). Another viable strategy to outcompete NOB in MABR is the control of the O_2:NH_4^+ surface loading ratio (Terada *et al.*, 2007). Gilmore and colleagues (2013) were able to maintain relatively low nitratation rates (14–19% of nitrite consumption) by applying a low O_2:NH_4^+ loading ratio (2.18 g O_2/g NH_4^+-N) in a continuously aerated MABR performing PNA.

Molecular characterization of MABR biofilms by fluorescence *in situ* hybridization, has shown the stratification of the different functional groups, with AerAOB growing in the vicinity of the membrane (aerobic region), and AnAOB growing in the outer region of the biofilm (anoxic) (Gong *et al.*, 2008). However, the aeration strategy further affects the spatial disposition of AnAOB. When intermittent aeration is provided, AnAOB grow closer to AerAOB (Pellicer-Nàcher *et al.*, 2010), than in a continuously aerated reactor, where the stratification is more evident (Gilmore *et al.*, 2013).

MABR provide an option for environments where the off-gas production is restricted, or where gas bubbles are to be avoided in the liquid (e.g. zero-gravity, space flight conditions).

Nitrous oxide emissions during autotrophic nitrogen removal

N_2O is an important greenhouse gas with a global warming potential 300 greater than CO_2. Anthropogenic N_2O emissions after the industrial revolution have triggered a 20% increase in the N_2O atmospheric concentration. About 40% of the total N_2O dumped into the atmosphere (17 Tg/year) originates from human activities (chemical: combustion, incineration, industrial production processes and atmospheric deposition; and biological: agriculture, livestock wastes) (IPCC 2007, Physical scientific basis). Greenhouse gas (GHG) emissions from wastewater treatment occur during biological nitrogen removal. Global N_2O emissions during sewage treatment have been estimated at 0.22 Tg N/year, which accounts for 3.2% of the total anthropogenic emissions (Mosier *et al.*, 1999), and up to 10.2% if nitrogen removal during manure treatment is included in the estimate (Scheehle *et al.*, 2006). Nonetheless, those figures may greatly underestimate the total N_2O emissions from wastewater treatment, since no data are available about N_2O emissions derived from nitrogen removal from industrial wastewaters or from landfill leachate, both of which may significantly contribute to the total GHGs emissions (Desloover *et al.*, 2012). Measurements made in full scale plants have revealed that N_2O can be produced in substantial quantities during wastewater treatment. The N_2O emissions in full-scale WWTPs range 0–14.6% of the total N-load of the plant (Kampschreur *et al.*, 2009).

N_2O is generated by both chemical and biological reactions (Fig. 11.6) (Desloover *et al.*, 2012). Chemical N_2O generation occurs by NH_2OH oxidation or nitrite and NO reduction by Fe^{2+}. The biological sources of N_2O are nitritation carried out by AerAOB, and by heterotrophic denitrifiers. During nitritation, two possible mechanisms lead to N_2O production. Firstly, in aerobic conditions, N_2O is generated from the HAO-catalysed oxidation of hydroxylamine, via the unstable intermediate HNO (Hooper and Terry, 1979; Poughon *et al.*, 2001). Secondly, under anoxic conditions, AerAOB are able to perform dissimilatory nitrite reduction to N_2O (using ammonium or H_2 as electron donor) (Bock *et al.*, 1995), with NIR and NOR reductases catalysing the reduction of nitrite and NO, respectively (nitrifier denitrification). Heterotrophic denitrification pathway includes N_2O and NO as intermediates during nitrite reduction to N_2. N_2O emissions arise when limiting electron donor does not allow for complete denitrification (Zumft, 1997). Finally, N_2O production has been detected from full scale anammox bioreactors (Kampschreur *et al.*, 2008). Although AnAOB do not have the genetic potential for N_2O production, one of the intermediates of anammox reaction, NO, is an important precursor of N_2O. Indeed, NO emission has been identified as one of the consequences of nitrite inhibition of AnAOB (Carvajal-Arroyo *et al.*, 2014). A N_2O production pathway in anammox reactors, has been recently hypothesized, consisting in the N_2O generation from chemical NO reduction, with Fe^{2+} as electron donor, and with anammox catalysing Fe^{2+} regeneration (Kampschreur *et al.*, 2011).

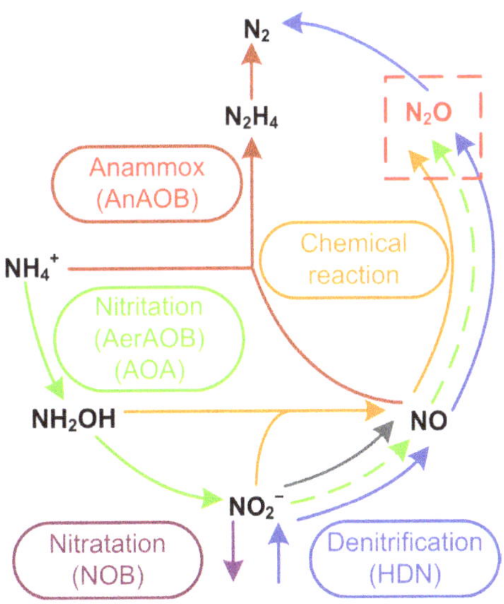

Figure 11.6 Routes of N_2O generation in partial nitritation–anammox reactors. Redrafted after Desloover *et al.* (2012). Green dashed line shows nitrifier-denitrification pathway for N_2O generation.

Different operational parameters have been found to play a major role in the production (and emission) of N_2O during autotrophic nitrogen removal. The DO concentration is one of the main factors controlling N_2O production. As NIR and NOR encoding genes are up-regulated during anoxia, oxygen limitation leads to increase N_2O production due to nitrifier denitrification (Tallec *et al.*, 2006). On the other hand, excess aeration may also enhance N_2O emissions (Kampschreur *et al.*, 2008). Therefore, aeration must be minimized to avoid aerobic N_2O production while maintaining enough DO to elude AerAOB denitrification. A bubbleless approach, in a MABR performing one-stage PNA, has been evaluated at laboratory scale, proving N_2O emissions to be 100-fold lower than in conventional systems, although no full scale confirmation exists (Pellicer-Nàcher *et al.*, 2010).

Other factors that affect N_2O emissions are nitrite and ammonia concentrations. Nitrite, as the main precursor of N_2O, is known to influence GHGs emissions during nitritation. Overall, higher nitrite concentrations (or sudden pulses in nitrite concentration) lead to greater N_2O

emissions (Colliver and Stephenson, 2000; Tallec *et al.*, 2006). In addition, ammonium concentration also influences N_2O emissions, since hydroxylamine production by HAO is regulated by ammonium. Accumulation of ammonium during the anoxic phase has been shown to trigger proportional N_2O peaks at the beginning of the following aerobic phase in a partial nitritation reactor (Yu *et al.*, 2010).

Although the reports on N_2O emissions in full-scale wastewater treatment plants are scarce, the available literature indicates that plants performing PNA emit a higher percentage of the nitrogen load as N_2O than plants designed for traditional nitrification-denitrification (by one to two orders of magnitude) (Desloover *et al.*, 2012). This can be partially explained by the fact that PNA plants treat more concentrated effluents (e.g. sludge centrate, industrial wastewaters), which results in highly variable conditions in the reactors, i.e. ammonium and nitrite concentrations, leading to higher N_2O emissions.

Overall, one-stage systems for autotrophic nitrogen removal have lower N_2O emissions than two-stage configurations. Measurements from one-stage full scale plants show N_2O emissions in the range of 1.2–1.3% of the total N load (Kampschreur *et al.*, 2009; Weissenbacher *et al.*, 2010), and maximum N_2O losses of 0.0015% were quantified in a lab-scale MABR (Pellicer-Nàcher *et al.*, 2010). N_2O emissions of 6.6% of the total nitrogen load were observed in a two-stage PNA system treating effluent of potato industry (Desloover *et al.*, 2011). Indeed, in one-stage biofilm systems, nitrite accumulation is generally 10–50 times lower than in two-stage systems (Desloover *et al.*, 2012), as it is immediately consumed by AnAOB. The preventative effect of AnAOB in the PNA biofilm can also be extended to the consumption of NO, which can be used by AnAOB when ammonium is available (Carvajal-Arroyo *et al.*, 2014; Kartal *et al.*, 2010b).

Application of PNA to the mainstream of sewage treatment

PNA technologies are a well-established alternative for the treatment of ammonium rich wastewaters, with influent concentrations higher than 500 mg NH_4^+-N l^{-1} (Lackner *et al.*, 2014). The fast implementation of the PNA process to the full scale has been driven by the significant savings brought about in terms of aeration, addition of chemicals and sludge management. Siegrist *et al.* (2008) estimated that about 40–50% of the total operation costs in a wastewater treatment plant could be saved by applying PNA to the treatment of sludge digester supernatant. For example, the implementation of this process allowed for net energy generation in a municipal wastewater treatment plant in Austria (Wett *et al.*, 2007). Nonetheless, the N load in sidestream is only about 15–25% of the total nitrogen load of the treatment plant. Therefore, even higher energy efficiency could be expected for future wastewater treatment plants, where the enhanced energy recovery from organic carbon is complemented with a minimum cost of nitrogen removal by application of PNA in the mainstream of sewage treatment (Kartal *et al.*, 2010a; Verstraete and Vlaeminck, 2011).

When compared to sludge liquor, sewage has lower N concentrations (20–60 mg N/l), lower temperatures (5–25°C) that are subjected to seasonal variations, and higher COD/N ratios. These differences pose additional challenges to the application of one-stage PNA to N removal in the mainstream. Low temperatures and low ammonium concentrations reduce both AnAOB and AerAOB rates. Dosta *et al.* (2008) showed almost complete loss of AnAOB activity when decreasing the temperature from 18 to 15°C in an anammox airlift reactor. Although AnAOB retrieved from low temperature (−1.7 to 4°C) natural environments have shown temperature optima at 15°C or even 12°C (Dalsgaard *et al.*, 2005; Rysgaard *et al.*, 2004), cultures obtained by adaptation of mesophilic reactors to low temperatures show higher optimum temperatures. A AnAOB rate of 39 mg N/g volatile suspended solids (VSS) d was observed in a laboratory scale anammox airlift at 10°C. Despite the difficulties, successful operation of one-stage PNA treating low strength synthetic wastewater, has been shown in MBBR at 13°C and 10°C (Gilbert *et al.*, 2014b; Persson *et al.*, 2014b), in a SBR at 12°C (Hu *et al.*, 2013) and in RBC at 15°C (De Clippeleir *et al.*, 2013).

Low N concentrations in the mainstream can further reduce AerAOB and AnAOB rates. AerAOB saturation constant for ammonium was estimated in 2.4 mg NH_4^+ -N/l (Wiesmann, 1994), which is already in the order of the ammonium concentrations expected in the reactor. While in suspended growth systems, AnAOB saturation constants for ammonium and nitrite are much lower than mainstream N concentrations, e.g.: $K_{s,}$ NO_2^- = 0.035 mg NO_2^- -N/l (Lotti *et al.*, 2014), $K_{s,}$ NH_4^+ = 0.07 mg NH_4^+ -N/l (Strous *et al.*, 1998), mass transfer limitation within biofilms generate high apparent saturation constants in the order of N concentrations in the water line, i.e.: high apparent saturation constants were calculated for AnAOB granules (2.4 ± 0.6 mm diameter) at 30°C, $K_{s,}$ NO_2^- = 4.90 mg NO_2^- -N/l, $K_{s,\ NH4+}$ = 8.96 mg NH_4^+ – N/l(Puyol *et al.*, 2013). Indeed, a decrease in N conversion rates due to low substrate availability was observed in a MBBR operated at 20°C, when treating sewage-like N concentrations (Gilbert *et al.*, 2014b).

The suppression of NOB activity is the main challenge brought by low temperature and low ammonium influent concentrations. High FA concentrations, usually present in reactors treating high strength ammoniacal wastewaters, pose inhibition to NOB, giving an advantage to AerAOB and AnAOB growth. Since NH_3 inhibition of NOB is unlikely during treatment of sewage, other strategies need to be followed to suppress aerobic nitrite oxidation. In suspended or hybrid growth systems and at moderate temperatures (25–30°C), maintaining low aerobic SRT favours NOB washout. This strategy is not possible in biofilm reactors, due to the impossibility to uncouple aerobic SRT from anoxic SRT. The application of low DO levels, a common strategy at moderate temperatures, cannot ensure NOB suppression at low temperatures, as *Nitrospira* spp. are K-strategists, with high affinity for oxygen, which facilitates NOB acclimation to low DO (Laanbroek and Gerards, 1993; Sliekers *et al.*, 2005). Some authors have put their efforts in the exploration of another strategy to suppress NOB activity, which is based on the application of alternating anoxic and aerobic conditions in the reactor. A delay or lag time in nitrate production by NOB, with respect to AerAOB oxidation of ammonia, has been observed when changing from anoxic to aerobic conditions (Alleman, 1984; Turk and Mavinic, 1986; Villaverde *et al.*, 2000). This phenomenon has been further studied with *Nitrospira* spp. at temperatures as low as 10°C, where short anoxic periods (5–20 min) were able to shut down NOB metabolism causing lag phases in nitrate production after restoration of aerobic conditions (Gilbert *et al.*, 2014a). This strategy has been successfully applied in full-scale hybrid growth PNA systems treating high strength wastewater at 30°C (Jardin and Hennerkes, 2011).

While application of one-stage PNA for mainstream N removal is a promising technology, important challenges still need to be solved. Stringent treatment standards cannot be met without effective suppression of NOB. Additionally, low removal rates, accompanied by the low growth rates of AnAOB at low temperature, require high catalyst concentrations, as well as very long SRT. This can only be achieved by efficient separation of the biomass, which is very much facilitated in biofilm systems.

Future trends

Existing applications of one stage PNA process are restricted to nitrogenous wastewaters from sludge digestion, landfill leachates and industry. Nonetheless, current research efforts aim to expand the field of application of PNA to other abundant nitrogen-containing wastewaters with low COD:N ratio in high volumetric loading rate processes. For example, the treatment of diluted source-separated urine has been demonstrated at the laboratory scale (Udert *et al.*, 2008) and technical scale trials have been done to treat digested black water (Meulman *et al.*, 2010; Vlaeminck *et al.*, 2009). The livestock industry generates waste manure with high N concentration which may be treated through PNA. Given the high concentration of N in manure (3–5 g N/l), treatment with PNA would bring considerable savings. The treatment of manure digestate with anammox based technologies has been explored at lab scale (Bernet and Béline, 2009; Hwang *et al.*, 2005; Molinuevo *et al.*, 2009; Scaglione *et al.*, 2013; Yamamoto *et al.*, 2011). However in these studies, the N removal is performed in two-stage systems, and the influent is diluted to mitigate the toxicity caused by the complex organic matrix. To date, the application of one stage systems for PNA in manure remains to be demonstrated.

References

Abma, W.R., Driessen, W., Haarhuis, R., and van Loosdrecht, M.C.M. (2010). Upgrading of sewage treatment plant by sustainable and cost-effective separate treatment of industrial wastewater. Water Sci. Technol. *61*, 1715–1722.

Abma, W.R., Schultz, C.E., Mulder, J.W., van der Star, W.R.L., Strous, M., Tokutomi, T., and van Loosdrecht, M.C.M. (2007). Full-scale granular sludge anammox process. Water Sci. Technol. *55*, 27–33.

Alleman, J.E. (1984). Elevated occurrence of nitrite in biological wastewater treatment sustems. Water Sci. Technol. *17*, 409–419.

Almstrand, R., Persson, F., Daims, H., Ekenberg, M., Christensson, M., Wilén, B.-M., Sörensson, F., and Hermansson, M. (2014). Three-dimensional stratification of bacterial biofilm populations in a moving bed biofilm reactor for nitritation-anammox. Int. J. Mol. Sci. *15*, 2191–2206.

Anthonisen, A.C., Loehr, R.C., Prakasam, T.B.S., and Srinath, E.G. (1976). Inhibition of nitrification by ammonia and nitrous-acid. J. Water Pollut. Con. F. *48*, 835–852.

Arp, D.J., and Stein, L.Y. (2003). Metabolism of inorganic N compounds by ammonia-oxidizing bacteria. Crit. Rev. Biochem. Mol. *38*, 471–495.

Aspegren, H., Nyberg, U., Andersson, B., Gotthardsson, S., and Jansen, J.l.C. (1998). Post denitrification in a moving bed biofilm reactor process. Water Sci. Technol. *38*, 31–38.

Barnes, D., and Bliss, P.J. (1983). Biological control of nitrogen in wastewater treatment (E., and FN. Spon., London, UK).

Barwal, A., and Chaudhary, R. (2014). To study the performance of biocarriers in moving bed biofilm reactor (MBBR) technology and kinetics of biofilm for retrofitting the existing aerobic treatment systems: a review. Rev. Environ. Sci. Biotechnol. *13*, 285–299.

Bernet, N., and Béline, F. (2009). Challenges and innovations on biological treatment of livestock effluents. Bioresource Technol. *100*, 5431–5436.

Bock, E., Schmidt, I., Stuven, R., and Zart, D. (1995). Nitrogen loss caused by denitrifying Nitrosomonas cells using ammonium or hydrogen as electron-donors and nitrite as electron-acceptor. Arch. Microbiol. *163*, 16–20.

Campos, J.L., Mendez, R., and Lema, J.M. (2000). Operation of a nitrifying activated sludge airlift (NASA) reactor without biomass carrier. Water Sci. Technol. *41*, 113–120.

Carvajal-Arroyo, J.M., Puyol, D., Li, G.B., Lucero-Acuna, A., Sierra-Alvarez, R., and Field, J.A. (2014). Pre-exposure to nitrite in the absence of ammonium strongly inhibits anammox. Water Res. *48*, 52–60.

Carvajal-Arroyo, J.M., Sun, W.J., Sierra-Alvarez, R., and Field, J.A. (2013). Inhibition of anaerobic ammonium oxidizing (anammox) enrichment cultures by substrates, metabolites and common wastewater constituents. Chemosphere *91*, 22–27.

Cema, G., Płaza, E., Trela, J., and Surmacz-Górska, J. (2011). Dissolved oxygen as a factor influencing nitrogen removal rates in a one-stage system with partial nitritation and Anammox process. Water Sci. Technol. *64*, 1009–1015.

Christensson, M., Ekström, S., Andersson Chan, A., Le Vaillant, E., and Lemaire, R. (2013). Experience from start-ups of the first ANITA Mox Plants. Water Sci. Technol. *67*, 2677–2684.

Colliver, B.B., and Stephenson, T. (2000). Production of nitrogen oxide and dinitrogen oxide by autotrophic nitrifiers. Biotechnol. Adv. *18*, 219–232.

Cortez, S., Teixeira, P., Oliveira, R., and Mota, M. (2008). Rotating biological contactors: a review on main factors affecting performance. Rev. Environ. Sci. Biotechnol. *7*, 155–172.

Courtens, E.N.P., Boon, N., De Clippeleir, H., Berckmoes, K., Mosquera, M., Seuntjens, D., and Vlaeminck, S.E. (2014). Control of nitratation in an oxygen-limited autotrophic nitrification/denitrification rotating biological contactor through disc immersion level variation. Bioresource Technol. *155*, 182–188.

Dalsgaard, T., Thamdrup, B., and Canfield, D.E. (2005). Anaerobic ammonium oxidation (anammox) in the marine environment. Res. Microbiol. *156*, 457–464.

Daverey, A., Su, S.-H., Huang, Y.-T., Chen, S.-S., Sung, S., and Lin, J.-G. (2013). Partial nitrification and anammox process: A method for high strength optoelectronic industrial wastewater treatment. Water Res. *47*, 2929–2937.

De Clippeleir, H., Defoirdt, T., Vanhaecke, L., Vlaeminck, S.E., Carballa, M., Verstraete, W., and Boon, N. (2011). Long-chain acylhomoserine lactones increase the anoxic ammonium oxidation rate in an OLAND biofilm. Appl. Microbiol. Biot. *90*, 1511–1519.

De Clippeleir, H., Vlaeminck, S.E., Carballa, M., and Verstraete, W. (2009). A low volumetric exchange ratio allows high autotrophic nitrogen removal in a sequencing batch reactor. Bioresource Technol. *100*, 5010–5015.

De Clippeleir, H., Vlaeminck, S.E., De Wilde, F., Daeninck, K., Mosquera, M., Boeckx, P., Verstraete, W., and Boon, N. (2013). One-stage partial nitritation/anammox at 15A degrees C on pretreated sewage: feasibility demonstration at lab-scale. Appl. Microbiol. Biot. *97*, 10199–10210.

Desloover, J., De Clippeleir, H., Boeckx, P., Du Laing, G., Colsen, J., Verstraete, W., and Vlaeminck, S.E. (2011). Floc-based sequential partial nitritation and anammox at full scale with contrasting N2O emissions. Water Res. *45*, 2811–2821.

Desloover, J., Vlaeminck, S.E., Clauwaert, P., Verstraete, W., and Boon, N. (2012). Strategies to mitigate N2O emissions from biological nitrogen removal systems. Curr. Opin. Biotechnol. 23, 474–482.

Di Trapani, D., Christensson, M., Torregrossa, M., Viviani, G., and Ødegaard, H. (2013). Performance of a hybrid activated sludge/biofilm process for wastewater treatment in a cold climate region: Influence of operating conditions. Biochem. Eng. J. *77*, 214–219.

Dosta, J., Fernandez, I., Vazquez-Padin, J.R., Mosquera-Corral, A., Campos, J.L., Mata-Alvarez, J., and Mendez, R. (2008). Short- and long-term effects of temperature on the Anammox process. J. Hazard. Mater. *154*, 688–693.

Egli, K., Bosshard, F., Werlen, C., Lais, P., Siegrist, H., Zehnder, A.J.B., and van der Meer, J.R. (2003). Microbial composition and structure of a rotating biological contactor biofilm treating ammonium-rich wastewater without organic carbon. Microb. Ecol. *45*, 419–432.

Egli, K., Fanger, U., Alvarez, P.J.J., Siegrist, H., van der Meer, J.R., and Zehnder, A.J.B. (2001). Enrichment and characterization of an anammox bacterium from a rotating biological contactor treating ammonium-rich leachate. Arch. Microbiol. *175*, 198–207.

Fuerst, J.A. (1995). The Planctomycetes – Emerging models for microbial ecology, evolution and cell biology. Microbiology UK *141*, 1493–1506.

Furukawa, K., Lieu, P.K., Tokitoh, H., and Fujii, T. (2006). Single-stage nitrogen removal using ANAMMOX and partial nitritation (SNAP). Water Sci. Technol. *53*, 83–90.

Fux, C., Huang, D., Monti, A., and Siegrist, H. (2004). Difficulties in maintaining long-term partial nitritation of ammonium-rich sludge digester liquids in a moving-bed biofilm reactor (MBBR). Water Sci. Technol. *49*, 53–60.

Gilbert, E.M., Agrawal, S., Brunner, F., Schwartz, T., Horn, H., and Lackner, S. (2014a). Response of different Nitrospira species to anoxic periods depends on operational DO. Environ. Sci. Technol. *48*, 2934–2941.

Gilbert, E.M., Agrawal, S., Karst, S.M., Horn, H., Nielsen, P.H., and Lackner, S. (2014b). Low temperature partial nitritation/anammox in a moving bed biofilm reactor treating low strength wastewater. Environ. Sci. Technol. *48*, 8784–8792.

Gilmore, K.R., Terada, A., Smets, B.F., Love, N.G., and Garland, J.L. (2013). Autotrophic nitrogen removal in a membrane-aerated biofilm reactor under continuous aeration: a demonstration. Environ. Eng. Sci. *30*, 38–45.

Gong, Z., Liu, S.T., Yang, F.L., Bao, H., and Furukawa, K. (2008). Characterization of functional microbial community in a membrane-aerated biofilm reactor operated for completely autotrophic nitrogen removal. Bioresource Technol. *99*, 2749–2756.

Gong, Z., Yang, F., Liu, S., Bao, H., Hu, S., and Furukawa, K. (2007). Feasibility of a membrane-aerated biofilm reactor to achieve single-stage autotrophic nitrogen removal based on Anammox. Chemosphere *69*, 776–784.

Hao, X., Heijnen, J.J., and van Loosdrecht, M.C.M. (2002). Sensitivity analysis of a biofilm model describing a one-stage completely autotrophic nitrogen removal (CANON) process. Biotechnol. Bioeng. *77*, 266–277.

Hellinga, C., Schellen, A., Mulder, J.W., van Loosdrecht, M.C.M., and Heijnen, J.J. (1998). The SHARON process: An innovative method for nitrogen removal from ammonium-rich waste water. Water Sci. Technol. *37*, 135–142.

Helmer, C., Kunst, S., Juretschko, S., Schmid, M.C., Schleifer, K.H., and Wagner, M. (1999). Nitrogen loss in a nitrifying biofilm system. Water Sci. Technol. *39*, 13–21.

Helmer, C., Schmid, M., Filipov, E., Gaul, T., Hippen, A., Rosenwinkel, K.-H., Seyfried, C.F., Wagner, M., and Kunst, S. (2002). Deammonification in biofilm systems: population structure and function. Water Sci. Technol. *46*, 223–231.

Hippen, A., Rosenwinkel, K.-H., Baumgarten, G., and Seyfried, C.F. (1997). Aerobic deammonification: A new experience in the treatment of waste waters. Water Sci. Technol. *35*, 111–120.

Hooper, A.B., and Terry, K.R. (1979). Hydroxylamine oxidoreductase of Nitrosomonas. Production of nitric-oxide from hydroxylamine. Biochim. Biophys. Acta *571*, 12–20.

Hu, Z., Lotti, T., de Kreuk, M., Kleerebezem, R., van Loosdrecht, M., Kruit, J., Jetten, M.S.M., and Kartal, B. (2013). Nitrogen removal by a nitritation-anammox bioreactor at low temperature. Applied and Environ. Microbiol. *79*, 2807–2812.

Hwang, I.S., Min, K.S., Choi, E., and Yun, Z. (2005). Nitrogen removal from piggery waste using the combined SHARON and ANAMMOX process. Water Sci. Technol. *52*, 487–494.

Jardin, N., and Hennerkes, J. (2011). Full scale experience with the deammonification process to treat high strength sludge water – A case study. Paper presented at: 11th IWA Specialised Conference on Design, Operation and Economics of Large Wastewater Treatment Plants (Budapest).

Jenni, S., Vlaeminck, S.E., Morgenroth, E., and Udert, K.M. (2014). Successful application of nitritation/anammox to wastewater with elevated organic carbon to ammonia ratios. Water Res. *49*, 316–326.

Jetten, M.S.M., Sliekers, O., Kuypers, M., Dalsgaard, T., van Niftrik, L., Cirpus, I., van de Pas-Schoonen, K., Lavik, G., Thamdrup, B., Le Paslier, D., *et al.* (2003). Anaerobic ammonium oxidation by marine and freshwater planctomycete-like bacteria. Appl. Microbiol. Biot. *63*, 107–114.

Jin, R.-C., Zheng, P., Mahmood, Q., and Zhang, L. (2008). Performance of a nitrifying airlift reactor using granular sludge. Sep. Purif. Technol. *63*, 670–675.

Joss, A., Salzgeber, D., Eugster, J., König, R., Rottermann, K., Burger, S., Fabijan, P., Leumann, S., Mohn, J., and Siegrist, H. (2009). Full-scale nitrogen removal from digester liquid with partial nitritation and anammox in one SBR. Environ. Sci. Technol. *43*, 5301–5306.

Kampschreur, M.J., Kleerebezem, R., de Vet, W.W.J.M., and van Loosdrecht, M.C.M. (2011). Reduced iron induced nitric oxide and nitrous oxide emission. Water Res. *45*, 5945–5952.

Kampschreur, M.J., Poldermans, R., Kleerebezem, R., van der Star, W.R.L., Haarhuis, R., Abma, W.R., Jetten, M.S.M., and van Loosdrecht, M.C.M. (2009). Emission of nitrous oxide and nitric oxide from a full-scale single-stage nitritation-anammox reactor. Water Sci. Technol. *60*, 3211–3217.

Kampschreur, M.J., van der Star, W.R.L., Wielders, H.A., Mulder, J.W., Jetten, M.S.M., and van Loosdrecht, M.C.M. (2008). Dynamics of nitric oxide and nitrous oxide emission during full-scale reject water treatment. Water Res. *42*, 812–826.

Kartal, B., Koleva, M., Arsov, R., van der Star, W., Jetten, M.S.M., and Strous, M. (2006). Adaptation of a freshwater anammox population to high salinity wastewater. J. Biotechnol. *126*, 546–553.

Kartal, B., Kuenen, J.G., and van Loosdrecht, M.C.M. (2010a). Sewage treatment with anammox. Science *328*, 702–703.

Kartal, B., Maalcke, W.J., de Almeida, N.M., Cirpus, I., Gloerich, J., Geerts, W., den Camp, H., Harhangi, H.R., Janssen-Megens, E.M., Francoijs, K.J., *et al.* (2011). Molecular mechanism of anaerobic ammonium oxidation. Nature *479*, 127-U159.

Kartal, B., Rattray, J., van Niftrik, L.A., van de Vossenberg, J., Schmid, M.C., Webb, R.I., Schouten, S., Fuerst, J.A., Damste, J.S.S., Jetten, M.S.M., *et al.* (2007). Candidatus 'Anammoxoglobus propionicus' a new propionate oxidizing species of anaerobic ammonium oxidizing bacteria. Syst. Appl. Microbiol. *30*, 39–49.

Kartal, B., Tan, N.C.G., Van de Biezen, E., Kampschreur, M.J., Van Loosdrecht, M.C.M., and Jetten, M.S.M. (2010b). Effect of nitric oxide on anammox bacteria. Applied and Environ. Microbiol. *76*, 6304–6306.

Kim, B.J., and Molof, A.H. (1982). The scale-up and limitation of physical oxygen-transfer in rotating biological contactors. Water Sci. Technol. *14*, 569–579.

Knowles, G., Downing, A.L., and Barrett, M.J. (1965). Determination of kinetic constants for nitrifying bacteria in mixed culture with aid of an electronic computer. J. Gen. Microbiol. *38*, 263–278.

Kuai, L.P., and Verstraete, W. (1998). Ammonium removal by the oxygen-limited autotrophic nitrification-denitrification system. Applied and Environ. Microbiol. *64*, 4500–4506.

Kuypers, M.M.M., Sliekers, A.O., Lavik, G., Schmid, M., Jorgensen, B.B., Kuenen, J.G., Damste, J.S.S., Strous, M., and Jetten, M.S.M. (2003). Anaerobic ammonium oxidation by anammox bacteria in the Black Sea. Nature *422*, 608–611.

Laanbroek, H.J., and Gerards, S. (1993). Competition for limiting amounts of oxygen between Nitrosomonas europaea and Nitrobacter winogradskyi grown in mixed continous cultures. Arch. Microbiol. *159*, 453–459.

Lackner, S., Gilbert, E.M., Vlaeminck, S.E., Joss, A., Horn, H., and van Loosdrecht, M.C.M. (2014). Full-scale partial nitritation/anammox experiences – An application survey. Water Res. *55*, 292–303.

Lackner, S., Terada, A., and Smets, B.F. (2008). Heterotrophic activity compromises autotrophic nitrogen removal in membrane-aerated biofilms: Results of a modeling study. Water Res. *42*, 1102–1112.

Lemaire, R., Webb, R.I., and Yuan, Z.G. (2008). Micro-scale observations of the structure of aerobic microbial granules used for the treatment of nutrient-rich industrial wastewater. ISME J. *2*, 528–541.

Lettinga, G., van Velsen, A.F.M., Hobma, S.W., de Zeeuw, W., and Klapwijk, A. (1980). Use of the upflow sludge blanket (USB) reactor concept for biological wastewater treatment, especially for anaerobic treatment. Biotechnol. Bioeng. *22*, 699–734.

Lotti, T., Kleerebezem, R., Lubello, C., and van Loosdrecht, M.C.M. (2014). Physiological and kinetic characterization of a suspended cell anammox culture. Water Res. *60*, 1–14.

Mathure, P., and Patwardhan, A. (2005). Comparison of mass transfer efficiency in horizontal rotating packed beds and rotating biological contactors. J. Chem. Technol. Biot. *80*, 413–419.

Mehrdad, M., Park, H., Ramalingam, K., Fillos, J., Beckmann, K., Deur, A., and Chandran, K. (2014). Anammox moving bed biofilm reactor pilot at the 26th Ward wastewater treatment plants in Brooklyn, New York: Start-up, biofilm population diversity and performance optimization. Water Sci. Technol. *Uncorrected Proof.*

Meulman, B., Elzinga, N., Gorter, K., Zeeman, G., Buisman, C.J.N., Vlaeminck, S.E., and Verstraete, W. (2010). Pilotscale demonstration of sustainable C and N removal from concentrated black water. In WA World Water Congress and Exhibition (Montréal, QC, Canada).

Mishima, K., and Nakamura, M. (1991). Self-immobilization of aerobic activated sludge—A pilot study of the aerobic upflow sludge blanket process in municipal sewage treatment. Water Sci. Technol. *23*, 981–990.

Molinuevo, B., Garcia, M.C., Karakashev, D., and Angelidaki, I. (2009). Anammox for ammonia removal from pig manure effluents: Effect of organic matter content on process performance. Bioresource Technol. *100*, 2171–2175.

Mosier, A., Kroeze, C., Nevison, C., Oenema, O., Sybil, S., and van Cleemput, O. (1999). An overview of the revised 1996 IPCC guidelines for national greenhouse gas inventory methodology for nitrous oxide from agriculture. Environ. Sci. Policy 2, 325–333.

Nielsen, M., Bollmann, A., Sliekers, O., Jetten, M., Schmid, M., Strous, M., Schmidt, I., Larsen, L.H., Nielsen, L.P., and Revsbech, N.P. (2005). Kinetics, diffusional limitation and microscale distribution of chemistry and organisms in a CANON reactor. FEMS Microbiol. Ecol. *51*, 247–256.

Ødegaard, H. (2006). Innovations in wastewater treatment: -the moving bed biofilm process. Water Sci. Technol. *53*, 17–33.

Paolini, A.E. (1986). Effect of biomass on oxygen-transfer in rbc systems. J. Water Pollut. Con. F. *58*, 306–311.

Pastorelli, G., Canziani, R., Pedrazzi, L., and Rozzi, A. (1999). Phosphorus and nitrogen removal in moving-bed sequencing batch biofilm reactors. Water Sci. Technol. *40*, 169–176.

Patwardhan, A.W. (2003). Rotating biological contactors: a review. Industrial and Engineering Chemistry Research *42*, 2035–2051.

Pellicer-Nacher, C., Franck, S., Gulay, A., Ruscalleda, M., Terada, A., Al-Soud, W.A., Hansen, M.A., Sorensen,

S.J., and Smets, B.F. (2014). Sequentially aerated membrane biofilm reactors for autotrophic nitrogen removal: microbial community composition and dynamics. Microb. Biotechnol. *7*, 32–43.

Pellicer-Nàcher, C., Sun, S., Lackner, S., Terada, A., Schreiber, F., Zhou, Q., and Smets, B.F. (2010). Sequential aeration of membrane-aerated biofilm reactors for high-rate autotrophic nitrogen removal: experimental demonstration. Environ. Sci. Technol. *44*, 7628–7634.

Persson, F., Sultana, R., Suarez, M., Hermansson, M., Plaza, E., and Wilén, B.-M. (2014a). Structure and composition of biofilm communities in a moving bed biofilm reactor for nitritation–anammox at low temperatures. Bioresource Technol. *154*, 267–273.

Persson, F., Sultana, R., Suarez, M., Hermansson, M., Plaza, E., and Wilen, B.M. (2014b). Structure and composition of biofilm communities in a moving bed biofilm reactor for nitritation-anammox at low temperatures. Bioresource Technol. *154*, 267–273.

Poughon, L., Dussap, C.-G., and Gros, J.-B. (2001). Energy model and metabolic flux analysis for autotrophic nitrifiers. Biotechnol. Bioeng. *72*, 416–433.

Puyol, D., Carvajal-Arroyo, J.M., Garcia, B., Sierra-Alvarez, R., and Field, J.A. (2013). Kinetic characterization of Brocadia spp.-dominated anammox cultures. Bioresource Technol. *139*, 94–100.

Puyol, D., Carvajal-Arroyo, J.M., Sierra-Alvarez, R., and Field, J.A. (2014). Nitrite (not free nitrous acid) is the main inhibitor of the anammox process at common pH conditions. Biotechnol. Lett. *36*, 547–551.

Pynaert, K., Smets, B.F., Wyffels, S., Beheydt, D., Siciliano, S.D., and Verstraete, W. (2003). Characterization of an autotrophic nitrogen-removing biofilm from a highly loaded lab-scale rotating biological contactor. Applied and Environ. Microbiol. *69*, 3626–3635.

Pynaert, K., Sprengers, R., Laenen, J., and Verstraete, W. (2002). Oxygen-limited nitrification and denitrification in a lab-scale rotating biological contactor. Environ. Technol. *23*, 353–362.

Quan, Z.X., Rhee, S.K., Zuo, J.E., Yang, Y., Bae, J.W., Park, J.R., Lee, S.T., and Park, Y.H. (2008). Diversity of ammonium-oxidizing bacteria in a granular sludge anaerobic ammonium-oxidizing (anammox) reactor. Environ. Microbiol. *10*, 3130–3139.

Rosenwinkel, K.H., and Cornelius, A. (2005). Deammonification in the moving-bed process for the treatment of wastewater with high ammonia content. Chem. Eng. Technol. *28*, 49–52.

Rusten, B., Ødegaard, H., and Lundar, A. (1992). Treatment of dairy wastewater in a novel moving bed biofilm reactor. Water Sci. Technol. *26*, 703–711.

Rysgaard, S., Glud, R.N., Risgaard-Petersen, N., and Dalsgaard, T. (2004). Denitrification and anammox activity in Arctic marine sediments. Limnol. Oceanogr. *49*, 1493–1502.

Scaglione, D., Tornotti, G., Teli, A., Ficara, E., R., C., and F., M. (2013). Advanced bioprocesses for N removal from the liquid fraction of co-digestated piggery/poultry manure and agro-wastes. In 13th World congress on anaerobic digestion (Santiago de Compostela, Spain).

Scheehle, E., Godwin, D., and Ottinger, D. (2006). Global anthropogenic non-CO2 greenhouse gas emissions: 1999–2020 (Washington, DC, U.S. Environmental Protection Agency).

Schmid, M., Twachtmann, U., Klein, M., Strous, M., Juretschko, S., Jetten, M., Metzger, J.W., Schleifer, K.H., and Wagner, M. (2000). Molecular evidence for genus level diversity of bacteria capable of catalyzing anaerobic ammonium oxidation. Syst. Appl. Microbiol. *23*, 93–106.

Schmid, M., Walsh, K., Webb, R., Rijpstra, W.I.C., van de Pas-Schoonen, K., Verbruggen, M.J., Hill, T., Moffett, B., Fuerst, J., Schouten, S., *et al.* (2003). Candidatus 'Scalindua brodae', sp nov., Candidatus 'Scalindua wagneri', sp nov., two new species of anaerobic ammonium oxidizing bacteria. Syst. Appl. Microbiol. *26*, 529–538.

Siegrist, H., Reithaar, S., Koch, G., and Lais, P. (1998). Nitrogen loss in a nitrifying rotating contactor treating ammonium-rich wastewater without organic carbon. Water Sci. Technol. *38*, 241–248.

Siegrist, H., Salzgeber, D., Eugster, J., and Joss, A. (2008). Anammox brings WWTP closer to energy autarky due to increased biogas production and reduced aeration energy for N-removal. Water Sci. Technol. *57*, 383–388.

Sliekers, A.O., Haaijer, S.M., Stafsnes, M., Kuenen, J.G., and Jetten, M.M. (2005). Competition and coexistence of aerobic ammonium- and nitrite-oxidizing bacteria at low oxygen concentrations. Appl. Microbiol. Biot. *68*, 808–817.

Solomon, S., Qin, D., Manning, M., Chen, Z., Marquis, M., Averyt, K.B., Tignor, M., and Miller, H.L. (2007). In Climate Change 2007: The Physical Science Basis, S. Solomon, D. Qin, M. Manning, Z. Chen, M. Marquis, A.B. Averyt, M. Tignor, and H.L. Miller, eds. (Cambridge University Press, Cambridge, UK), pp. 21–92

van der Star, W.R.L., Abma, W.R., Blommers, D., Mulder, J.W., Tokutomi, T., Strous, M., Picioreanu, C., and Van Loosdrecht, M.C.M. (2007). Startup of reactors for anoxic ammonium oxidation: Experiences from the first full-scale anammox reactor in Rotterdam. Water Res. *41*, 4149–4163.

Strous, M., Heijnen, J.J., Kuenen, J.G., and Jetten, M.S.M. (1998). The sequencing batch reactor as a powerful tool for the study of slowly growing anaerobic ammonium-oxidizing microorganisms. Appl. Microbiol. Biot. *50*, 589–596.

Strous, M., Fuerst, J.A., Kramer, E.H.M., Logemann, S., Muyzer, G., van de Pas-Schoonen, K.T., Webb, R., Kuenen, J.G., and Jetten, M.S.M. (1999a). Missing lithotroph identified as new planctomycete. Nature *400*, 446–449.

Strous, M., Kuenen, J.G., and Jetten, M.S.M. (1999b). Key physiology of anaerobic ammonium oxidation. Appl. Environ. Microb. *65*, 3248–3250.

Szatkowska, B., Cema, G., Plaza, E., Trela, J., and Hultman, B. (2007). A one-stage system with partial nitritation and Anammox processes in the moving-bed biofilm reactor. Water Sci. Technol. *55*, 19–26.

Tallec, G., Garnier, J., Billen, G., and Gousailles, M. (2006). Nitrous oxide emissions from secondary activated sludge in nitrifying conditions of urban wastewater treatment plants: Effect of oxygenation level. Water Res. *40*, 2972–2980.

Terada, A., Lackner, S., Tsuneda, S., and Smets, B.F. (2007). Redox-stratification controlled biofilm (ReSCoBi) for completely autotrophic nitrogen removal: The effect of co- versus counter-diffusion on reactor performance. Biotechnol. Bioeng. *97*, 40–51.

Third, K.A., Sliekers, A.O., Kuenen, J.G., and Jetten, M.S.M. (2001). The CANON system (completely autotrophic nitrogen-removal over nitrite) under ammonium limitation: Interaction and competition between three groups of bacteria. Syst. Appl. Microbiol. *24*, 588–596.

Turk, O., and Mavinic, D.S. (1986). Preliminary assessment of a shortcut in nitrogen removal from wastewater. Can. J. Civil Eng. *13*, 600–605.

Udert, K.M., Kind, E., Teunissen, M., Jenni, S., and Larsen, T.A. (2008). Effect of heterotrophic growth on nitritation/anammox in a single sequencing batch reactor. Water Sci. Technol. *58*, 277–284.

Vázquez-Padín, J., Mosquera-Corral, A., Campos, J.L., Méndez, R., and Revsbech, N.P. (2010). Microbial community distribution and activity dynamics of granular biomass in a CANON reactor. Water Res. *44*, 4359–4370.

Vázquez-Padín, J.R., Figueroa, M., Fernández, I., Mosquera-Corral, A., Campos, J.L., and Méndez, R. (2009). Post-treatment of effluents from anaerobic digesters by the Anammox process. Water Sci. Technol. *60*, 1135–1143.

Verstraete, W., and Vlaeminck, S.E. (2011). ZeroWasteWater: short-cycling of wastewater resources for sustainable cities of the future. Int. J. Sust. Dev. World *18*, 253–264.

Villaverde, S., Encina, P.A.G., Lacalle, M.L., and Fdz-Polanco, F. (2000). New operational strategy for SBR technology for total nitrogen removal from industrial wastewaters highly loaded with nitrogen. Water Sci. Technol. *41*, 85–93.

Vlaeminck, S.E., Cloetens, L.F., Carballa, M., Boon, N., and Verstraete, W. (2009). Granular biomass capable of partial nitritation and anammox. Water Sci. Technol. *58*, 1113–1120.

Vlaeminck, S.E., De Clippeleir, H., and Verstraete, W. (2012). Microbial resource management of one-stage partial nitritation/anammox. Microb. Biotechnol. *5*, 433–448.

Vlaeminck, S.E., Terada, A., Smets, B.F., De Clippeleir, H., Schaubroeck, T., Bolca, S., Demeestere, L., Mast, J., Boon, N., Carballa, M., *et al.* (2010). Aggregate size and architecture determine microbial activity balance for one-stage partial nitritation and anammox. Applied and Environ. Microbiol. *76*, 900–909.

Vlaeminck, S.E., Terada, A., Smets, B.F., Van der Linden, D., Boon, N., Verstraete, W., and Carballa, M. (2009). Nitrogen removal from digested black water by one-stage partial nitritation and anammox. Environ. Sci. Technol. *43*, 5035–5041.

Volcke, E.I.P., Picioreanu, C., De Baets, B., and van Loosdrecht, M.C.M. (2012). The granule size distribution in an anammox-based granular sludge reactor affects the conversion-Implications for modeling. Biotechnol. Bioeng. *109*, 1629–1636.

Wang, L., Zheng, P., Xing, Y., Li, W., Yang, J., Abbas, G., Liu, S., He, Z., Zhang, J., Zhang, H., *et al.* (2014). Effect of particle size on the performance of autotrophic nitrogen removal in the granular sludge bed reactor and microbiological mechanisms. Bioresour. Technol. *157*, 240–246.

Weissenbacher, N., Takacs, I., Murthy, S., Fuerhacker, M., and Wett, B. (2010). Gaseous nitrogen and carbon emissions from a full-scale deammonification plant. Water Environ. Res. *82*, 169–175.

Wett, B., Buchauer, K., and Fimml, C. (2007). Energy self sufficiency as a feasible concept for wastewater treatment systems. Paper presented at: IWA Leading Edge Technology Conference (Singapore).

Wiesmann, U. (1994). Biological nitrogen removal from wastewater. In Biotechnics/Wastewater, Fiechter, A., ed. (Springer, Berlin/Heidelberg, Germany), pp. 113–154.

Winkler, M.K., Kleerebezem, R., and van Loosdrecht, M.C. (2012). Integration of anammox into the aerobic granular sludge process for main stream wastewater treatment at ambient temperatures. Water Res. *46*, 136–144.

Wyffels, S., Boeckx, P., Pynaert, K., Verstraete, W., and Van Cleemput, O. (2003). Sustained nitrite accumulation in a membrane-assisted bioreactor (MBR) for the treatment of ammonium-rich wastewater. J. Chem. Technol. Biot. *78*, 412–419.

Yamamoto, T., Wakamatsu, S., Qiao, S., Hira, D., Fujii, T., and Furukawa, K. (2011). Partial nitritation and anammox of a livestock manure digester liquor and analysis of its microbial community. Bioresource Technol. *102*, 2342–2347.

Yu, R., Kampschreur, M.J., van Loosdrecht, M.C.M., and Chandran, K. (2010). Mechanisms and specific directionality of autotrophic nitrous oxide and nitric oxide generation during transient anoxia. Environ. Sci. Technol. *44*, 1313–1319.

Zhang, L., Jiang, J., Yang, J., Hira, D., and Furukawa, K. (2012). High rate nitrogen removal by the CANON process at ambient temperature. Water Sci. Technol. *65*, 1826–1833.

Zumft, W.G. (1997). Cell biology and molecular basis of denitrification. Microbiology and Molecular Biology Reviews *61*, 533–616.

Glossary

The following glossary includes the definition of key concepts used among the different chapters throughout the entire book.

Alpha-diversity: the local diversity of a community contained in a habitat patch.

Aquatic biofilms: complex microbial communities attached and growing on living and non-living surfaces found in marine or freshwater or man-made environment. In biofilms microbes are attached to each other and embedded in the matrix of self-produced extra polymeric substances.

Autotroph: organism that is capable of producing nutritive organic molecules from inorganic sources, using different energy sources (light, chemical reactions).

Benthos: community of organisms living in close relationship with their substrate, in general permanently attached.

Beta-diversity: the variability of species identities among communities across space and time. Beta-diversity can occur as directional turnover along a gradient or as non-directional variation.

Bioaccumulation: general term referring to the accumulation of chemical substances, such as metals or organic pollutants, in the biota. Bioaccumulation generally provides a good proxy of the bioavailability of compounds. The term *accumulation* is often used for biofilms when the methodology does not discern between the portion of bioaccumulated chemical from the portion remaining outside of the cells, mainly adsorbed to EPS and/or inorganic particles.

Biofilm biobarriers: are physical structures where biofilm develops that are constructed to enhance bioremediation of pollutants in aqueous environments such as groundwater and surface water.

Biodiversity/Diversity: the total variety and variability among living organisms, the communities and ecosystems they are part of. Diversity measures may consider the number of species (or some other biological units), i.e. richness, their relative abundances, and the varying dissimilarities between species.

Bioelectrochemical system (BES): biological technology capable to produce electrical power or chemical compounds by the action of a biocatalyst (exoelectrogenic microorganims) generally using wastewater as anodic fuel.

Biofouling: attachment and growth of microorganisms, algae, and invertebrates on submerged surfaces. The term is specially used for man-made surfaces, such as in water distribution systems. It is also called *microbial fouling*. Such accumulation is also referred to as *epibiosis* when the host surface is another organism and the relationship is not parasitic.

Community: a group of interacting species that overlap in time and space.

Community ecotoxicology: is the study of the effects of toxicants on ecological systems focusing on the effects at community level.

Community level physiological profile (CLPP): is a rapid community-level culture approach which is used to characterize the metabolic profile of microbial communities by measuring the utilization of a range of different carbon sources.

Confocal laser scanning microscopy (CLSM): microscopy technique which allows the visualization of images at different depths of the observed sample (such as a biofilm) to compile a three-dimension final image. Confocal laser scanning microscopy may be used to detect reflection, autofluorescence signals (such as pigments of phototrophs) or the emission signals of specific fluorochromes most commonly targeting nucleic acids, lipids, carbohydrates, or proteins.

Cooperation: beneficial interactions between organisms.

Coulombic efficiency (CE): parameter used to evaluate the electrical efficiency in Bioelectrochemical Systems (BES). CE is the ratio between electrons actually transferred and the electrons potentially transferred to the electrical circuit considering the oxidation-reduction reaction occurring at the bioelectrode.

Co-diffusion and Counter-diffusion: related to the direction of diffusion of substances in biofilms. Transport of substances occurs by diffusion within the biofilm layer and is driven by a concentration gradient. In co-diffusive systems, the gradients of the electron donor and electron acceptor are parallel. On the other hand, in counter-diffusive systems, both gradients are opposed and thus electron donor and acceptor diffuse in opposite directions.

Dispersal: the movement of individuals from one habitat patch to another. Microorganisms are assumed to have higher dispersal rates due to their high population sizes, high transportability and short generation times, which should increase their chance to reach new habitats and establish populations therein.

Dissolved organic matter (DOM): the organic fraction of the dissolved material in water (defined as those passing a 0.5 μm filter). DOM contains a mix of organic compounds (including humic substances, polysaccharides, peptides, lipids). DOM is usually quantified in carbon units and then referred as DOC (dissolved organic carbon).

Electron transfer mechanisms: the microbial mechanisms to transport and release electrons inside the membranes (intracellular electron transfer (ICET)) and to/from an electrode material (extracellular electron transfer (EET)). Different EET mechanisms are defined, such as direct (direct contact, conductive biofilm and nanowire) or indirect (mediators) extracellular electron transfer. Both mechanisms can take place simultaneously in order to maximize the microbial benefits.

Electrotrophic microorganism: group of microorganisms capable to directly receive electrons from an electrode to grow. These microorganisms are able to reduce compounds taking electrons from the electrode surface.

Epilithic biofilm: biofilm attached on rocks or cobbles. It is also named *epilithon.*

Epiphytic biofilm: biofilm growing on living plants such as that developing on macrophytes. Significant interactions with plant (which might be both synergic and antagonistic) typically occur in this biofilm type.

Epipelic biofilm: biofilm attached to the particles of cohesive sediments (clay and silt). Typical epipelic biofilms occur in intertidal areas where biofilms are mainly colonized by diatoms. It is also named *epipelon.*

Epipsammic biofilm: Biofilm attached to the particles of sandy sediments (sand and gravel). It is also called *epipsammon.*

Epixylic biofilm: Biofilm growing on dead plant material such as wood and leaves. Epixylic biofilm is mainly formed by fungi due to their ability to degrade lignocellulose compounds which are the main constituent of plant tissues. It is also called *epixylon.*

Evenness: a measure for the variation in species abundances in a community and reaches its maximum value when all species are equally abundant.

Exoelectrogenic microorganism: group of microorganisms that have the ability to transfer electrons extracellularly to the electrode material. In practical applications, exoelectrogenic microorganisms are able to oxidize organic matter from wastewater transferring electrons to the electrode material and generating electricity.

Extracellular enzymes: enzymes bound to the cell surface of microorganisms or in the periplasmatic space in gram-negative bacteria, acting outside the cell. In biofilms, free enzymes are also found within the EPS matrix. Extracellular enzymes play a key role in the decomposition of dissolved organic matter in aquatic environments since they degrade polymeric and macromolecular organic matter into low molecular weight molecules which can cross the bacterial cell membranes.

Extracellular polymeric substances (EPS): mainly composed by polysaccharides (but also by proteins, lipids, particulate material and detritus), EPS provide the mechanical stability of biofilms, mediate their adhesion to surfaces and form a cohesive three-dimensional polymer network that interconnects and transiently immobilizes biofilm cells. This matrix provides a protection from predation, toxic substances and physical perturbations.

Functional diversity: diversity of physiological or ecological units in a community. It is also a diversity measure (see Biodiversity/Diversity definition) which includes a functional notion of dissimilarity between species.

Gamma-diversity: is the diversity of a region holding multiple habitat patches with contained communities.

Granular biomass: compact granules and dense aggregates with an approximately spherical external appearance that do not coagulate under decreased hydrodynamic shear conditions and settle faster than flocs, allowing for better biomass retention and high volumetric conversion in a reactor.

Heterotroph: organism that obtains carbon from organic compounds.

Macrophytes: aquatic plants that grow in or near water. They can be floating, submersed or emerged. Aquatic macrophytes provide a substrate for algae and epiphytic biofilms.

Mass effects perspective: a meta-community perspective which emphasizes that spatial dynamics, such as source-sink effects or rescue effects, affect local community structures.

Meta-community: a set of local communities which are linked by dispersal of potentially interacting species. Several perspectives within the framework of meta-community ecology emphasize different mechanisms as potential drivers of local community composition (see the mass effect, the neutral, and the species sorting perspective definitions).

Methanogenesis: biogenic formation of methane (CH_4) as a form of anaerobic respiration in which the terminal electron acceptors are carbon compounds of low molecular weight. Methanogenesis occurring in sewers is carried out by Methanogenic Archaea (MA).

Microbially induced concrete corrosion (MICC): biological process occurring in biofilms growing on the crown of sewer pipes, surface exposed to the gas phase, that leads to the corrosion of sewers, cracking of the pipes and ultimately, structural collapse. MICC is caused by the biological production of sulfuric acid (H_2SO_4) from oxidation of hydrogen sulfide (H_2S) with oxygen (O_2) in sewers atmosphere.

Microphytobentos: synonym of the term "biofilm" but usually referred to populations of photoautotrophic microorganisms such as diatoms, euglenids, crysophyceans, dinoflagelates that colonize benthic substrata in marine systems, especially in intertidal and lower supra-tidal sediments where the light arrives.

Multichannel imaging: application of up to five separate excitation–detection combinations either simultaneously or sequentially in conventional CLSM.

Neutral perspective: a meta-community perspective, in which all species are similar in their competitive abilities. Dynamics are derived from dispersal and stochastic demographic processes.

Nestedness: degree of order/organization of a community, in which the number (due to gain and loss) of species is related to site-specific factors.

Next-generation sequencing (NGS): also known as high-throughput sequencing, is the catch-all term used to describe recent technologies (including, among others, Illumina (Solexa) sequencing and Roche 454 sequencing), which allow us to sequence relatively short DNA and RNA sequences along the entire genome much more quickly and cheaply than the previously used Sanger sequencing.

Nutrient uptake length: is the physical distance a molecule of a nutrient is transported in the flowing water until it is uptaken by microorganisms. In streams, it can be measured by adding a known content of nutrient (usually performed with phosphate, ammonia or nitrate) and follow its disappearance downstream. Nutrient uptake length is an important parameter for quantifying nutrient cycling in streams.

Nutrient stoichiometry: is the molar ratio between major nutrients (i.e. C, N and P) in organisms and in their food. The first author defining stoichiometry of aquatic organisms was Redfield who defined the molar C:N:P ratio of 106:16:1 for marine plankton.

Periphyton: synonym of the term "biofilm" and specially used in studies focusing on photosynthetic organisms.

Persistent organic pollutants (POPs): are chemical compounds that are carbon based and due to their structure possess chemical and physical characteristics that make them persistent to biodegradation in the environment for many years. They are distributed as a result of natural processes involving soil/sediment, water and air. The consequence is accumulation in fatty tissues and biomagnification in the food web, where they can cause diverse harmful effects due to their toxicity.

Phylogenetic diversity: a concept of diversity, which includes the phylogenetic dissimilarity between species, additionally to their presence/absence or abundance.

Pollusensitivity: sensitivity value of species towards levels of pollution (of various natures). This value (or the sensitivity profile of a species) is usually calculated from large datasets and describes species probability of presence in relationship with water quality.

Polychlorinated biphenyls (PCBs): are organic compounds that consist of two biphenyl rings onto which 1–10 chlorines are attached. All congeners are hydrophobic with a high biomagnification potential in the food chain, exhibit toxicity to a varying degree and a significant proportion display dioxin-like toxicity.

Quorum sensing: a form of bacterial population density depended chemical cell-to-cell communication and gene regulation.

Recovery: ability of a community under perturbation (e.g. a chemical) to restore its functional and structural attributes to initial values (measured before the perturbation) after the perturbation has ended.

Resistance: ability of a community to remain unchanged under perturbation (e.g. by the presence of a contaminant).

Sensitivity: in ecotoxicological studies it refers to the degree of modification in the function and structure of a community in response to a perturbation, such as the presence of a contaminant.

Sewers: underground network of physical structure-installations composed of pipelines, pump stations, manholes and channels that convey the wastewater from its source to the point where it is treated and discharged.

Solid retention time (SRT): is defined as the ratio between total biomass contained in a bioreactor and the rate at which biomass is washed out in the effluent. If the solid retention time is too short, slow growing microorganisms are not able to maintain their populations in the reactor and will eventually be completely washed out of the reactor.

Species sorting perspective: a meta-community perspective, which emphasizes spatial niche separation and local interactions between species and their abiotic environment.

Stable isotope probing (SIP): is a technique that is used to identify the microorganisms in environmental samples that use a particular growth substrate (i. e. organic pesticides). SIP is based on the incorporation of ^{13}C-labelled substrate into cellular biomarkers such as nucleic acids (DNA and rRNA), the separation of labelled from unlabeled nucleic acids by density gradient centrifugation, and molecular identification of active populations carrying labelled nucleic acid.

Stromatolites: ancient microbial mats which were abundant and diverse in the shallow zone of the oceans in the Proterozoic. Usually they are formed by precipitation and microbial carbonate sedimentation, resulting in a layered structure where typically cyanobacteria are present. Stromatolites played a crucial role for the early establishment of life since they consumed CO_2 and produced free O_2 and H_2.

Sulfate-reducing bacteria (SRB): group of bacteria predominant in anaerobic sewers that uses sulfate respiration to obtain energy while oxidizing organic compounds and reducing sulfate (SO_4^{-2}) to hydrogen sulfide (H_2S).

Suspended aggregates: highly fragile structures suspended in fresh and seawater made of microorganisms, organic and inorganic particles. Suspended aggregates typically occur during bloom periods after an increased input of nutrients. They are also called *lake snow*, *river snow* or *marine snow*.

Synergy: cooperation, not obligatory between microorganisms, in different processes that result in a greater benefit or production than if microorganisms were individual. In communities, cooperation helps in degradation because some strains produce the metabolites that are used for other species to complete degradation. In BESs, syntrophic interactions have been described on exoelectrogenic and non-exoelectrogenic bacteria in anode biofilms.

Syntrophic interactions: the use of one organism's metabolic intermediates and by-products by another organism as a metabolic substrate. It serves to increase internal nutrient cycling, reduce energy expenditures on resource acquisition, and can create favourable environmental conditions for novel niche development.

Taxonomic diversity: diversity based on the presence/absence or the abundance of taxonomic units, such as species, and is insensitive towards phylogenetic or functional differences among taxa.

Tolerance: the capacity of a community to support an alteration of its environment (e.g. exposure to toxicants), with no significant modifications in its structure and function.

Toxicity endpoint: is the measurement of a biological effect caused by exposure to a toxicant, usually measured as the 50% lost of the specific endpoint (i.e. metabolic activity such as organism growth, survival or reproduction, or biomass).

Xenobiotics: group of chemicals including for instance pesticides, polycyclic aromatic hydrocarbons (PAHs), polychlorinated biphenyls (PCBs) and antibiotics which is foreign to biological systems, but is commonly detected in the environment.

Index

A

Activated sludge 39, 100–101, 173, 175–180, 192, 211–212
Aerobic 10, 15, 39, 93, 102, 147–148, 154, 155, 160, 168, 169, 171–176, 194, 205, 207, 208, 210–217
Aerobic ammonium-oxidizing bacteria (AerAOB) 205–208, 210–217
Aggregate/suspended aggregates 3, 5, 6, 9–11, 16, 18–19, 29–31, 34–35, 39, 47–48, 55, 63–64, 70–73, 76, 89, 205, 207, 212, 213, 225
Algae 3–10, 12–13, 15, 19, 29, 31–34, 47–49, 52–53, 63, 65, 70–74, 91, 93, 96–97, 100, 111–112, 121, 126, 135, 137–138, 157, 167
Alpha-diversity 63, 68, 69, 223
Ammonium 93, 96–97, 134, 160, 205–208, 210–217
Anaerobic 8, 10, 15, 53, 93, 145, 147, 148–152, 155, 156, 160, 168, 169, 171–176, 183, 194, 211, 213
Anammox 205–207, 210, 213, 215–217
Anoxic 8, 10, 12, 15, 71, 147, 152, 153, 172, 205, 207, 210, 211–217
Anoxic ammonium oxidizing bacteria (anaob) 205–216
Antifouling 49–50, 54, 56
Archaea 5, 9, 19, 29, 39, 47, 50, 63, 71, 97–99, 145, 150, 151, 153, 154
Architecture 32, 34, 36, 47, 49–50, 53, 93, 100, 127, 192
Attachment 3, 9–11, 13, 16–17, 48–49, 53, 56, 66–67, 157, 211
Autotroph/autotrophic 4, 8, 65, 76, 93–96, 126–128, 133, 136, 158, 205, 209, 210, 212–216, 223

B

Bacteroidetes 10, 35, 39, 50, 63, 70–71, 176
Benthos 4–5, 11, 18, 90, 223
Beta-diversity 63, 67–70, 72, 75–76, 78, 223
Bioaccumulation 130, 135, 171, 223
Bioanode 186, 192
Bioassessment 111–113
Biobarrier 172–176, 223
Biocathode 189–192
Biocidal 49, 50, 154
Biodegradation 3, 167, 170–173, 175–177
Biodiversity/diversity 8, 13, 15–16, 33, 53, 63–80, 92, 99, 112, 114, 116, 119–120, 128, 135, 138, 147, 176, 210, 223
Bioelectrochemical system (BES) 183–190, 192, 194–196, 198, 223
Bioelectrogenic biofilm 183
Biofouling 3, 5–6, 11, 18, 47–49, 55, 56, 223
Bioindicator 111–113, 121
Biomonitoring 111–117, 120, 121
Bioremediation 3, 53, 98, 129, 167–169, 171, 173–177, 185, 190, 223
Biotic interaction 63–66, 68, 89, 118–119
Brominated flame retardants (BFRs) 167, 169, 170, 172

C

Carrier material 185, 205, 208, 211, 213
Chlorinated ethenes 167–170, 172
Ciliate 53, 71, 129
Co-difusion/counter-diffusion 208, 211, 213–214, 223
Colonization 8–11, 15–16, 35, 49, 51, 53, 64–65, 71, 73, 79, 91, 119, 133, 157
Community ecotoxicology 125, 137, 138, 223
Community level physiological profile (CLPP) 128, 223
Competition 7, 47, 49, 50, 64, 68, 72, 77, 117, 119, 152, 213
Conductive biofilm 187, 193, 194, 224
Confocal laser scanning microscopy (CLSM) 4, 30–38, 40, 90, 99–101, 127, 223
Cooperation 7, 47, 50, 53, 223
Corrosion 3, 48, 145, 148, 149, 155–160, 224
Coulombic efficiency (CE) 188, 190, 223
Cyanobacteria 5–6, 8, 10, 16, 19, 29, 31–34, 52, 54, 63, 70–74, 76, 93, 96, 100, 111, 121, 225

D

Decay/decaying 9, 64–65, 70–71, 75, 101, 210, 213, 214
Degradation/biodegradation 3, 14–15, 33–35, 52–53, 69, 71, 73, 79, 91, 95, 100, 117, 127, 129, 152, 160, 167, 169–177, 185, 225, 226
Denitrification 13, 15–16, 34, 96–97, 101–102, 127–128, 153, 154, 175, 189–190, 205, 206, 209, 210, 213–216
Diatom 4–5, 8, 10–11, 17, 34–35, 47, 50, 53, 70, 73, 75, 111–121, 126–128, 132, 138
Dioxin 167–169, 171
Dispersion/dispersal 53, 55–56, 63–68, 74–75, 119, 224
Dissolved organic carbon (DOC) 34, 78, 92, 96, 99, 224
Dissolved organic matter (DOM) 13, 74–75, 95–96, 135, 147, 152, 155, 224

E

Electron transfer mechanisms 183–187, 189, 192, 193, 195–197, 224
Electrosynthesis 184–185, 190, 195
Electrotrophic 189–191, 224
Endpoint/toxicity endpoint 125–130, 134, 138, 226
Epilithic biofilm 5, 7–8, 13–15, 64, 67, 72, 76, 78, 97, 224
Epipelic biofilm 5, 116, 224
Epiphytic biofilm 5, 10, 13, 14, 18, 66, 70, 72–73, 75, 78, 79, 224
Epipsammic biofilm 5, 14, 69, 72, 78, 224
Epixylic biofilm 5, 9, 13, 14, 224
Evenness 68, 72, 74, 76, 79, 99, 120, 224
Exoelectrogen/exoelectrogenic 183, 185–189, 193, 194, 197, 198, 224
Extracellular enzymes 6, 13, 79, 89–90, 94–97, 99–101, 127–129, 224
Extracellular polymeric substances (EPS) 4, 6–7, 13, 15, 29, 32, 34, 37–39, 47–49, 72, 89, 93, 100, 130, 147, 158, 167, 184–186, 194, 224

F

Fermentation 15, 145, 148, 152, 189, 191–192
Firmicutes 50–51, 71, 176
Flagellates 50, 53, 71
Flow 9, 11–13, 15–16, 19, 32, 34, 36–37, 50, 67, 75–76, 78–79, 98, 92, 96, 101, 136, 147, 159, 188
Fluorescence *in situ* hybridization (FISH) 34–35, 49, 99–100, 127–128, 186, 212, 214
Fluorochromes 31, 36–38, 39, 223
Functional diversity 63, 78–80, 224
Fungi 3–9, 13, 17, 29, 47, 50, 54–55, 63, 71–72, 91, 95, 126–128, 157–159, 167

G

Gamma-diversity 69, 224
Glycoconjugates 31, 35, 39
Granules/granular 5, 6, 10, 18, 32, 34, 39, 171, 185, 205, 207–209, 211–213, 217, 224
Grazing 3, 6–8, 12, 17–19, 29, 36, 37, 50, 53, 70, 72, 118, 127, 129, 134, 135, 136, 137
Greenhouse gas 5, 145, 150, 205, 214

H

Heterotroph 4, 5, 7–9, 13–15, 53, 65, 72, 74, 76–77, 91, 93–96, 133, 153–155, 157, 158, 210, 213, 215, 224
Hydraulic retention time (HRT) 148, 150, 159, 213
Hyporheic zone 5, 9, 63–64, 69–73, 78, 93

I

Inhibition 13, 49, 54, 128, 131, 135, 138, 154, 207–208, 210, 214, 215, 217
Inoculum 53, 171–172, 212
Invertebrates 18, 29, 36, 47–48, 51, 53, 75, 112

L

Larvae 18, 47–48, 51, 53, 56
Lectin 31–35, 38–40
Light 12–15, 18, 29, 47, 50, 52, 55–56, 68, 71–72, 74–75, 79, 93, 95–97, 100, 116, 118, 126, 131–132, 136

M

Macrophyte 7, 10, 73, 111–112, 224
Marine sediments 15, 63, 66, 70, 76–77
Marine snow 5–6, 10–11, 18, 34, 52, 53, 226
Mass effects perspective 66–67, 224
Mediator 183, 187, 193, 195–196
Membrane-aerated biofilm reactors (MABR) 205, 208, 209, 213–216
Membrane biological reactor (MBR) 175
Metacommunity 65–68, 75, 119, 224
Methane 10, 15, 93, 145, 148, 150, 153–155, 175, 184, 189–191, 198
Methanogenic/methanogenesis 15, 71, 145, 150, 153–155, 175, 224
Microbial electrolysis cells (MEC) 183–184, 188, 191, 192
Microbial fuel cell (MFC) 176, 183–188, 191, 192, 197
Microbial mat 5, 16, 18, 31, 52, 73, 75–76, 96, 225
Microbially induced concrete corrosion 145, 155–156, 158, 160, 224
Microphytobenthos 4, 225
Motility 31, 49, 54, 71, 114, 120
Moving bed biofilm reactors (MBBR) 172, 205, 209, 211, 215–217
Multichannel imaging 29, 225
Multimetric index 120

N

N-Acyl homoserine (AHL) signals 51–52, 54–55
Nanoparticle 56, 101, 129, 130, 132, 135, 138
Nanowires 56, 187, 193–195, 197
Nestedness 119, 225
Neuston 35
Neutral perspective 67, 225
Next-generation sequencing 120, 127–128, 225
Nitrate 77, 93–94, 96–97, 134, 147–148, 152–154, 172, 174–175, 184, 189–190, 198, 205–207, 210–213, 217
Nitrification 15, 17, 53, 93, 96–97, 101–102, 205–206, 213, 216
Nitritation 205–208, 210–213, 215–216
Nitrite 55, 93, 147, 153–154, 175, 189–190, 205–207, 211–217
Nitrous oxide 201, 214–215
Nutrient stoichiometry 91, 94, 225
Nutrient uptake/nutrient uptake length 8–10, 89, 96, 102, 132, 134, 138, 225

O

One-stage PNA (partial nitritation–anammox) 205–209, 213, 216–217
Optical coherence tomography (OCT) 31, 37, 40
Organic carbon 14, 34, 65, 72, 77–78, 80, 99, 155, 206, 209, 210, 213, 216
  see also Dissolved organic carbon
Oxidative stress 130–132
Oxygen 11–12, 15, 19, 34, 36, 53, 55–56, 77, 91, 93, 97, 101, 132, 145, 147–150, 153–157, 174, 184
  see also Oxygen transfer *and* Oxygen uptake
Oxygen transfer 205, 207, 209–210
Oxygen uptake 155, 201, 209–210

P

Partial nitritation 205, 211–213, 215–216
Perchloroethene (PCE) 169, 174, 176, 179
Periphyton 4, 5, 18, 96, 116, 121, 225
Persistent organic pollutants (POPs) 167–171, 173, 175–177, 225
Phosphorus 9, 14, 34, 77, 89, 90, 97–98, 116, 128, 145
Photosynthesis 10, 15–16, 94, 96, 126–128, 136, 138
Phylogenetic diversity 66, 69, 225
Pollutants/pollution 3, 77, 111–116, 120, 125–126, 128–135, 137–138, 145, 147, 160, 183, 185, 189, 190
  see also Persistent organic pollutants
Polychlorinated biphenyls (PCBs) 167–171, 175–176, 225
Predation 6, 18, 47, 49–50, 89
Proteobacteria 10–11, 33, 50–52, 63, 70–71, 73, 175–176, 186, 206
Protozoa 3–7, 10, 18–19, 47, 63, 126–127, 129, 130, 167

Q

Quorum sensing (QS) 36, 47, 51–56, 94, 207, 225

R

Reactive oxygen species (ROS) 55, 131
Reactor 3, 32–34, 37–39, 92, 147, 154, 172, 174–176, 191–192, 205–217
Recovery 76, 95, 114, 117, 133, 136, 225
Resistance 4, 7, 113, 127–128, 160, 192, 225
Richness 14, 64–65, 72, 74–78, 114, 116–117, 120, 129, 223
Rotating biological contactors (RBC) 205, 209, 210, 216

S

Salinity 12, 17–19, 74, 76, 111, 113
Sensitivity/pollusensitivity 111–113, 119–120, 126, 129, 136, 137, 225
Sewage 15, 114, 145–150, 155, 159, 172, 175–176, 205, 212, 214, 216–217
Sewer 145–161, 225
Sludge 39, 100–101, 145, 147, 172–176, 192, 205–208, 211–213, 216–217
Sludge/solid retention time (SRT) 205, 207–208, 213, 217, 225
Species sorting 34, 66, 67, 74, 225
Sponge 52, 54
Spores 18, 47, 48, 53
Stable isotopes/stable isotope probing technique 90, 99, 102, 129, 225
Stromatolite 5, 52, 73, 225
Succession 15, 17, 33, 50, 64, 92–93, 97, 119–120, 129, 157
Sulfate-reducing bacteria (SRB) 5, 15, 52, 98, 145, 148, 151, 153–156, 169–170, 225
Sulfide 98, 145, 148–159
Synergy 185, 187, 226
Syntrophic interactions 93, 150, 152, 185, 187, 226

T

Taxonomic diversity 63, 74, 76, 78–80, 120, 226
Temperature 4, 8–9, 11–14, 17–19, 35, 47, 49, 69, 71, 74–76, 79, 91–92, 95, 101, 118, 130–132, 136, 147, 155, 158–159, 186, 191–192, 207, 216, 217
Thickness 8, 100, 130, 211, 214
Tidal sediments 4, 70
Tolerance 13, 17, 19, 92, 111, 126, 127, 226
Toxic/toxicity 6, 47, 49–50, 100–101, 111, 114–115, 126–132, 135–137, 145, 149, 152–155, 167–173, 177, 195, 207
Trichloroethene (TCE) 169, 172–174, 196

V

Viability 36–37, 129
Virus 4, 6, 19, 35, 50, 63, 71

W

Wastewater 3, 55, 98, 145–152, 155–156, 159, 167, 173–177, 183–195, 198, 205, 208–217

X

Xenobiotics 176, 226

www.ingramcontent.com/pod-product-compliance
Lightning Source LLC
LaVergne TN
LVHW071411210926
848486LV00004B/246
* 9 7 8 1 9 1 0 1 9 0 1 7 3 *